CONSTRUCTION PROJECT PLANNING

CONSTRUCTION PROJECT PLANNING

By

Prof. (Dr.). R.K. Khitoliya

Former Professor and Head
Civil Engineering Department and
Former Member-Board of Management
Punjab Engineering College
Chandigarh - 160012
(India)

DPH

DISCOVERY PUBLISHING HOUSE
INDIA

Published by:

DISCOVERY PUBLISHING HOUSE
4383/4B, Ansari Road, Darya Ganj
New Delhi-110 002 (India)
Phone : +91-11-23279245; 23253475; 43596065
E-mail : discoverybooksindia@gmail.com
discoverypublishinghouse@gmail.com
namitwasan9@gmail.com
web : www.discoverypublishinggroup.com

***First Edition:* 2023**

ISBN: 978-93-94917-00-2

Construction Project Planning

(All are Very Sweet & Loving)
my Granddaughters
Ayesha, Ananya & Aditi

Preface

Planning, an important and essential phase in any project, is more or less lacking in our country. Due to lack of planning, money, manpower and time are lost. The concept of planning should be emphasised in the curriculum at all levels of education. This is an attempt to tell the young friends the necessity of the planning and planning methods.

Project is a mission, undertaken to create a unique facility, product or service within the specified scope, quality, time, and costs. With the emerging global opportunities, projects cross geographical boundaries, corporate channels, traditional systems and cultural diversities. The knowledge areas needed to manage such projects comprise project management techniques, general management practices and technology-related subjects. The project management techniques of planning, scheduling and controlling are the tools and devices that bind the subject's knowledge areas. These techniques can be applied to all types of projects. This book covers their application in the field of construction.

In spite of all these facts, construction is the most unorganised sector in the country. In major projects, most modern technology is tried to be employed. But, lack of planning and management has the effect on these projects resulting in long delays, cost increases, disputes, quality at stakes on almost all the projects.

There is a vast scope for improving performance through knowledge in the construction industry, where men, materials, machinery, money work together to build a facility. Perhaps, it is one of the rare industries that can provide a vast scope for cost and time reduction at micro-level.

Construction Project Planning is a unique subject introduced in the study-curriculum of B.Tech/M.Tech in Civil Engineering. The subject is based upon the modem technology of construction planning, construction contracts and universal management of building construction in Indian context which can be considered as the prerequisite for the survival in the present scenario of the civilization. Relevant topics of discussion include objectives and concerns of network techniques, contract management, selection of contractors, tenders and conditions of contract, legal aspects of contract, quality control, cost effectiveness and lastly the increasing role of computers in all spheres of design, construction and management.

The author will greatly appreciate being informed about errors and receiving constructive criticism for improvement of the book. To achieve the goal of understanding, the author has extensively used publications/website of BIS, Various Departments/Ministries of Government of India, and other International agencies. The help received from these publications/website is therefore gratefully acknowledged. The material for this book has been drawn freely from various books, technical publications, websites etc., as sources and the author express his sincere thanks to them. In spite of his sincere efforts, some contributions might not have been acknowledged. Such omissions are not intentional and the author regret the same. Information contained in this book has been obtained by author from sources believed to be reliable and are correct to the best of his knowledge. Every effort has been made to avoid errors and omissions and ensure accuracy. Any error or omission noted may be brought to the notice of the publisher/author which shall be taken care of in the forthcoming edition of this book. However, neither the publisher nor the author guarantee the accuracy or completeness of any information published herein.

The author of the book has taken all possible care to ensure that the contents of the book do not violate any existing copyright or other intellectual property rights of any person in any manner whatsoever. In the event, the author has been unable to track any source and if any copyright has been inadvertently infringed, the facts may be brought to the notice of the publisher in writing.

–Author

Contents

1

Planning for Construction Project

Construction Management Plan

The project manager's prime activity is planning which is one of the key functions of the management process. Planning is selecting objectives and then establishing programs and procedures for achieving the objective; it is decision making for the future; it is looking ahead. Decisions are required because a choice must be made from among the alternatives that are available. It may be a choice of systems, equipment or contract strategy to mention a few. Eventually the plan is accomplished through a structured sequence of events which will lead to a desired set of objectives.

A proactive management style sets a plan and makes it happen. This is the opposite of reactive management which results from lack of planning. Reactive management is often referred to as crisis management or as a fire-fighting mode. Planning allows leadership to control whereas reactive management is a lack of leadership and control.

An engineer who is looking for challenging work will soon find himself or herself responsible for a project that demands the utmost in planning skills. In fact. it may be necessary for the engineer to carry out all the planning; to visualize all the operations of the project; to arrange these operations in their proper sequence; to achieve confidence that each operation is understood; to acquire the know-how and means necessary to perform each operation; and to feel convinced that the method thought out for performing each operation is the most economical. Confidence is achieved only through systematic planning. Unfortunately, because of overconfidence many people do not make the necessary effort to develop a plan, they resort to "seal of the pants" management which produces one panic after another and below par performance. Their targets keep shifting from day to day. For our purposes this picture presents a vivid example of how a project should not be carried out.

If, however, one is confident that one knows a project extremely well and it is so small that its progress can be plotted mentally without the need of any formal planning, that individual is justified in saving planning time and using it for other efforts such as organization and coordination. But how many projects are that simple?

Planning is not only scheduling, which is but one of the many plan that can be put in place. Planning results in setting out schedule milestones, which are time objectives. It clearly establishes the work to be done to achieve certain contract scope and cost objectives.

The main purpose of planning is that it reduces the uncertainty that exists before a project or portion of a project is launched. Besides avoiding the crisis management, planning improves the efficiency of execution; it clarifies the objectives.

Another important product of planning is that it provides a basis for executing, monitoring, and controlling. Thus during the early overall planning the PM must be a key participant in the planning process. This planning responsibility should not be delegated; to do so would mean that the PM is abdicating responsibility for planning, requiring that someone else carry this load. This someone else becomes the true project manager.

A construction management plan is a document that entails detailed information on mechanisms, tools, resources, and the knowledge transfer necessary to deliver complex construction projects. The plan enables project managers to determine project feasibility and develop resource schedules that align with the goals and objectives. It further facilitates efficient project resource planning and management.

Construction management plans bridge the communication gap between cross-disciplinary teams which is critical before beginning the projects.

With an organized plan and schedule, managers can be held accountable and ensure that the workforce is committed towards timely and quality accomplishment of the construction work.

Overall, it is a handbook for the project managers that can help them in managing construction projects to succeed.

Define the Project Scope

The project's scope is inclusive of the project's goals and objectives, deliverables, features, financials, tasks, and deadlines. Defining these variables builds the foundation of your construction management plan.

When managers have detailed documentation of every project attribute, it becomes feasible to plan construction projects and allocate tasks with accuracy.

Resource Capacity Planning

Once the project scope is in place, the construction management plan should have a resource schedule ready. Building the right team to accomplish varied tasks is crucial to get a head start on the construction project.

Managers must therefore assemble the right workforce and allocate the right tasks to them to avoid any potential bottlenecks in the future. Capacity planning before project execution also provides enough window to the manager to recruit the resources well in advance and eliminate delays.

Phase Design and Budgeting

The overall project plan must be divided into small phases or small projects. This will enable decision-makers to track the progress at every stage conveniently. In addition, designing each phase facilitates accurate budgeting as the management practices and methods to be employed are explained explicitly.

Phase designing and budgeting is vital to get an estimate of the total revenue that will be generated. It will also facilitate managers to assess the feasibility of the project.

Stating the Interdependencies

Enunciating tasks and their interdependencies on other departments is a must in construction projects. That's because construction projects operate with cross-functional teams and coordination is required between every team for the project's success.

Thus the plan created by the project manager helps them supervise the construction planning and ensure that projects are completed on time.

These are the essentials of a construction project management plan. Of course, managers can configure the plan as per their projects and requirements.

Let us now understand the steps that will help you create an effective and systematic construction management plan.

Importance of Planning

Today construction is a project-oriented industry. Most work is carried out as a project which means that facilities to be constructed or objectives to be achieved are defined and that an effort is then made to achieve these within certain time and cost parameters.

One characteristic of projects is their continuous growth in size and complexity as technology advances. Complexity generates the necessity for specialization, and since each specialist has his own jargon, specialization may lead to a breakdown of communications. For example, in a project for the construction of a certain plant, many specialists-structural, construction, mechanical, electrical, heating and ventilating, sanitary, production and system engineers-may be involved. Each is responsible for a specialized technology. Each has a viewpoint and thinks differently about the project. Some method must be found to facilitate communication so that each person is working toward the same set of project objectives: time, cost, quality and goodwill. The emphasis that the project manager puts on the project objectives may be different from that of the project team. It is only by using a project plan that the project manager can effectively communicate to the team his or her emphasis on the various project objectives.

Another characteristic of projects is the increasing importance of timely completion. If a company is trying to develop a certain product before a competitor can introduce a similar product into the market, the company must meet its own schedule or it may lose even its present market share. Today engineering design can be completed by modern computers much faster than could be done only a decade ago. Therefore more organizations can quickly place more designs on the market, reducing the life of a design and increasing the necessity to plan and schedule. The need for an early debut to enhance the potential of a design to capture the market makes a sound plan imperative.

In view of present-day resource limitation, optimum use of resources is essential. A project must be efficiently planned to make optimal use of the limited resource. It is thus possible to economize on the present project and to go that much further with other projects in an organization.

The cost of capital is another factor that places a high emphasis on establishing a schedule and working to achieve its targets. No one can afford to lose interest on the money investment in a project while it is unproductive. It is important to make the investment productive as early as possible.

Because of increased outlay, technical complexity, price escalation and longer time spans the element of risk in construction projects has increased. Financiers of large projects demand not only technical feasibility reports but also documented schedules and reliable estimate to substantiate the management's ability to execute and control the operations economically. In effect, the financiers require a workable plan.

Management must absorb information and make decisions. With the growth in complexity and size of projects, the volume of information has increased tremendously; consequently in practice it is necessary to manage by exception. A prerequisite for management by exception is the existence of a plan selling the targets against which performance can be measured. When performance matches the standard set by the plan, no exception report to the management is required: only the factors requiring management action are reported. Along with exception reports, accomplishments should be reported also.

The efficiency of a group depends on each individual contributing his share. Planning provides a unity of purpose to the members of a group. The increased efficiency of each individual which results from this unity leads to increased production by the group as a whole. Planning is therefore essential for achieving an increase in production.

Another factor that makes planning absolutely essential in projects is the transiency of construction personnel. Individuals familiar with the job often leave before the project is completed. Breaking in a new person is always more costly than working with someone who has knowledge of the job. Basically a newcomer works much more slowly while he is acquiring the knowledge and experience necessary to work at a proficient rate. The existence of a plan provides continuity that reduces the learning period for a newcomer.

The need for planning can be considered from another point of view. A project is executed only once. Every project manager likes to claim that it was done most efficiently but there should be evidence to validate this claim. To substantiate a claim, there must not only be a plan but it must be the best plan possible. If work is done according to the plan, the assertion that the project was done most efficiently can be upheld.

Projects have four common objectives: time, cost, quality and goodwill.

If efficient communications are maintained and the project is built on time is within cost limits, conforms to the specified quality, and meets the utility, flexibility, service, and aesthetic needs as defined during the design phase, there is no problem maintaining goodwill. Thus goodwill is a variable dependent on the first three variables and the design objectives. Of these, however, the third variable-quality-is controlled by the design engineers through project drawings and specifications. So the only variables amenable to managerial control are time and cost. In this text, cost planning and control is dealt with in a limited sense that is, as it is affected by planning and control of progress and not in its broad sense, which would encompass cost reduction modeling, optimization, and analysis. In physical terms a system is a set, of things or parts forming a whole. The time and cost plan comprising a schedule and cost estimates is a system in the physical sense.

Planning for the sake of planning is of no use. The value of a plan lies in its implementation. Progress is measured against the planned target and the schedule or cost deviations are corrected. If the corrective action cannot bring the project within the limits, the plan is modified. The project is kept on schedule and within the budget through the control function which is as important as if not more important than the initial plan.

STEPS FOR CONSTRUCTION PROJECT PLANNING

Document the Scope

A comprehensible scope outlines the construction project's goal and purpose in terms of its value and benefits. It is a compilation of both the stakeholder's and client's expectations and inputs. After diligent analysis of every input, a scheduling plan for the construction is developed.

In addition to this, a project's scope explains each and every team member's roles and responsibilities in detail. It also includes the project's timeframe, financials, methods and techniques to be employed. Moreover, it consists of factors that may adversely affect the project's progress.

In a nutshell, the project's scope is an overview of all the project attributes. Documenting the scope is therefore the most important step while drawing out the plan.

Draw out a Work Breakdown Structure (WBS)

A work breakdown structure (WBS), as the name suggests, is a deliverable-oriented breakdown or bifurcation of projects into smaller tasks. Drawing out a WBS allows project managers to categorize the projects into manageable chunks.

They can schedule and allocate the right task to the right resource at ease. One of the major benefits of designing a WBS is that it enhances task management efficiency and helps managers ensure uniform allocation of workload. A WBS is, therefore, necessary to help project managers track the project's progress at every stage. This will help them implement corrective measures in case of discrepancies.

Work Breakdown Structure in Project Management

Work breakdown structure (WBS) in project management is a method for completing a complex, multi-step project. It's a way to divide and conquer large projects to get things done faster and more efficiently.

The goal of a WBS is to make a large project more manageable. Breaking it down into smaller chunks means work can be done simultaneously by different team members, leading to better team productivity and easier project management.

How to Create a Work Breakdown Structure

Before you create a work breakdown structure, it's essential to first assess the project scope by talking to all stakeholders and key team members involved.

As the project manager, you want to ensure that all critical input and deliverables are gathered and transparently prioritized. We may use Gantt charts, flow charts, spreadsheets, or lists to show the hierarchical outline of importance and connectivity between the tasks needed to complete the project.

After outlining the deliverables and tasks in order of completion, we can then assign each task to a project team member. Ensure no team member carries the majority of the project's weight by spreading duties and responsibilities across the team.

Characteristics of a Work Breakdown Structure

The Project Management Institute (PMI) defines WBS as "a deliverable-oriented hierarchical decomposition of the work to be executed by the project team to accomplish the project objectives and create the required deliverables."

Each WBS level represents a new and increasingly detailed definition of work needed to complete the project.

PMI's definition adds that a WBS structure must be constructed in a way that each new level in the hierarchy includes all the work needed to complete its parent task. This means that every parent task element must have more than one child task within it to consider the parent task element complete.

Examples of Work Breakdown Structure

The work breakdown structure for each project can be different.

As a project manager, we may have to experiment to see which WBS works best for you and your team. The goal is to show the hierarchy of your projects and make progress clear to everyone involved — whether they are a team member or an external stakeholder.

Here are some work breakdown structure examples. We can use any of these to outline our WBS.

1. **WBS spreadsheet:** We can structure your WBS efficiently in a spreadsheet, noting the different phases, tasks, or deliverables in the columns and rows.
2. **WBS flowchart:** We can structure your WBS in a diagrammatic workflow. Most WBS examples and templates you may find are flowcharts.
3. **WBS list:** We can structure your WBS as a simple list of tasks or deliverables and subtasks. This is the most straightforward approach to make a WBS.
4. **Work breakdown structure Gantt chart.** We can structure your WBS as a Gantt chart that represents both a spreadsheet and a timeline. With a Gantt chart-structured WBS, you can link task dependencies and show project milestones.

Difference between WBS and a Work Breakdown Schedule

Various detailed project documents support the WBS. Amongst them are a risk management plan, quality plan, procurement plan, communications plan, staffing plan, and a work breakdown schedule plan.

The work breakdown schedule includes the start and completion dates for all tasks, activities, and deliverables defined in the WBS.

Benefits of Work Breakdown Structure

The WBS provides the foundation for integrating the work package and intermediate deliverables with all other aspects of project initiation, planning, execution, monitoring and controlling, and closing.

A deliverable-oriented WBS provides many benefits to the project, including the following:

- Better communication to project sponsors, stakeholders and team members.
- More accurate estimation of tasks, risks, timelines and costs.
- Separating project deliverable into component parts to ensure the project plan matches the approved project scope and will fulfill the overall objectives of the project.
- Increase confidence that 100% of the work is identified and included.
- A foundation for the control process within the project.
- Decomposes (or disassembles) the overall project scope into deliverables and supports the definition of the work effort required for effective management.

The 100% Rule

The 100% rule is a core characteristic of the WBS. This rule states that the WBS includes 100% of the work defined by the project scope and captures. All deliverables-internal, external, and interim in terms of work to be completed, including project related management. The 100% rule is one of the most important principles guiding the development, decomposition and evaluation of the WBS. The role applies at all levels within the hierarchy.

The WBS should not include any work that falls outside the actual scope of the project that is it cannot include more than 100% of the work.

Representation of the WBS

The WBS can be represented in a variety of ways including graphical, textual, or tabular views. Regardless of the representation used, the WBS enables the project team to predict and forecast costs, schedules, resource requirements, and allocations more accurately.

Problem Diagnostic Checklist

The following are representative examples of major project problems resulting from key WBS defects:

- There are frequently missed deadlines and an extended schedule:
  - Have all major and minor deliverables been included? Failure to include all deliverables within the initial WBS can increase project schedules when missed deliverables are identified.
  - Have deliverables been defined specifically enough to allow for appropriate work packages to be developed?
  - Does the WBS facilitate the use of Earned Value Management techniques?
- Project is over budget:
  - Does the WBS provide logical summary points for assessing accomplishments, as well as for measuring costs and schedule performance?
  - Does the WBS facilitate the use of Earned Value Management techniques?
- Individuals are unable to use the new product or feature:
  - Are deliverables decomposed into smaller, more specific deliverables? For example, a deliverable of training might not be decomposed thoroughly enough to cover all of the people who need training to use the new product, process, or service.
  - Are the WBS elements deliverable-focused?
  - Were appropriate assembly or integration deliverables and testing activities present?
  - Were training and implementation deliverables defined?
- The project scope has changed and is unmanageable: Has a WBS been created for the project?

  - Does the WBS decompose the overall project scope into deliverables?
  - Does the WBS provide a level of flexibility for change?
  - Has the WBS been updated when necessary changes are approved by the change control process?
  - Has the WBS been placed under change control?
- The project has become an ongoing project with no end in sight:
  - Has a maintenance plan been developed for post implementation if needed?
  - Does the project have a specific end point?
  - Does the WBS include a close-out phase or plan?
  - Is the endeavor actually a project or is it an ongoing operation:
- Project team members are confused about their individual responsibilities:
  - Do the WBS elements define overlapping responsibilities for the creation of a deliverable?
  - Is the information within the WBS at the appropriate level of detail and in communication processes and decision authorities agreed upon beforehand?
  - Do the WBS elements reflect work with specific, tangible deliverables?
  - Have all key stakeholders, including subject matter experts, contributed to the creation and validation of the WBS?
- Some planned work does not get done:
  - Has all required work been included in the WBS?
  - Are the WBS elements deliverable-focused?
  - Has the WBS organized around deliverable rather than process steps?
  - Was decomposition completed before dependencies and durations were defined?

Things to Consider when Preparing a WBS

A high level WBS can often be developed early in the conceptual stage of the project. Once the project is defined and specifications are prepared, a more detailed WBS can then be developed. Syed Ali Dilawer,. (2010).

The following list of questions should be considered when developing a WBS to manage a project:

- Is the project scope statement defined and issued?
- Has the project scope statement been defined and issued?
- Has the Project Manager and the team formulated a vision of the final product(s), services, or results?
- Has personnel who will do the work been assigned to develop the WBS?

- What project are the project's component parts?
- How do the pieces work together?
- What needs to be done?
- Have the project's intended business objectives been defined? What is required to achieve the business value?
- Has the entire project been thought through? Have the high-level deliverables been progressively decomposed?
- Have all deliverables, both interim and final, been identified? What is to be provided?
  What is required?
- Has the relationship of each component to the end product been defined? What methods will be employed? What special processes will be needed? What are the quality requirements? What kinds of inspections need to be done?
- Have the activities that are needed to support the deliverables been identified, including those that directly or indirectly facilitate their creation?
- Does the project require any external sources to contribute to the project and have they been identified?
- Has all work associated with risk management been identified? Have risks associated with project assumptions been identified?

These thoughts and questions are intended to help the Project Manager develop a clear statement of what the product(s) of the project are. Once completed, all of the work packages (i.e., the lowest-level WBS elements) should together comprise the complete list of deliverables for the project.

Guidance in Choosing a Method for Preparing a WBS

In developing a WBS, the project management team needs to decide first which development method to use. Some guidelines and explanations for which approach might be more appropriate are as follows:

Top-Down

Use the top-down in this situation:

- The Project Manager and Project Management Team have little to experience m developing Work Breakdown Structures. Top-down development allows for progressive understanding and elaboration on of the WBS. Syed Ali Dilawer,. (2010).

Bottom-Up

Use the bottom-up approach in these situations:

- The nature of the projects, the project team might already have a very good understanding of all interim deliverables required for the new project.
- The nature of the project life cycle is well known. If the organization always uses the same project life cycle, the interim deliverables for that life cycle are well known and can be used to begin bottom-up WBS development. Syed Ali Dilawer,. (2010).

- Appropriate WBS templates are available. If the organization has WBS from projects with similar products or services, and these can be reused, a bottom-up approach enhances the team's ability to customize the WBS template.

WBS Standards and Templates

There are plenty of sample WBS available in the literature, but the choice to use sample WBS as templates must be made carefully. The organization can have WBS templates for very similar projects, and the use of these templates is highly encouraged. However, if the project significantly differs from other projects in the organization, and no template seems to apply develop the WBS from scratch with a top-down approach.

Iterations

Construction of the WBS is an iterative process and may rely on more than one method to produce the final high quality WBS. For example, a WBS template and the top-down approach may be used initially to determine the overall structure of the WBS, while it might be more appropriate to use the bottom-up method to verify that all the elements are present to achieve a particular deliverable.

General Factors to be Considered

In developing a WBS, the following basic tenets should be considered:

- Each WBS element represents a single tangible or intangible deliverable.
- Deliverables include both final and interim deliverables that are required to create the final results.
- Deliverables include intangible items, such as information/ communication, integration, administration, training, process management, and procurement.
- All deliverables are explicitly included in the WBS.
- Deliverables are unique and distinct.
- All significant reporting mechanisms, such as review meetings, monthly reports and test reports are included and identified in the WBS.
- Clearly defining the project deliverables, so that each is unique, ensures there will be no duplication in the outcomes of the project or of the work performed to produce the end-products.
- Accountability for each work package can be assigned to a single project team member or subcontractor. If this is not possible, then reconsider whether or not the work package can be further decomposed.
- Each element in the WBS representing subcontracted or externally committed deliverables directly corresponds to matching elements in the subcontractor's WBS.
- The deliverables are logically decomposed to the level that represents how they will be produced and managed (e.g. designed, purchased, subcontracted, or fabricated).

- All WBS elements are compatible with organizational and accounting structures.

The following basic guidelines should be considered when organizing WBS elements into the WBS hierarchy: Syed Ali Dilawer,. (2010).

- Each WBS element belongs to only one parent WBS element.
- The set of child elements into which a parent element is decomposed includes all of the work contained in the parent, such that the 100% Rule applies.
- A coding scheme is used for WBS elements that clearly represent the hierarchical structure when viewed in text format.
- All "legs" of the WBS need not be to the same depths. Some areas of the WBS will need to show more detail than others.
- There is no need to have all work packages at the same level.
- The WBS development process should:
  - Be iterative.
  - Be reviewed and revised as the rest of the project planning process progresses.
  - Provide a vehicle for flexibility, particularly when the scope of the project effort might change.

Selection of the Type of WBS Element to be included

- A WBS organizes and defines the total work scope of the project. Not every WBS, however, needs to include all types of work. Rather, the kinds of work included in a WBS should be dictated by the scope and nature of the project for which the WBS is being developed.
- Structuring the Logic of the Decomposition:
- An essential feature of a WBS is that it clearly and comprehensively defines the scope of the project work through decomposition of deliverables into a hierarchy of simpler components, thereby providing one of the primary methods for managing complex projects. The way that the Project Manager decomposes the project (i.e., the logic used for decomposing the work) can vary depending on the needs and requirements of the performing organization and the use to which the WBS will be put.
- It is important that the WBS remain deliverable-oriented, rather than process-oriented, and explicitly contains all intermediate deliverables.

WBS Structure Guidelines

- The contents of this section reflect a view of recognised standards for the development of Work Breakdown Structures customised to meet the needs of the projects.
- These are, as stated, guidelines only, and the WBS for a specific should be structured to meet the needs of that project, but should still aim to align with these guidelines. In addition WBS designs must be reviewed and agreed with the head of planning and Project Manager.

WBS Guidelines for Multiphase Projects

The proposed WBS guidelines for a multiphase consists of 6 levels as follows:

- Level 1: Program Name
  Example: XXX, etc.
- Level 2: Project
  Example: Infrastructure, or Zones
- Level 3: Area of Work
  Example: Plot no., phase
- Level 4: Type of Work
  Example: Engineering, Interfaces, Procurement, NOCs, Construction
- Level 5: Area Details
  Example: Golf, Marina, etc. for area detail
- Level 6: Work Details
  Example: Scheme design, detail design, civil work, finishing work, etc.

WBS guidelines for single phase projects

- Level 1: Project
  Example: Infrastructure, or Zones
- Level 2: Area of Work
  Example: Plot no., phase
  Level 3: Project Stages
  Stage 0-Master Planning
  Stage 1-Concept Design Stage
  Stage 2-Schematic Design Stage
  Stage 3-Detail Design Stage
  Stage 4-Final Design and Tendering Docs
  Stage 5, 6-Tendering/Award
  Stage 7-Construction

OR

Type of Work
Example: Engineering, Interfaces, Procurement, NOCs, Construction
Level 4: Area Details
Example: Golf, Marina, etc. for area detail
Level 5: Work Details
Example: Scheme design, detail design, civil work, finishing work, etc.

Feasibility Study

A construction project depends on environmental, social, and financial factors. A shift in even one of these can impact the project's progress. Before beginning the execution, project managers must meticulously assess every factor. They should also analyze if the project is prone to environmental hazards or if the project will be profitable enough.

This careful evaluation constitutes the feasibility study. Before investing the resources, their time, and the project costs, conducting the feasibility study is necessary to ensure the project's success.

Cost and Time Analysis

Major construction projects cost thousands of dollars to your firm every hour. A slight delay in the timeline can eat into your organization's profitability. Managers can prevent these setbacks from happening by running an extensive cost and time analysis. This is only possible when they have centralized visibility of their project tasks and resources booked to complete them.

With a unified view, project managers can calculate the resource and equipment costs which will help them calculate approximate total revenue. Thus cost and time analysis is crucial to ensure that projects are delivered within time and budget. Note that employee leaves, holidays, buffer time must be taken into consideration while analysing the timeline.

Quality Control and Risk Management

Stakeholders and clients have to be specific with the project quality they want. This quality is clarified in terms of material and the final output. The onus is on project managers to ensure that this quality is maintained throughout the project lifecycle.

More importantly, a risk management plan has to be in place when you are dealing with construction projects. This is because these projects are not only prone to time and cost overruns but also pose safety risks. A construction management plan therefore must have a contingency plan and quality control audits.

Documentation of Communication Plan

When a project pivots on cross-functional departments, transparent communication and coordination are imperative to ensure smooth functioning.

The project manager's role is to document the collaboration channels beforehand to make sure no department or team is left out of the loop.

If one team faces a challenge, it can communicate and notify other departments who are dependent on them. This way, curative steps can be taken in time to avoid delays and further roadblocks.

These are the steps to be followed to document a systematic and well-organized construction management plan.

In conclusion, mentioned are some tips and tactics for project managers to effectively plan construction projects,

The construction industry projects demand a detailed plan to ascertain that no factor or attribute is overlooked. A shift in one factor can cause a ripple effect restricting the project's progress. Thus, managers must make a diligent effort to develop this plan. The above-mentioned constituents and steps will help you formulate a systematic plan and lead your project to succeed.

Below are some additional tips to add glory to our plan:

1. Open all channels of communication and collaboration. Ensure that all the documents are uploaded on the drive to make them accessible from anywhere.
2. Use effective resource management or project management software to gain 360-degree visibility of the enterprise. This will help you track progress at all times.
3. Time and task management are two important components to ensure project profitability. Plan and workforce smartly and efficiently to avoid over-utilization of resources and delays in meeting deadlines.
4. Safeguard your project data by investing in secure software and creating a backup at regular intervals. Losing this data can cause a catastrophic effect on the project's progress.
5. Review and improve your plan every time. A construction management plan is always a Work in Progress. Based on the changing factors, keep updating the plan to enhance efficiency.
6. Project managers must keep in mind the lessons learned from past experiences and improve their project planning process.

Construction Financing and Cost Control

The procedure of distinguishing and lowering the expenditure to maximize profit is known as Cost Control. The first stage of cost control is the budgeting process. A business analyst or company compares its actual results with the projected expectations, and if actual costs are higher than planned or projected expenditure then management takes action to rectify the issue causing the undesired changes. For instance, a company can call upon the various seller of the same project and compare their cited cost, which will lower costs. The cost control aims to manage the project within the approved budget. Continuous and timely cost reporting will have the best possible estimate of:

- Established project cost
- The anticipated final cost of the project.
- Future cash flow.

Cost Management

The process of planning and controlling the budget of a business is termed as cost management. It is a form of management and accounting that allows a business to predict the expenditures and to help reduce the chance of going over budget.

- Planning of Resources
- Estimation of Cost Incurred
- Budgeting
- Cost Control

The money spent can be categorised into three broad groups i.e. Manpower, Machine and Materials. All the three mentioned modes of expense are and can be further sub-divided in basic categories to monitor them effectively. Description of what types of resources are required and in what quantities for each element of the work break down the structure. Projects need to be guided throughout the project duration right from the tender bidding until the defect liability period of the project in order to achieve the required output at the end of the project. It helps in predicting the expenses of the business so that one can avoid going over budget.

Cost Control Techniques

Following are some of the valuable and essential techniques used for efficient project cost control:

- **Planning the Project Budget:** Initially the budget is planned using approximations at the start of project. It is this budget that you would have to help you for all payments that need to be made and costs that you will incur during the project life cycle. The making of this budget therefore entails a lot of research and critical thinking. Like any other budget, you would always have to leave room for adjustments, as the costs may not remain the same right through the period of the project. Adhering to the project budget at all times is key to the profit from project.
- **Keeping a Track of Costs:** Keeping track of all actual costs is also equally important as any other technique. Here, it is best to prepare a budget that is time-based. This will help you keep track of the budget of a project in each of its phases. The actual costs will have to be tracked against the periodic targets that have been set out in the budget. These targets could be on a monthly or weekly basis or even yearly if the project will go on for long. This is much easier to work with rather than having one complete budget for the entire period of the project.
- **Effective Time Management:** Another effective technique would be effective time management. Although this technique does apply to various management areas, it is very important with regard to project cost control. The reason for this is that the cost of your project could keep rising if you are unable to meet the project deadlines; the longer the project is dragged on for, the higher the costs incurred which effectively means that the budget will be exceeded. The project manager would need to constantly remind his/her team of the important deadlines in order to ensure that work is completed on time.
- **Project Change Control:** Project change control is yet another vital technique. Change control systems are essential to take into account any potential changes that could occur during the course of the project. This is because each change to the scope of the project will have an impact on the deadlines of the deliverables, so the changes may increase project cost by increasing the effort needed for the project.
- **Use of Earned Value:** Similarly, in order to identify the value of the work that has been carried out thus far, it is very helpful to use the

accounting technique commonly known as 'Earned Value'. This is particularly helpful for large projects and will help you make any quick changes that are essential for the success of the project.

Cost Control

The process that is supposed to be continued throughout the construction period to ensure that the cost of the building is kept within the agreed cost limits is the cost control. The cost control can be divided into two major areas:

- The control of cost during design stages
- The cost control by the contractors when the construction of the project has begun.

Cost control is looked upon as a managerial process, which produces information in support of the decision-making, to improvise value improvement, cost reduction, and continuous improvement in the organization. For every project, the management of cost is an extremely vital portion of project management to run it successfully. The cost management is commonly reflected in company's strategic goals, mission statements and business plans of a project organization in many ways. The three main functions to support the cost system performance of any project are discussed below. First, financial reporting function which is the system that present production expenses in reports covering each period of interest such as cost of goods sold and inventory. Secondly, cost system gives the producers and consumers economic feedback for instance process efficiency and expense control. Finally, according to cost information, companies can estimate the costs of products, services, activities and customers for better control. Cost management is the process of effectively planning and controlling the costs involved in a business. It is considered one of the more challenging tasks in business management. Generally, the costs or the expenses in a business are recorded by a team of experts using expense forms. The process includes various activities for instance collecting, analysing, evaluating and reporting cost statistics for budgeting. A company's overall budgeting can be brought under control by implementing an effective cost management system. Cost management is employed by many businesses as a vital part of business management. Cost management is also considered a form of management accounting that helps to identify future expenditures in a business to reduce budget overages. When cost management is being applied to a specific project, analyse the expected costs in the business. This is to be done in the beginning phase of the planning period. The project manager then approves the predicted expenses in purchasing the materials required for the project. The costs and expenses are recorded and monitored during the project execution period to ensure that the cost is in line with the actual cost management plan. Once the project is complete, the actual costs are compared with the predicted costs, which will help in predicting future expenses.

Advantages of Cost Control

Cost control has the following advantages:

(i) It helps the firm to improve its profitability and competitiveness.

(ii) In the absence of cost control, profits may be drastically reduced despite a large and increasing sales volume.

(iii) It is indispensable for achieving greater productivity.

Cost Control Techniques

Project managers are responsible for overseeing project goals and carrying out effective cost control. The following are valuable cost control techniques used to boost your bottom line:

- Budget planning
- Cost tracking
- Time management
- Project change controlling
- Earned value use

Direct and Indirect Cost

Costs can be sub divided into two main categories: direct costs and indirect costs. For budgeting and accounting purposes, it's crucial not only to document your business expenses but to classify them properly. The difference between "direct" and "indirect" costs is self-explanatory and indicative. While it's true neither word takes on a dramatically different meaning when categorizing the various costs of doing business, there's a bit more nuance in practice. Direct costs are traceable to the production of goods or services as required by the consumer. The operative word is "specific." Indirect costs are sometimes necessary to production, but they are not traceable. Indirect costs are the ones which are necessary to keep your business in operation. Think of them as the prerequisites for the production of any specific good or service.

Direct Costs

Direct costs are directly generated by the project. The cost of labour and materials could be the best examples of direct costs. Generally, labour costs are viewed as direct cost because of its relevance to the workers who actually are involved in the project. However, some labour costs might not be considered as direct costs for the project such as the costs of support personnel. For instance the accountant costs or consultant costs that come from other organizations may not be allocated directly and integrated as a part of the project. Specifically when their duties are servicing and monitoring something that could be viewed as extra in any sense. If the types of material required to accomplish the project are known direct cost of materials is easy to calculate. For instance, the direct costs of building a library or holding a party for 100 people can be estimated with fair accuracy, meaning these costs can be pinpointed in the project directly in a systematic way. Historically, to some industries like manufacturing, the costs of two resources were applied to projects directly and those were material costs and labour costs. There are few reasons behind assigning the resources directly. First of all, these kinds of costs accounted for a great number of the costs of the final product. Second, the perspective of scientific management believed that the usage and consumption of these expensive resources under full control was viewed as very important step. Finally, it was simple to measure the consumption of these kinds of resources by observation. These three characteristics can be justified the measurement costs for estimating how much materials and how much labour will be spent on each product.

Indirect Cost

Indirect costs are generally linked to two features, overhead cost and selling and general administration cost. The former refers to all sources of indirect materials, utilities, taxes, insurances, properties and repairs, depreciation on equipment, and health and retirement benefits for the labour force. They could be quite complicated to estimate as the costs could derive from different projects and different works within a company. Selling and general administration costs usually refer to advertising, shipping, salaries, sales and secretarial support, sales commissions and similar costs. Tracking and relating these costs to projects is not nearly as forthright compared with the direct costs, so the procedures applied are different from organization to organization. Measurement approaches could include flat rate (such as a percentage multiplier from 20% to over 50% on top of direct costs) which is a way to estimate the cost relative to the direct costs of a project framed as overhead costs, or a project-by-project-basis-measurement which is based on individual analysis. Organizing business expenses as either direct costs or indirect costs is a matter that goes beyond simple product pricing – it affects your tax payments, too. Overhead expenses, such as the utilities that power equipment and the inventory needed to manage the office, are tax-deductible.

Cost Management

Almost all the projects got to be target-hunting right throughout so as to receive the specified and expected output at the top of the project. It is the team that is liable for the project and most significantly the project manager that must to be able to perform out effective dominant of the prices. There are, however, many techniques that may be used for this purpose. Additionally to the project goals that the project manager has got to administer, the management of varied prices is additionally very necessary task for any project. Project management would not be effective the least bit if a project manager fails during in this respect because it would primarily confirm whether or not your organization would create a profit or loss. It Includes processes required to ensure that the project is completed within the approved budget processes involved are:

- Resource Planning
- Cost Estimating
- Cost Budgeting
- Cost Control

Resource Planning

Resource planning is a process of allocating tasks to human and non-human resources in a way that would maximize the efficiency of the resources. It can be both of those things and it can be something in between. Some organizations need to get a general overview and others manage the everyday work. Some organizations only need to know if there are enough resources to take on a new project. Some need to plan the work with 15-minute accuracy with all the details they can get. Some need to find a balance in between. There are no rights and wrongs. There's only what suits you. Resource planning tools are the tools that allow you to do the kind of planning that suits you (or at least that's the case if you have chosen a good one, here). At the core, there are three types of resource management. There's scheduling, there's planning, and there's management.

Simple Scheduling

Simple scheduling is for those that need the overview to forecast resource availability and nothing else. It can be a tool to provide a link between budget and the plan. It is described as dragging and dropping tasks to available resources with the right skills. They use all-day tasks and they don't adjust the utilization. Scheduling isn't necessarily done for better communication inside the project team or for collaboration which might sound like an illegal thing to say in today's project management world. However, resource scheduling is done to give those that plan the work a better understanding of what's going on. If done right, it will improve team communication and collaboration. Simple scheduling should be done as often as needed. That's the rule with any kind of resource planning. Plans only make sense if they are up to date. If there are changes, the changes should be apparent in the plans. Otherwise, the resource schedule has no point. That being said, resource scheduling is most often done once a week or even less than that. Resource scheduling is for those that need to visualize the data using Gantt charts. It's usually not the only resource planning that's done in the organization.

Detailed Planning

More detailed resource planning is for those that want to optimize the work that's being done. What makes it more detailed is usually the capacity planning? Default view is of 9 months and they plan one week at a time. Resource planning should be done several times per week. The rule that all changes should be apparent in the plans as they happen and not ad-hoc still holds. Resource planning means constant resource allocation and task prioritization. The people that the plans are made for should be able to see the plans whenever they want and sync them with their calendar. The project managers and the sales team should be present. That way it's clear how many resources are needed and where there is room for a new project.

Therefore, resource planning is for team leaders that need to plan the daily tasks of their team. Since it has a lot to do with capacity planning, the necessary features are resource utilization calculations, task utilization, conditional task colouring, and integrated reports.

Full on Management

Full on resource management is daily task planning. It's the type of resource planning that calls for live updates, a mobile app and calendar sync. It's taking resource planning to the next level. Resource management is that the most elaborate out of the 3 and you would like a lot of details if you wish to try to do detailed resource management. Resource management is concern having the resource management software system at your sight at all times. Whether it's for planning and reallocating or if it's for checking what's your next task. It's for those that need more than high-level planning. It's for field workers that are constantly on the go and need to know what's next on the to-do list through a mobile app.

Cost Estimsating Processes

The aim of this process is to estimate the cost of products and processes involved in the production. This means, a whole understanding of the design, contacts and production so as to properly model the use of materials by transformation and flow activities. Once the price estimating method is enforced in the construction, sometimes begins by making a budget within the early stages of the project. This budget part is extremely necessary, thanks to its part within the main project furthermore a reference for production and control. Project cost estimating is a very complex task due to the inherent uncertainty and variability in construction. This can be why feedback from the cost management is of crucial importance, that the cost estimating method can increase the standard of the knowledge available within the price set up, which can be in further use by the cost estimating method.

Cost estimation is the art of assigning value. It is also a science making use of a wide range of techniques to predict the costs of activities and assets. There exist a wide range of methods, applications and names for estimates. A cost estimate is that the approximation of the value of a program, project, or operation. The cost estimate features a single total value and will have recognisable component values. A drag with a price overrun are often avoided with a reputable, reliable, and correct cost estimate. Different professionals such as quantity surveyors and cost control department may prepare cost estimates or contribute to cost estimates. Building Costs can vary between builders/developers. This may flow from to the scale or buying skills of an organisation or the discount that it receives from suppliers. Professional fees must be added to the cost of construction and these will vary according to the project. In the past, fees were supported on a group scale that was a share of the building price, although within the previous few years a part of aggressiveness has emerged which may end in the professions negotiating their fees. This has resulted in a decrease in the amount paid for skilled fees although if a scale is employed these are often very little as 8% of the value of a basic industrial unit to 20% or additional for more refined developments with advanced mechanical, electrical and communication services installations. The preparation of prices are going to be done by the Quantity Surveyor so as that the client will get an appropriate building within the value constraints obligatory.

- Cost estimation is part of the cost engineering profession. A project might be any process that is started to perform work activities and/or create assets. The accuracy of the estimate depends heavily on the level of project scope definition: as the design and conditions of the project become better defined, so do the estimated values.
- Cost estimation is needed to provide decision makers with the means to make investment decisions, choose between alternatives and to set up the budget during the front end of projects. For this, estimates made by vendors and contractors need to be validated by clients as well. In later phases of the project, the budget estimate is used as a baseline to assess the performance of a project.
- Estimating is done by breaking down the total scope of a project in manageable parts, to which resources can be assigned and cost estimated. There are standardised ways of breaking down a project, like the Work Breakdown Structure (WBS) and the Cost Breakdown Structure (CBS), but depending on the needs of the project team and external party's multiple structures are often implemented to align reporting and sharing of cost data.
- A cost estimate is more than a list of costs. It also includes a detailed Basis of Estimate (BOE) report that describes the assumptions, inclusions, exclusions, accuracy and other aspects that are needed to interpret the total project cost. Otherwise, it would be a meaningless number. The BOE is required to communicate the estimate to the various parties involved in the decision making, but is also handy during close out, when the performance of the project is compared with other projects. It is the vital part often overlooked, that allows you to learn from your experience and mistakes.

Cost Budgeting

Cost budgeting may be a tool to estimate the price or necessary efforts for projects, work packages or activities in project management. Cost budgeting includes the estimation of prices, setting a hard and fast budget, and managing and dominant the particular prices (compared to the calculated ones). The prices then have to be compelled to be allotted to the activities or work packages in a project. A fastidiously enforced schedule and resource arrange allow an additional precise cost budgeting. Keeping a keen eye on the prices of operational your business will assist you build choices like once to renegotiate contracts, expand to alternative markets and increase or lower your cost. Cost budgets detail the expenses related with operating your business, running a project or developing a product. It tells the quantity of cash you expect to pay out for a given period and includes things like labour and utility costs. A budget is an accounting or plan for successive amount, which may be a month, quarter, year or may be the period of a project. A cost budget may be plan concerning your company's known expenses for the next period. You will be able to build budgets for varied prices like project costs or development costs. For instance, the cost budget for a project can embody all expenses needed to run the project, as well as participants' salaries and project supplies, whereas a manufacturing cost budget may embody raw materials and overhead prices. In most cases, you will produce cost budgets for expenses that are vital to your operations.

Costs or expenses are components of your business that cut your profits, however these prices change you to produce services or manufacture things. The crucial issue to contemplate is that the kind and level of costs you must incur. The budgeting process enables you to identify and clarify your expenses. It forces you to think of ways to control these expenses and identify the expenses you need to prioritize to meet your objectives. It helps you establish what your work prices and what limitations there are so you create realistic plans. Cost budgets additionally helps you avoid waste as a result you can see wherever you spend your money. To set your budget, you'll first have to be compelled to determine your costs. Your bank statements are an honest place to begin as a result you can see specifically what you paid out monthly. For example, when you set a budget for electricity overhead, analyse past bills and assess the units you consumed for each month. Also, consider the expected changes in rates charged per unit. Based on this, get the average amount you spent and apply that expense to each month over the entire period to set your budget.

Cost Planning and Control Processes

Cost planning suggests that purification the initial cost estimate and providing a project income, build on extra information that is generated along the project (like the deadlines of payments for the most material supplier) that is made on the assembly set up. These additional processes should support the decision-making in a dynamic way, in order to raise the possibility to accomplish the business goals. The cost management involves watching actual cost performance and identifies business opportunities. The aim of this control is not just comparing the current and estimated performance, but also concentrates on value generation. Having all this in mind, would deliver greater quality in the cost management by becoming proactive and enabling the company to face the dynamic, uncertain and complex construction environment that exists in many projects. Finally, you may agree that, cost management should be seen as a method and its implementation needs innovation, automation, learning, involvement and commitment from folks. The construction prices for a project can depend on the dimension, form of building, standard of finish required, and location, the economic climate of the construction industry i.e. if there is a shortage of construction work available firms will reduce the amount of their tender in order to try and attract work. If the opposite is the case and there is a lot of work available, firms will increase their tenders, as they will not be too keen to get the contract which can stretch their resources, unless it is worth their while financially. In an exceedingly recession, construction corporation can virtually get add order to stay their workforce and to confirm some cash flow. Whether or not the rate of inflation must be thought as it will depend on the duration of the contract and also the rate of inflation.

PDCA

PDCA, typically known as PDSA, the "Deming Wheel," or "Deming Cycle," was developed by far-framed management advisor Dr William Edwards Deming within the fifties. Deming himself named it the "Shewhart Cycle," as his model was supported a concept from his mentor, Walter Shewhart. Deming wanted to form some way of distinctive what caused merchandise to fail to fulfil customers' expectations. His resolution helps businesses to develop hypotheses regarding what has to amendment, and then take a look at these in a very continuous feedback circuit. PDCA helps in following objective:

- Helping to implement Total Quality Management or Six Sigma initiatives, and generally helping to improve processes.
- Exploring a variety of solutions to issues, and piloting them in a very controlled manner before choosing one for implementation.
- Avoiding wastage of resources by rolling out an ineffective solution on a good scale.

PDCA encourages an engaged, problem-solving workforce – the method is not restricted to managers, however are often used across the organizational structure, exploiting combined knowledge and information. This helps business to innovate through creative problem solving. Though it stimulates organizational problem finding, PDCA is not a problem solving approach as such. It is more of a coming up with and implementation methodology. It is often used as part of alternative problem solving techniques like Six-Step Problem Solving or Simplex. The easiest way to think of it is a prototyping model. But it can be used for larger projects, where testing is not feasible – for example, if a business has already bought new, expensive equipment to solve a problem, then the implementation has to be on a large scale. Usually the Problem Solving Group (PS) has already identified what they want to solve and the changes they would like to make.

- **Plan:** This initial stage clarifies the objectives of the chosen solution, and can establish that processes got to amendment (the downside will have already been outlined, and an answer projected exploitation alternative problem finding techniques). The PS group will plan how the solution is tested, and forecast what is expected to happen during the PDCA the cycle. This will result in a set of outputs, and usually set a baseline for improvement. Focused work at this stage includes agreeing what data is to be collected, what resources are needed, and the actions that will take place and when. The PS group will then decide where the trial run will be held. In our case we have choose to bring the cost down of the project using PDCA cycle with special focus on Billing department works like rate- analysis and converting supply work to task work.
- **DO:** This involves physically implementing the answer, and collection of information for analysis in the next step, observation and project management.

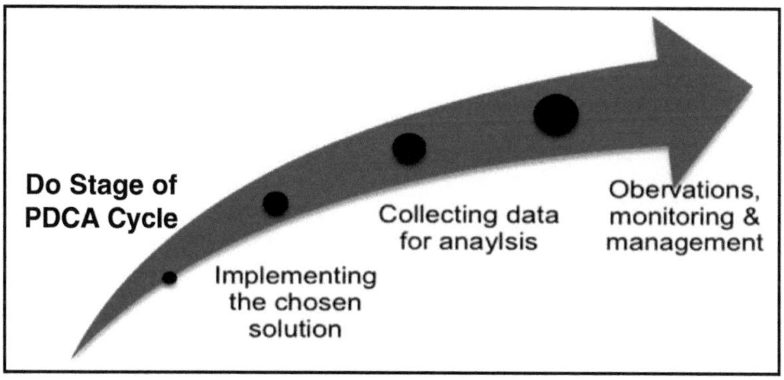

Fig. 1.1: Different Stages of DO

This step should also include oversight to ensure that the solution is implemented as per the specification. The new assessment package is designed, and 3 nurses are trained in it. They then conduct the assessments as planned, with associate degree observer recording knowledge of the new method. In DO part it was successfully tried conversion of few contractors from supply work to task work and it proved much cost-saving and profitable. Also, the work order prepared previously without properly performing rate analysis was also rechecked and the cost down was achieved.

Check: This step is about analysing 'before', 'during', and 'after' data to see what can be learned. (Another name for this step is 'Study'). 'During' data is easy to overlook, but any plan is implemented in phases, and if issues are to be found and identified quickly, these actions need to be measured. Throughout the check part, the info and results gathered from the do part is evaluated. Information is then compared to the expected outcomes to envision any similarities and variations. If the data is placed in a chart it can make it easier to see any trends if the PDCA cycle is conducted multiple times. This helps to envision what changes work higher than others and if the aforementioned changes can be improved.

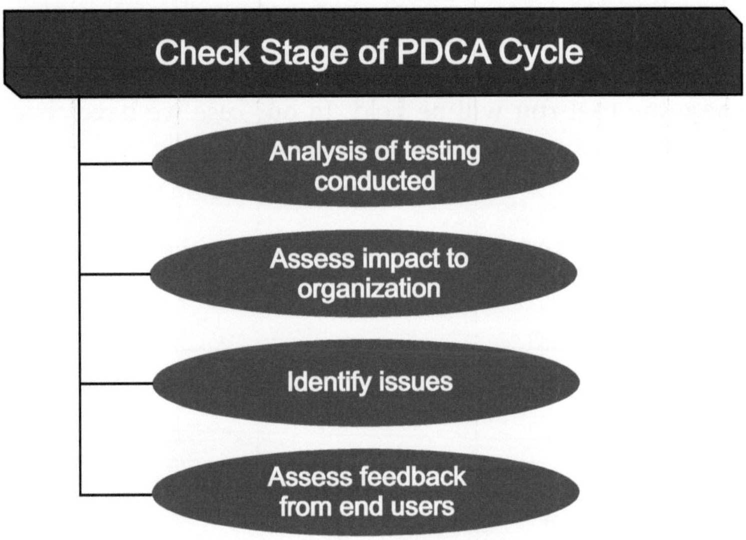

Fig. 1.2: Different Stages of Check

Not only will the process be observed qualitatively as well as quantitatively, feedback can be gathered from the end users of the solution. The data from the observations, and feedback from staff is analysed, and compared to the baseline. Issues and successes are identified. After plan and Do part, there comes CHECK part in which we ensure that the changes which were planned and then implemented are followed or not. Also in check part we investigate for desired result of our prior implemented decisions.

- **Action:** If the 'Check' demonstrates that the 'Plan' phase was implemented effectively and improvements can be seen on the baseline, then a new baseline is created and also the cycle returns to 'Plan', treating the new baseline. If 'Check' results infer that there has been no improvement from the 'Do' part, then the prevailing baseline continues in situation. In either scenario, if 'Check' reveals something different from expected (whether it has out-performed or underperformed), then it identifies that more learning is needed.

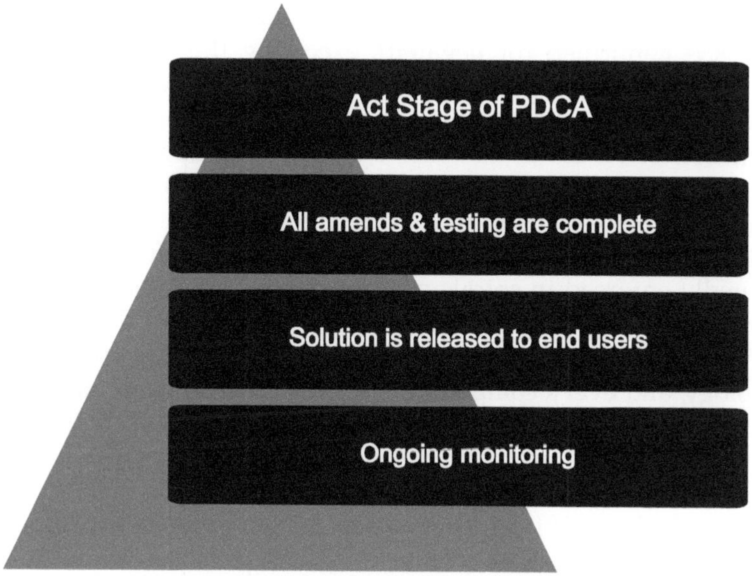

Fig. 1.3: Different Stages of Action

This may even result in the returning to the overarching problem solving technique, and the process of problem identification and definition, using the chosen problem-solving model. Changes or corrective actions must not be undertaken at this stage, and will never be done evaded a proper 'Plan' part. The baseline must not be modified from verified one, discovered in 'Act'. The PS group, based on the data, decide if the assessment package has helped or hindered the admissions process. Whatever is decided, the method needs any modifications, supported on the learning from 'Do' and 'check'. The cycle then integrates, on a wider scale, as it is refined and improved.

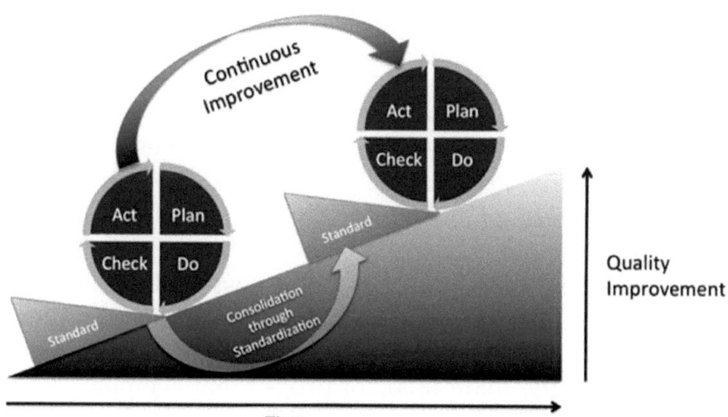

Fig. 1.4: Standardization through PDCA process

A basic principle of the methodology and PDCA is iteration—once a hypothesis is confirmed (or negated), executing the cycle once more will extend the information more. Repetition of the PDCA cycle will bring its users nearer to the goal, sometimes an ideal operation and output. Another basic function of PDCA is that the "hygienic" separation of every section, for if not properly separated measurements of effects because of varied synchronous actions (causes) risk turning confounded. Deming frequently stressed iterating towards the improved system, therefore PDCA ought to be repeatedly enforced in spirals of skyrocketing information of the system that converge on the final goal, every cycle nearer than the previous. One will envision an open spring, with every loop being one cycle of the methodology, and every complete cycle indicating a rise in our information of the system underneath study. This approach relies on the idea that our knowledge and skills are restricted, however rising. Particularly at the beginning of a project, key information might not be known; the PDCA—scientific method—provides feedback to justify guesses (hypotheses) and increase information. Instead of enter "analysis paralysis" to urge it perfect the first time, it is better to be roughly right than specifically wrong. With improved information, one might like better to refine or alter the goal (ideal state). The aim of the PDCA cycle is to bring its users nearer to whatever goal they opt for.

In cases of government grants or other forms of external funding, identifying direct and indirect costs becomes doubly important. Such programs are often strict about what constitutes a direct or an indirect cost and will allocate different amounts of funding to each classification. Often funding is for a specific project and will largely support direct costs. Certain funders will allow companies the opportunity to justify the necessity of indirect costs for specific projects, but they will expect honesty and rigor, and the decision to grant funding as requested is still at their ultimate discretion. For many companies, costs such as consultants, travel, communication, postage and printing, and computers may fall into a grey area. It may be difficult for a company to quantify their use of these resources for a designated purpose.

The limited objective of project control deserves emphasis. Project management procedures are primarily supposed to spot deviations from the project set up instead of to counsel doable areas for cost savings. This characteristic reflects the advanced stage at that project management becomes necessary. The time at which major cost savings will be achieved is during planning and design for the project. During the actual construction, changes are probably to delay the project and result in immediate price increases. As a result, the main focus of project management is on fulfilling the original design plans or indicating deviations from these plans, instead of checking out important improvements and cost savings. It is only when an operation is needed that major changes will usually occur within the construction setup. Throughout the execution of a project, procedures for project management and record keeping become indispensable tools to managers and different participants within the construction process. The problems related to the integration of would force some discussion. Project management activities and functional concerns are intimately linked, yet the techniques used in many instances do not facilitate comprehensive or integrated consideration of project activities. Especially, managers are typically forced to infer the value of schedule changes, instead of being given with aids for this method. Communication or integration of varied types of data can serve variety of helpful functions, though it will need special attention within the establishment of project management procedures. These tools serve the twin purpose of recording the monetary transactions that occur furthermore giving managers a sign of the progress and issues related to the project. The importance of the cost control is also understood by knowing its advantages in implementing within the project.

(a) **Lower Expenses:** The main benefit of putting cost controls in place is lowering your company's overall expenses. This will allow your company to keep more cash on hand, or to invest larger amounts of money in other ways, such as in capital expenses or paying down debt.

(b) **Gain Operational Efficiency:** Cost-control management can help you clearly identify activities running smoothly and staying within budget from the ones constantly breaking down and consuming extra money.

(c) **Realize Procurement Effectiveness:** Maybe you've bought the same types of materials from the same vendors for many years. Are you really getting the most for your money? A cost-control management review can allow you to look for better bargains on the goods and services that make your business run.

(d) **Streamline Technology:** At this time, regarding each business method has been tormented by technology in a way or another but, in some cases, corporations overspend on technology that is not extremely necessary and, in different cases, they are missing out on ways that to automatize or augment an activity with the proper technical tool. Cost-control management will assist you decide whether or not to change or upgrade.

Budget Preparation

A budget is a projection of revenue and expenses over a specific future amount of time; it is compiled and re-evaluated on a periodic basis. Budgets can be created for an individual, a family, a bunch of individuals, a business, a government, a country, and an international organization or simply concerning anything that produces and spends money. At firms and organizations, a budget is an internal tool employed by management and is usually not needed for coverage by external parties. Once the price of each component is known and fixed then we go for preparation of budget. A budget is telling your money where to go instead of wondering about your expenditure. Budget is used as a baseline against the actual expenditure. Budget consists of all the quantified data of materials, salary of staffs, maintenance cost, taxes and royalties. A budget is like a roadmap for business — it provides an overview of what you will be spending and making over a future time period. A proper budget will include educated estimates as to what the company will make (revenues), and a precise plan for your spending. Following the budget successfully can ensure your business is profitable and achieves its goals. A budget will outline estimated revenues, and then include a plan for expenses that is less than those revenues, so that a profit is expected. A balanced budget means the revenues are equal to expenses. A surplus means revenues exceed expenses, and a deficit means expenses exceed revenues. As a business, budget should always strive to be in a surplus state.

The process starts with assuming various factors to be considered in the upcoming budget period. These assumptions are related to project sales trends, cost trends and the overall economic outlook of the market, industry or sector. Specific factors affecting potential expenses are addressed and monitored. The budget is published in a packet that outlines the standards and procedures used to develop it, including the assumptions about the markets, key relationships with vendors that provide discounts, and explanations of how certain calculations were made. A well-formed budget is essential to the success of the business because it allows to match what the company spends to what it earn. Without a clear plan for spending, it is very easy to outspend revenues over time, which can lead to losses, increases in debt, and the potential closure of business. A budget should guide every single business expenditure. For example, if the management realizes midway through a year that business desperately needs updated computers, they can consult budget to see how much estimated surplus revenue it will generate for the remainder of the year, then explore costs for computer upgrades and see if that fits within the surplus figure while allowing to earn a profit, or alternatively, if budget have the additional revenue to support taking out a loan for the computers. A budget can also help to visualize if we are spending too much, and need to make cuts midway through the year.

The purposes of budgeting are for resource allocation, planning, coordination, management and motivation. It is conjointly very important tool for deciding, observing business performance and foretelling financial gain and expenditure. With proper budgeting, limited resources are managed efficiently. Budgeting is critical in the business planning process. A business owner has got to predict whether or not the corporation will be profitable. Budgeting conjointly enables a model of the potential monetary performance of a business, only if specific strategies and plans are followed. It provides a financial framework for making important decisions. To manage a business effectively, expenditure must be properly controlled. Budgeting conjointly helps measure the forecast business performance against the particular business performance. It permits a business owner or manager to work out whether or not the business lives up to expectations through variation between budgeted and actual expenditure. A particular budget provides data on what proportions a business will pay every month. The usefulness of budgeting depends on the accuracy of available information.

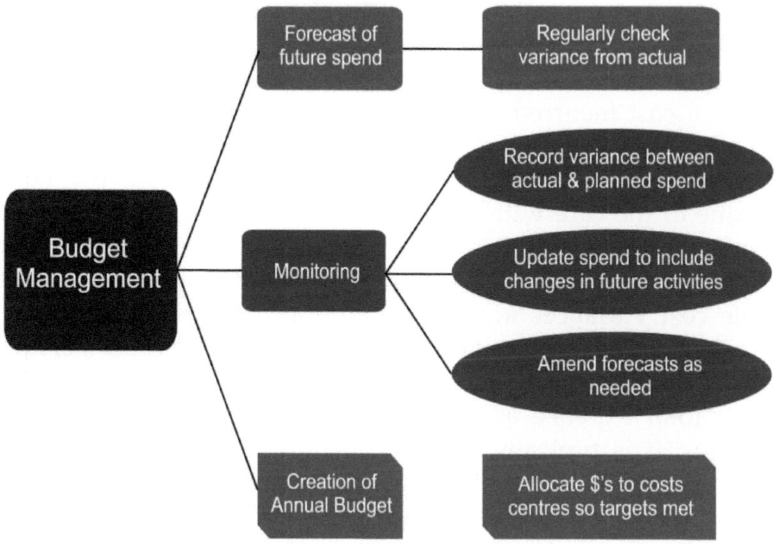

Fig. 1.5: Budget Management in a nutshell

Preparation of Reconciliation

Reconciliation is the process of comparing the estimated or theoretical data before performing work with actual data after considering data relating to variances. Subcontractor Reconciliation as well as Bulk Material Reconciliation is prepared so that the amount of work done by each sub-contractor and usage of bulk material may be properly maintained. Reconciliation is used to ensure that the theoretical value and the actual amount spent or used at the site. Also Reconciliation ensures that we have not paid for the same work twice either to same contractor or two different contractors. A master copy of reconciliation for each type of work like Fabrication, Erection, Anchoring & Rock Bolting so that even if another person is allotted to certify the next bill they have proper and exact info about what work has already been done and paid to which contractor.

Preparation of Budget Control Sheet and Cost Variance

Budget Control sheet is used to show the actual cost of work done, budgeted cost for that work and the estimated cost at the project completion. Making a budget management setup is one of the most effective ways in which to stay to outlay goals. Budget control is a process of budget planning and implementation that is designed to ensure that resources are allocated in an efficient, transparent and secure way.

Cost variance is the distinction between the cost really incurred and therefore the budgeted or planned quantity of cost that ought to incur. Cost variances are most ordinarily tracked for expense line things, however, can also be tracked at the task or project level, as long as there is a budget or standard against which it can be calculated. Some cost variances are formalized into standard calculations. There is an unfavourable cost variance when the actual cost incurred is greater than the budgeted amount. There is a favourable cost variance when the actual cost incurred is lower than the budgeted amount. Whether a variance ends up being positive or negative is partially due to the care with which the original budget was assembled. If there is no reasonable foundation for a budgeted cost, then the resulting variance may be irrelevant from a management perspective. Not all unfavourable cost variances are bad. Spending more money in one area may create a favourable cost variance somewhere else. For example, it might be necessary to spend twice as much on preventive maintenance to avoid a much greater total expense associated with replacing fixed assets more frequently. Thus, it is typically better to review price variances from the amount of an entire department, facility, or single product, instead of at an additional elaborated level. This higher level of research offers managers room during which to portion funds in a manner to boost total profits.

ACWP = AC + UNDER MEASURE – OVER MEASURE

Where ACWP = Actual Cost for Work Performed

AC = Actual Cost Incurred

Under Measure = Work has been done but not paid till now

Over Measure = Work has not been done yet but paid

ETC = BAC – BCWP

Where ETC = Estimated Cost to Complete

BAC = Budgeted Actual Cost

BCWP = Budgeted Cost for Work Performed

EAC = ACWP + ETC

Where EAC = Estimated cost at Completion

CV = BCWP – ACWP

Where CV = Cost Variance

If $CV > 0$, Under Budget

If $CV < 0$, Over Budget

If $CV = 0$, On Budget

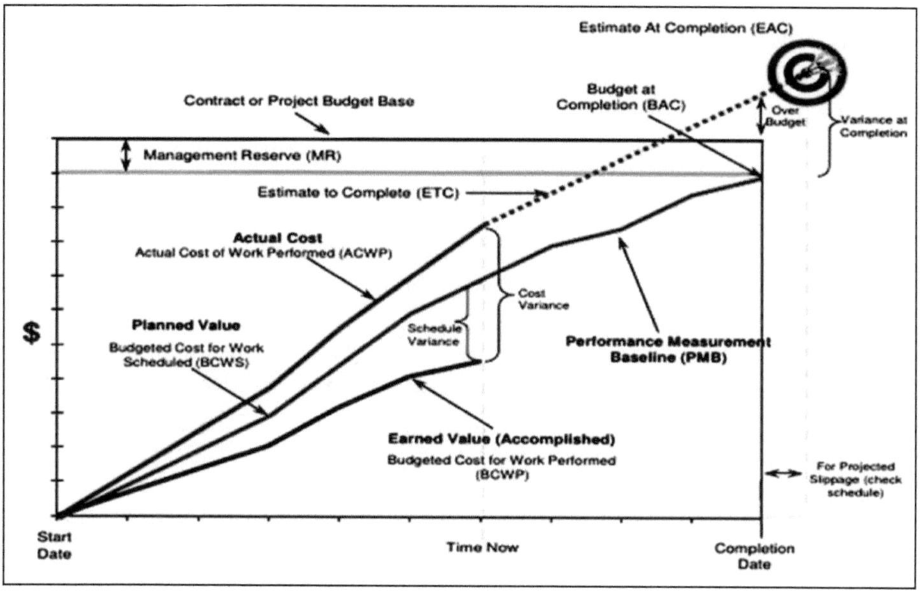

Fig. 1.6: Budgeting terms and various types of cost

Record Maintenance

In cost control part of the project, a no. of paper works is needed to be taken care of so as to get the desired results. Advantages of records management embrace simpler effective management of your current records (both paper and electronic); a reduced/eliminated level of record-keeping redundancies; reduced prices for records storage instrumentation and supplies; and inflated usable workplace area through the elimination of superfluous file storage. Additionally, Records Management provides institutional responsibility and timely access to information. This portion deals with maintaining the papers like Inward/Outward slips, Bill movement slip, Work done abstracts, invoice verification, backup verification etc.

- **Bill movement slip** insures the timely verification of the bill and proper progress of the bill. There is master sheet for bill movement and the delay and current position of bill is intimated by official mail to the responsible person so that they may take required action and process the bill as soon as possible. Ttrack the bill which is dispatched to the Head Office for further processing and another level of check applied on the quantity certified and they then cross check with their reconciliation prepared. Also the debit from various departments are checked like stores, P&M, Manpower Debit and Safety Penalty.
- **Invoice Verification:** In process of maintaining records to verify the invoice and properly maintain the record is one of the important part in this stage we check the sub-contractor name, invoice no., Invoice date, Bill No., proper work order no and its related ammendment if

any, Exact name and address of the civil contractor and GST details. After verifying the above mentioned matter we then move to the verification part of the money which are the previous amount of this bill should match with the total amount of the last bill.

- **Debit Records:** Debit is the sum saved for the company which we recover from the various sub-contractors for our machinery and manpower which they utilise for their designated works. The debit may be of either Plant and Machinery, Manpower Debit, Material Debit or any type of penalty imposed by safety team for not following safety regulations properly.

The very first step in order to prepare a machinery debit is to have a detailed log sheet of that machine which is prepared by the operator himself on a shift basis. This log book has to be counter-signed by respective authorized personnel like section In-charge and Plant In-charge.

Then after we decide whether the work performed by the machine is debit-able to that contractor or not which is confirmed by going through the different clauses of the latest amendment of work order for the concerned contractor. Once the work is confirmed to be debit-able then we get the hiring rate of that machinery from the schedule rates issued by government or the market rate or as discussed with the contractor. The log sheet is then maintained in both hard copy and scanned format, also excel sheet is prepared which is then prepared and attached along with the bill as a backup for deduction made.

Similarly, if any manpower is employed in either cleaning or disposal of muck then it is debited from the excavation contractor at a rate of minimum wages of unskilled labour. The record of manpower employed is gathered from the labour report sheet which is prepared by timekeepers at every site and then sent to the billing department.

- **Daily labour report** is prepared by dedicated timekeepers on shift basis. A timekeeper is responsible for accurately recording the detail of labour along with the work performed by them which is then verified and cross checked by higher authority. These time keeper sheets are consolidated and then sent to the billing department for calculation of debit.

Cost Control Through Field Supervision

The primary task in field supervision work in any construction project is to raise an efficient organisational team to look into the following aspects during implementation of the project.

(a) Inspection of supplies and control of inventory.
(b) Inspection of day-to-day construction activities to control time, workmanship and quality.
(c) Reporting: Weekly, monthly, quarterly as needed.
(d) Scrutinise the reports and take corrective actions.

These are the main items of supervision work, which are to be diligently carried out for an effective control over time as well as cost.

In addition, the field management staff have the onerous duty of looking after the safety aspects to prevent accidents at site, as far as possible, so that work is not interrupted. Any serious accident causes disruption in labour deployment, humanitarian and legal problems and a carry-over effect on the activity performances for a considerable period of time. As a result, work gets delayed and costs escalate. It is therefore the duty of the field management staff to have a sharp eye over machinery operations and movement of materials and labour to prevent accidents. A good idea of safety management and labour rules would help the field management staff also in this respect.

Cost Control Through Communication Network

An efficient communication network at a construction site can overcome locational disadvantage of the site (particularly in hilly, remote areas), ensures speedy procurement of materials from suppliers, ensures control over many sections/site at a time, reduces travelling of field staff and a good co-ordination can be achieved between the authorities and field staff. In effect, time is saved and hence money. Additionally, if computers are used, they can be used profitably as labour-saving devices in preparing data-base site accounting, preparation of wage bills, inventory records, material reconciliation and reporting. Satellite communication facilities, coupled with e-mail and internet, have now made it possible for a construction manager to control and monitor construction activities in far-flung areas inside the country and also sites located abroad.

Estimation

Most of the construction activities are deterministic in nature and their performances can be assessed from records in similar jobs elsewhere. One such source of information is the PWD Schedule of rates and specifications in which the material costs, labour charges and specifications for various items of work may be found. These schedules cover building work, road work, sanitary and plumbing work and river-training work also. Similar schedules are also obtainable from oil companies, port trust authorities, municipal bodies and railways. These should be consulted by an estimator while preparing estimates of any project.

The type and format of the estimates are tailored to suit the requirement of the owner, contractor and supplier as the case may be. Broadly the estimates for construction projects are of the following types:

1. Budgetary estimate

The budgetary estimate is prepared in the planning stage and gives the forecast of the project cost without going into much details. For buildings, costs are generally calculated on plinth-area basis, cost of machinery may be on list price plus cost of operation, maintenance, repair, depreciation as a percentage on list price. For road construction projects, cost may be calculated on 'per km' basis for different types of roads, excluding cost of major bridges. The budgetary estimate is extremely helpful in assessing project finance and also for project appraisal.

2. Detailed item-rate estimate

Detailed item-rate estimates are prepared after the design is done, working drawings prepared and all activities or work-packages are listed and scheduled. The detailed estimates give the quantities of different items of work (like excavation, brick work, concreting, plaster, steel work etc.) with a consolidated rate covering cost of materials, labour, hire charges of equipment, consumables, taxes and contractor's margin of profit. These estimates form the Bill of Quantities (BOQ) for tender purpose. Even when the job is awarded on lump-sum basis, the lump-sum figure is arrived at from the detailed estimates. If the items in the detailed estimate are the same as the activities/work package of the network schedules, these estimates can be conveniently used subsequently during periodic cost monitoring stage.

3. Material estimate

Material estimates show break-up of quantities of materials like cement, brick, reinforcement rods of different diameter, structural steel sections, aggregates etc. for different items of work for procurement/supply purpose. Phase-wise requirement of materials are prepared from the material estimates following the network programmes with a view to finding out cash-flow requirements during implementation of the project.

4. Estimates by detailed costing

For un-scheduled non-stereotype specific jobs like repair, rehabilitation or strengthening of an existing structure, it is advisable to study the sequence of work and calculate materials, labour, hire-charges of equipments and other over-head costs including contractor's margin at each stage of the work. These costs are then added up to find out the total cost.

Rate Analysis For Construction Work

The process of determining rate per unit of any work in Civil Engineering project like earthwork, concrete work, brickwork, plastering, painting etc. is known as Analysis of Rates or simply Rate Analysis. The rates of materials and labour vary from place to place and hence the rates of different items of works also vary from place to place. The rates of these works further help in determining cost of particular work and in turn cost of the project.

Objective of Rate Analysis

- With the help of rate analysis actual cost of items can be determined on the location.
- It helps to cross check the rates offered by contractor.
- To find out the exact quantity of materials and labours required for specific work.
- To finalise the labour contract rates.

Necessity of Rate Analysis

- To determine the actual cost per unit of the items.
- To work out the economical use of materials and processes in completing the particulars item.
- To calculate the cost of extra items which are not provided in the contract bond, but are to be executed as per the directions of the department.
- To revise the schedule of rates due to increase in the cost of material and labour or due to change in technique.

Factors Deciding Rate of Items

The various factors that are involved in determining rate of any item, process or work are mentioned below:

- Specifications of works and material about their quality, proportion and constructional operation method.
- Quantity of materials and their costs.
- Cost of labour and their wages.
- Location of site of work and the distances from source and conveyance charges.
- Overhead and establishment charges.
- Profit and miscellaneous expenses of the contractor.

Determination of the rate of any specified item of work from the known quantities of materials and the labour required, cost of the materials and wages of the labour employed for the completion of the given item of work, is called analysis of rates.

The cost of different materials required and the wages of different labour to be employed for the execution of the project are usually taken from the prevailing market rates. As the cost of the materials and wages of the labour differ from place to place in a large country like India, the rates of different items of same specifications vary considerably. Contractor's profit @10% over the cost of the project is generally allowed and included in the analysis of rates. The rates of materials are taken at the delivery point at the project site. It includes cost as origin, cost of transport and railway freight if any.

Each estimate is submitted along with the analysis of rates for the various items of the work. The analysis of rates includes the following:

1. Cost of material. The cost of the materials is taken at the project site.

2. Wages of the labour. The wages of the different types of labour to be employed for the execution of the various items of the work.
3. T and P (Tool and Plant) and sundries. Generally, 3% of the labour cost is taken as T and P and sundries charges for that item.

Factors Affecting Rate Analysis

1. Materials required for Rate Analysis

Almost in every type of construction materials are the major part. About 50% of the total cost of any work is due to the materials itself. Hence it is very much important that the cost of the materials should be exactly found out. Market surveys should be conducted to analyse the exact cost of each and every material required for the construction.

The quantity of the materials either more or less, both effects the rate analysis. Hence one should maintain proper quantity of the materials on the site.

2. Labour

The number of the labours required for the execution of the work should be properly calculated. Selection of the labours should be done according to their categories like skilled, semi-skilled and unskilled and the payment of the labours should also be decided according to their class. About 30% of the total cost of the work is due to the cost of labours.

3. Item Specification

The specification of the items indicates the proportion of materials, the process and the method of the work. If superior and high quality material is issued then definitely the cost of the materials will be higher.

4. Tools and Plants

In every type of construction work different types of tools and plants are required. It is the duty of the estimator to decide whether buying or hiring of tools and plants will be more economical. It is decided on various factors such as duration of work, number of labours, budget of work, type of work etc. If special type of instruments are used in the execution of work then the cost of the construction will be increased.

5. Overhead Charges

Overhead charges mainly includes expenses of office, labour welfare, material testing, water supply, electrification, depreciation amount etc. This is usually considered as 2-5% of the total cost of the work.

6. Contractor's Profit

Normally 10% of the overall cost of the work is considered to be contractor's profit.

7. Location of the Work

If the location of the site is in the remote areas then it will be very difficult to execute the work. Transportation charges increases as labours have to be transported to the location and the cost of labours also varies as the location is far away from their residence.

8. Miscellaneous

Adverse climatic conditions, completion period of work are also considered to effect the rate analysis. As the effects caused by this directly reflects into rate analysis.

DATA

The process of working out the cost or rate per unit of each item is called as Data. In preparation of Data, the rates of materials and labor are obtained from current standard scheduled of rates and while the quantities of materials and labor required for one unit of item are taken from Standard Data Book.

Schedule of Rates

In order to determine the rate of a particular item, the factors affecting the rate of that item are studied carefully and then finally a rate is decided for that item.

This process of determining the rates of an item is termed as analysis of rates or rate analysis. The rate of particular item of work depends on the following.

1. Specifications of works and material about their quality, proportion and constructional Operation method.
2. Quantity of materials and their costs.
3. Cost of labours and their wages.
4. Location of site of work and the distances from source and conveyance charges.
5. Overhead and establishment charges
6. Profit

Procedure of Rate Analysis

The analysis of rates is worked out for the unit payment of the particular item of work under two heads: Materials and Labour.

- The cost of items of work = Material cost + Labour cost
- Other costs included to the above cost of items of work are:
  - Tools and Plants (T & P) = 2.5 to 3% of the labour cost
  - Transportation cost (if conveyance more than 8 km is considered.)
  - Water charges = 1.5 to 2% 0f total cost
  - Contractor's profit = 10%

The determination of rate of particular item of work per unit from the cost of quantities of materials, cost of labours and other miscellaneous petty expenses required for the completion of the work is known as 'Rate Analysis'. A reasonable profit nearly 10 to 15% for contractor is included in rate analysis.

If the materials are carried from the distant place more than 8 kilometers then the cost of transport is also added. Since rates of materials and labours varies from place to place. The rates of different items of work also vary from place to place.

Cost of materials at source and at site of construction. The costs of materials are taken as delivered at site inclusive of the transport local taxes and other charges.

Requirement of Labour and Materials

Cost of labour-types of labour, standard schedule of rates. The labour can be classified into 1) Skilled 1st class 2) Skilled IInd Class 3) un skilled. The labour charges can be obtained from the standard schedule of rates 30% of the skilled labour provided in the data may be taken as Ist class, remaining 70% as II class. The rates of materials for Government works are fixed by the Superintendent Engineer for his circle every year and approved by the Board of Chief Engineers. These rates are incorporated in the standard schedule of rates. Lead statement: The distance between the source of availability of material and construction site is known as "Lead" and is expected in Km. The cost of convenayce of material depends on lead. This statement will give the total cost of materials per unit item. It includes first cost, convenayce loading, unloading stacking, charges etc. The rate shown in the lead statement are for mettalled road and include loading and staking charges. The environment lead on the metalled roads are arrived by multiplying by a factor (a) for metal tracks - lead × 1.0 (b) For cartze tracks - Lead × 1.1 (c) For Sandy tracks - lead × 1.4 Note: For $1m^3$ wet concrete = $1.52m^3$ dry concrete approximately SP.Wt of concrete= 1440 kg/m^3 (or) 1.44 t/m^3, 1 bag of cement = 50 Kg

Abstract Estimate

This is the third and final stage in a detailed estimate. The quantities and rates of each item of work, arrived in the first two stages, are now entered in an abstract form. The total cost of each item of work is now calculated by multiplying the quantities and respective rates.

Abstract of Estimate Form

Item No.	Description Particular	Quantity	Unit	Rate	Per Unit	Amount

Cash Flow

A cash flow statement provides information about the historical changes in cash and cash equivalents of an enterprise by classifying cash flows into operating, investing and financing activities. It requires that an enterprise should prepare a cash flow statement and should present it for each accounting period for which financial statements are presented.

Objectives of Cash Flow

Cash flow statement shows inflow and outflow of cash and cash equivalents from various activities of a company during a specific period. The primary objective of cash flow statement is to provide useful information about cash flows (inflows and outflows) of an enterprise during a particular period under various heads, i.e., operating activities, investing activities and financing activities. This information is useful in providing users of financial statements with a basis to assess the ability of the enterprise to generate cash and cash equivalents and the needs of the enterprise to utilize those cash flows. The economic decisions that are taken by users require an evaluation of the ability of an enterprise to generate cash and cash equivalents and the timing and certainty of their generation.

Benefits of Cash Flow Statement

Cash flow statement provides the following benefits:

- Cash flow statement when used along with other financial statements provides information that enables users to evaluate changes in net assets of an enterprise, its financial structure (including its liquidity and solvency) and its ability to affect the amounts and timings of cash flows in order to adapt to changing circumstances and opportunities.
- Cash flow information is useful in assessing the ability of the enterprise to generate cash and cash equivalents and enables users to develop models to assess and compare the present value of the future cash flows of different enterprises. It also enhances the comparability of the reporting of operating performance by different enterprises because it eliminates the effects of using different accounting treatments for the same transactions and events.
- It also helps in balancing its cash inflow and cash outflow, keeping in response to changing condition. It is also helpful in checking the accuracy of past assessments of future cash flows and in examining the relationship between profitability and net cash flow and impact of changing prices.

Cash from Operating Activities

Operating activities are the activities that constitute the primary or main activities of an enterprise. For example, for a company manufacturing garments, operating activities are procurement of raw material, incurrence of manufacturing expenses, sale of garments, etc. These are the principal revenue generating activities (or the main activities) of the enterprise and these activities are not investing or financing activities. The amount of cash from operations' indicates the internal solvency level of the company, and is regarded as the key indicator of the extent to which the operations of the enterprise have generated sufficient cash flows to maintain the operating capability of the enterprise, paying dividends, making of new investments and repaying of loans without recourse to external source of financing.

Cash flows from operating activities are primarily derived from the main activities of the enterprise. They generally result from the transactions and other events that enter into the determination of net profit or loss. Examples of cash flows from operating activities are:

Cash Inflows from operating activities

- cash receipts from sale of goods and the rendering of services.
- cash receipts from royalties, fees, commissions and other revenues.

Cash Outflows from operating activities

- Cash payments to suppliers for goods and services.
- Cash payments to and on behalf of the employees.
- Cash payments to an insurance enterprise for premiums and claims, annuities, and other policy benefits.
- Cash payments of income taxes unless they can be specifically identified with financing and investing activities.

Break Even Point (BEP)

In units, BEP = Fixed costs Contribution per unit Contribution per unit/ Total contribution.

Break even charts

The type of graph which might be used to show the effects of changes in costs and revenues (or perhaps more appropriately changes in unit costs and selling prices) is a break even chart (though a PV graph would have the same effect).

The Limitations of Breakeven Analysis

The use of simple break even analysis assumes that all costs can be split into fixed and variable components and furthermore that one activity base (units produced, units sold, hours worked) will be suitable for describing all variable costs and revenues. Linear relationships are assumed. This means that fixed costs are assumed to be fixed at all levels of activity, and that variable cost per unit and sales price per unit are constants and independent of the level of activity. In practice, fixed costs will only be constant over the relevant range. Over a larger range, most fixed costs are stepped. If necessary, this can be adjusted for on the break even chart. It is more difficult to show how variable costs will behave at various activity levels, but the cost accountants linear model is bound to be incorrect over a wide range. For example, as activity increases variable cost may fall because it is possible to buy raw materials cheaper in bulk, or because of increases in worker efficiency. On the other hand, scarcity of resources at high level of activity may lead to higher variable costs. The linear relationship used for revenue totally ignores the sales price/demand relationship for a product. The market may not be perfect, and in order to increase sales of the product it may be necessary to reduce its price. Simple break even analysis also assumes that there is a constant sales mix, or that sales of only one product are being considered. Furthermore, it assumes that there are no changes in stock levels and that units produced equals units sold. This last point will not matter if stock is valued at variable cost, but if an absorption costing system is in use it becomes more difficult to predict profits at different levels of activity if stock levels are changing. Simple break even analysis is therefore most useful when predictions are made within the range of the company normal activity and when there is no significant building up or running down of stocks.

Cost Control

Cost control is therefore concerned with:

- influencing the factors that create changes to the cost plan and ensuring that changes are agreed upon;
- determining that the cost plan has changed;
- managing the actual cost changes as and when they occur;
- monitoring cost performance to detect and understand variances from the cost plan;
- ensuring that all appropriate changes are recorded accurately in the cost plan;
- preventing incorrect, inappropriate or any unauthorized changes from being included in the cost plan;
- taking actions to bring expected costs within acceptable limits.

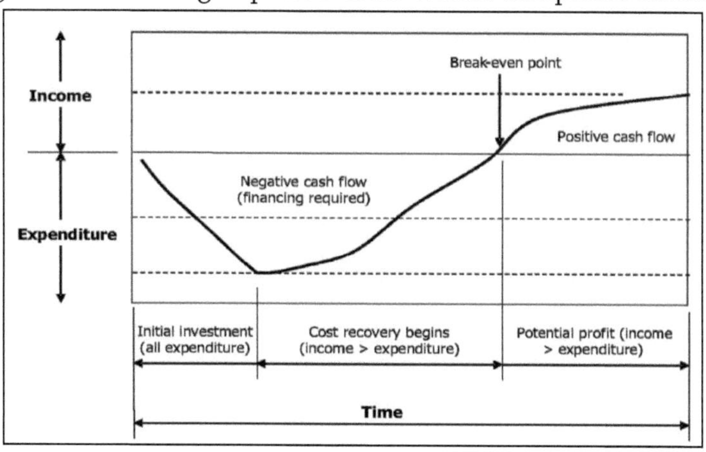

Fig. 1.7

Cost control can be seen to include examining and understanding the reasons for both positive and negative cost variances. It is often integrated with other control processes such as scope change control, schedule control, quality control and so on and requires the following tools:

- a cost plan for the project agreed with the client and key stakeholders;
- performance reports which provide information on the project scope and cost performance such as which budgets have been met and which have not;
- a change request process to capture any changes which occur in many forms: oral and written, direct and indirect, externally or internally initiated, and legally mandated or optional. The changes may require increasing the budget or may even allow it to be reduced in the event of a revised scope of work. A cost change control system defines the procedures by which the cost baseline may be changed and typically includes the paperwork, tracking systems and approval levels necessary for authorizing budget or expenditure changes. The cost change system should also be integrated with the functional or physical change control system.

Tools and Techniques for Cost Control

The available techniques of cost management and forecasting differ with regard to the level of detail and preciseness of the assessment of the actual state and forecasts. In reality, many factors operating during the project lifecycle, such as the time delays, exceeding of costs, changes of orders (and the associated cost changes) influence the real cash flow in the project.

The basic problem of forecasting and managing cash flow is the issue of taking into account these factors in order to obtain accurate control. Furthermore, a second challenge is assessing the actual monetary values associated with the project implementation, such as time delays, cost deviations and so on in the context of the work value. Often assimilating all of these factors is very difficult and, sometimes it has to be said, even impossible. Consequently, cost control typically uses detailed plans and schedules which were devised at the early stage of the project life cycle.

Typical 'S' type budget-planned cost curves in figure 1.8 shows the 'time now' position and the cost value of the work done compared to the budgeted expenditure and the actual costs incurred.

At the 'time now' point, the planned costs are not exceeded by the actual project costs. This position would suggest that while the project is not exceeding the planned rate of spend, it is also not achieving the expected progress for the costs incurred at this time. By conducting a series of regular project reviews it is possible to predict the project cost overrun in time and overspend in monetary terms and then examine options to rectify the situation. The variance – the difference between the planned and the actual performance – may be significant and if so will need to be addressed.

Typical cost/time 'S' curve

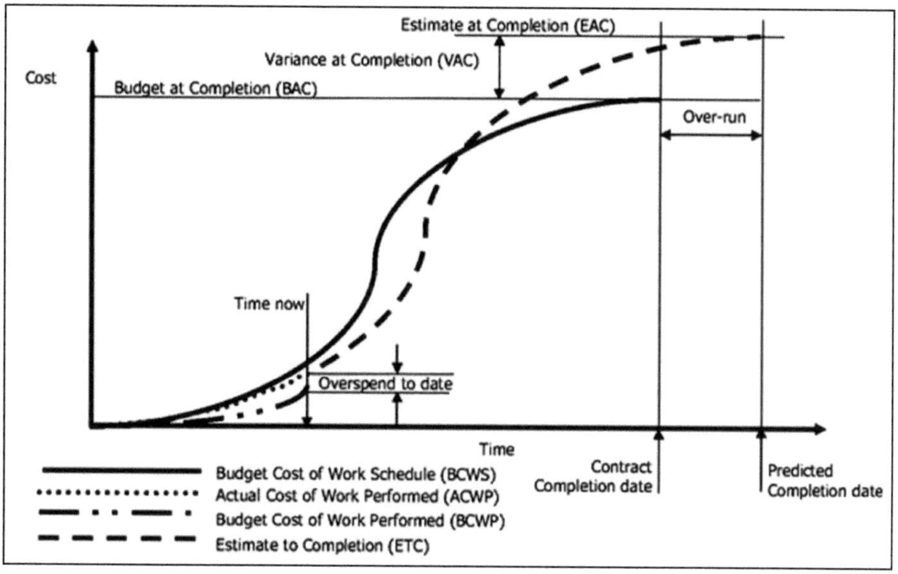

Fig. 1.8

Cost Control in Construction Projects

For cost control on a project, the construction plan and the associated cash flow estimates can provide the baseline reference for subsequent project monitoring and control. For schedules, progress on individual activities and the achievement of milestone completions can be compared with the project schedule to monitor the progress of activities. Contract and job specifications provide the criteria by which to assess and assure the required quality of construction. The final or detailed cost estimate provides a baseline for the assessment of financial performance during the project. To the extent that costs are within the detailed cost estimate, then the project is thought to be under *financial control*. Overruns in particular cost categories signal the possibility of problems and give an indication of exactly what problems are being encountered. Expense oriented construction planning and control focuses upon the categories included in the final cost estimation. This focus is particular relevant for projects with few activities and considerable repetition such as grading and paving roadways.

For control and monitoring purposes, the original detailed cost estimate is typically converted to a *project budget*, and the project budget is used subsequently as a guide for management. Specific items in the detailed cost estimate become job cost elements. Expenses incurred during the course of a project are recorded in specific job cost accounts to be compared with the original cost estimates in each category. Thus, individual job cost accounts generally represent the basic unit for cost control. Alternatively, job cost accounts may be disaggregated or divided into *work elements* which are related both to particular scheduled activities and to particular cost accounts.

In addition to cost amounts, information on material quantities and labor inputs within each job account is also typically retained in the project budget. With this information, actual materials usage and labor employed can be compared to the expected requirements. As a result, cost overruns or savings on particular items can be identified as due to changes in unit prices, labor productivity or in the amount of material consumed.

Material Cost

The rate of various materials as per specifications for the items under consideration can be chalked out from market survey. The costs of materials are taken as delivered at site of work. This is inclusive of:

- The first cost (cost at origin),
- Cost of transport, railway freight (if any), etc.
- Local taxes and other charges.

Lead Statement

The distance between the source of availability of material and construction site is known as "Lead" and is expressed in Km. The cost of conveyance of material depends on lead. This statement is required when a material is transported from a distant place, more than 8 kms (5 miles). The lead statement will give the total cost of materials per unit item including first cost, conveyance loading-unloading, stacking charges etc.

A typical lead statement is provided as follows:

Sl. No.	Materials	Unit	Cost at Source (per unit)	Lead (in Km)	Conveyance Charges (Per Km/ Per Unit)	Total Conveyance Charges (Per Unit)	Total Cost (In Rs./ Per unit)
1.	Rough Stone	Cum	250.00	25	5.00	125.00	375.00
2.	Sand	Cum	12.00	20	4.00	80.00	92.00
3.	Cement	Bag	370.00	Local	-	-	-

Labour Cost

To obtain labour cost the number and wages of different categories of labourers, skilled (Skilled 1st Class), semi-skilled (Skilled 2nd Class) and unskilled, required for each unit of work should be known and this number is multiplied by the respective wage per day. The labour charges can be obtained from the standard schedule of rates. 30% of the skilled labour provided in the data may be taken as 1st class, remaining 70% as 2nd class.

The length of time required to do a certain piece of the work may vary according to the skill and mental development of the workmen and working conditions to the particular job.

(a) Task or out-turn work

This is the quantity of work which can be done by an artisan or skilled labour (with the help of semiskilled and unskilled labours) of the trade working for 8 hours a day. The out-turn of work per artisan varies according to the nature, size, height, situation, location etc. Out-turn is more in larger cities, as the more specialized and experienced labours are available, than the small cities and country sides.

Out turn of Artisans

The out turn of work per artisan varies according to their nature, size, height, situations, etc. The following out turns are generally accepted for an average artisan per day for determining the labour cost.

Table 1.1

S. No.	Item	Quantity	Per day
1.	Brickwork in lime or cement mortar in foundation and plinth	1.25 m^3	45 ft^3 per mason
2.	Brickwork in cement mortar in superstructure	1.00 m^3	35 ft^3 per mason
3.	Brickwork in mud mortar in foundation and plinth	1.50 m^3	55 ft^3 per mason
4.	Brickwork in mud mortar in super structure	1.25 m^3	45 ft^3 per mason
5.	Half brick wall in partition	5.00 m^2	50 ft^2 per mason
6.	Lime concrete in foundation or floor	8.50 m^3	300 ft^3 per mason
7.	Lime concrete in roof terracing	6.00 m^3	200 ft^3 per mason
8.	Cement concrete (1:2:4)	5.00 m^3	175 ft^3 per mason
9.	R.B. work	1.00 m^3	35 ft^3 per mason
10.	RCC work	3.00 m^3	125 ft^3 per mason
11.	12mm plastering with cement	8.00 m^2	80 ft^2 per mason
12.	Earthwork in excavation (ordinary soil)	3.00 m^3	100 ft^3 per beldar
13.	Earthwork in excavation (hard soil)	1.00 m^3	35 ft^3 per beldar
14.	White washing	70.00 m^2	700 ft^2 per washer
15.	Numbers of bricks laid per mason in brickwork up to a height of 3 m (10ft)	500 bricks per mason	

Table 1.2: Materials for different items of work

1.	Bricks (9" × 4.5" × 3") Bricks (20cm × 10 cm × 10 cm)	1350 for 100 cu ft 500 for one cu m
2.	Dry mortar for brickwork 30%	30 cu m for100 cu m
3.	Bricks for brick ballast lime concrete	1050 for 100 cu ft 37000 for 100 cu m
4.	Dry mortar for lime concrete in foundation and floor 35%	35 cu ft for 100 cu ft 35 cu m for 100 cu m
5.	Materials for cement concrete (1:2:4)	Cement 22 cu ft. For 100 cu ft sand 44 cu ft for 100 cu ft ballast 88 cu ft for 100 cu ft

(Table Contd...)

6.	Bricks for R.B work	1200 for 100 cu ft
7.	Dry material for 12 mm plastering	6 cu ft for 100 sq ft 2.0 cu m for 100 sq ft
8.	Lime for white washing one coat	1 kf for 100 sq m 10 kg for 100 sq ft
9.	Bricks (20×10×10 cm) for brick flooring or half brick wall	500 for 100 sq ft 5000 for 100 sq m
10.	Bricks for brick flat floor	350 for 100 sq ft 3500 for 100 sq m

Out-turn or Task

Particulars of items	Quantity of work per day (8 hrs. a day)
(1)	(2)
1. Earthwork in excavation in foundation in ordinary soil, lead up to 50m and lift up to 1.5 m	3.00 cum per mazdoor/Beldar
2. Earthwork in excavation in hard soil for 100 m lead and 1.5 m lift	2.00 cum per mazdoor/Beldar
3. Excavation in rock	1.00 cum per mazdoor
4. Sand filling in plinth	4.00 cum per mazdoor
5. Breaking of brick ballast 40 mm gauge	0.75 cum per labour/breaker
6. Breaking of stone ballast 40 mm gauge	0.40 cum per labour
7. Breaking of stone ballast 20 mm gauge	0.25 cum per labour
8. Brickwork in cement mortar in foundation and plinth	1.25 cum per mason
9. Brickwork in cement mortar in superstructure	1.00 cum per mason
10. Half brick wall in partition	5.00 square meter per mason
11. Brick work in cement mortar in arches	0.55 cum per mason
12. Lime concert in foundation/flour	8.50 cum per mason
13. Lime concreating in roof terracing	6.00 cum per mason
14. Cement concrete (1:2:4)	5.00 cum per mason
15. R.C.C. work	3.00 cum per mason
16. 12 mm plastering with cement morter	8.00 square meter per mason
17. Pointing with cement/line mortar	10.00 sq.m per mason
18. 25 mm I.P.S. (cement concrete) floor	7.50 sqm per mason
19. Terrazo floor 6 mm thick mosaic work over 20 mm cement concrete (1:2:4)	5.00 sq.m. per mason
20. Brick flat floor in cement or lime mortar	8.00 sq. m per mason
21. Timber framing sal or Teak wood	0.07 cum per carpenter
22. Timber framing in country wood	0.15 cum per carpenter
23. Door and window shutters panelled or glazed 24. White washing or colour washing one coat 25. White washing or colour 3 coats	0.15 sq.m per carpenter 200 sq.m white washer 70 sq.m per white washer

(Contd…)

(1)	(2)
26. Painting or varnishing doors or windows one coat	25 sq. m per painter
27. Distempering one coat	35 sq.m. per painter
28. Amount of work done by a mazdoor (helper) per day. (i) Mix (ii) Delivery bricks (iii) Delivery mortar	 3 cum per mazdoor 4000 to a distance of 15 m per mazdoor 5.5 cum of brick work

The recommendation of All India Standard Schedule of Rates and various other govt. reports are used to work out approximate quantity of labour required to prepare the analysis of rates. IS: 7272 (part 1)-1974, provides recommendations for labour output constants for building work which can be used to fix up the labour cost.

A typical labour output constant issued by National Building Organization is provided below:

Labour Requirements

Description of work	Quantity	Labour
(1)	(2)	(3)
1. Earthwork in excavation in foundation, trenches etc. in ordinary soil including disposal up to 30m and lift of 1.5 m	28.30 m^3 (1000 cft)	Beldar - 5 nos. Mazdoor - 4 nos.
2. Refilling of excavated earth in foundation, plinth etc. including consolidation in 150 mm layer	28.30 m^3 (1000 cft)	Beldar-3 nos. Mazdoor-2 nos. Bhisti-0.5 nos.
3. Laying cement concrete	2.83 m^3 (100 cft)	Beldar-2 nos. Mazdoor-3 nos. Bhisti-3/4 nos. Mason/¼ nos.
4. Laying of R.C.C. work	2.83 m^3 (100 cft)	Beldar-3 nos. Mazdoor-3 nos. Bhisti-1.5 nos. Mason-0.5 no.
5. Reinforcement work for R.C.C	1 quintal	Blacksmith-1 no. Beldar-1 no.
6. First class Brickwork in 1:4 cement morter in superstructure	2.83 m^3 (100 cft)	Mason-2.25 nos. Mazdoor-4.5 nos. Bhisti-0.5 no.
7. Wood work in door/window fromes	0.18 m^3	Carpenter-2 nos. Beldar-1 nos.
8. Wood work in panelled, glazed shutters etc.	0.30 m^3	Carpenter-15 nos. Beldar-4 nos.
9. 40 mm cement concrete flooring	40 m^2	Mason-5 nos. Beldar-4 nos. Mazdoor-3 nos. Bhisti-1 no.

(Contd...)

(1)	(2)	(3)
10. 12 mm cement mortar plastering	40 m^2	Mason-3 nos. Mazdoor-3 nos. Bhisti-1 no.
11. Three coats white washing/colour washing	60 m^2	White washer-1 no. Mazdoor-1 nos.
12. Two coats painting on wood or steel	10 m^2	Painter-3 nos. Mazdoor-2 nos.

Miscellaneous cost

(a) Cost of equipment, Tools and Plants (T & P)

The cost of equipment and ordinary tools and plats and miscellaneous petty items (sundries) are added to the specific item rate as lump-sum. A provision of 2.5 to 3% of the labour cost is made for such items. In certain tools and plants if it is difficult to allocate their use for a particular item of rate; then the cost of such tools or plants may be allocated to the over-head expenditure.

For big works and projects where it becomes necessary to use special types of equipment like batching plants or WMM plant or dumpers or cranes for transportation of concrete mix, provisions of an amount 1% to 1.5% of the estimated cost is provided in the estimate under the head "special tools and plants".

(b) Water charges

For drinking purpose of the workers and for the work, arrangement of water is made sinking tube well; bore well or from temporary connection from municipality. For this purpose a provision of 1.5 to 2% of total cost (Material + Labour + Sundries) is made in the estimate.

(c) Over head charges

Overhead charges include general office expanses, rents, taxes, supervision and other cost which are indirect expanses on the job. Expanses for small tools such as planks, ladders, ropes and other hand tools are also included in the over-head charges. A provision of 2.5% to 5% is made in the rate analysis as overhead charge. Overhead charges can be divided under two categories: General Overhead and job overhead.

General overhead

These are the expanses made throughout the year irrespective to running works in hand. These include:

- Establishment charge including rent of office space and taxes
- Salaries to office staff
- Purchase of stationary, Printing, postage etc.
- Electricity, telephone and water bills
- Travelling expanses

Job overhead

These are the expanses indirectly incurred for the job or the project. These include:

- Salaries of personnel engaged for the work (Site engineers, Surveyors or site office staff)
- Rent of temporary site office space, electricity, telephone and water bills
- Handling of materials
- Repairs, carriage and depreciation of T & P.
- Labour welfare, safety measures and insurance etc.
- Interest on investment
- Theft and other losses.

(c) Contractor's profit

Generally a provision of 10% is made in the rate analysis as contractor's profit for ordinary contracts. For small jobs 15% profit and for large jobs 8% profit may be considered as reasonable. Contractors profit is not included in rate analysis if material is supplied by the department.

Illustration: Rate Analysis of Some Important Items

Earthwork in excavation in foundation including filling in trenches up to 30 m lead and 1.5 m lift

Assume volume of excavation = 100 cu m

Particulars	Qnty/Nos.	Rate (Rs.)	Cost (Rs.)
Material Charges	-	-	-
Labour Charges			
1. Head Mason	D Nos.	450.00 per day	225.00
2. Beldar	18 Nos.	250.00 per day	4500.00
3. Mazdoor	14 Nos.	220.0 per day	3080.00
T&P, Sundries, etc.	LS	240.00 LS	240.00
		Total Materials and Labour	**8045.00**
		Add 1.5% water charges	120.67
		Add 10% Contractors profit	804.50
		Grand Total	8970.17
		Rate per cu m	Rs. 89.70

First class brickwork in super structure with cement mortar (1:6)

(a) Estimation of Materials

Assume volume of brickwork = 10 cu m

Nominal size of modular brick = 10 cm×10 cm× 20 cm Hence, the number of bricks required =

Actual size of modular brick = 9 cm×9 cm×19 cm

The remaining space is filled by mortar, hence the volume of mortar required for 10 cum

= 10 – (5000 × 0.09 ×0.09 × 0.19) = 2.3 cu m.

Additional mortar required for frog filling, brick bonding and wastages @ 15%. Thus volume of set mortar = 2.3 + 2.3 × 15/100 = 2.64 cum.

But, 1.25 cu m of dry volume of mortar materials produces 1.0 cu m set mortar.

Hence, volume of dry materials required for 2.64 cu m of set mortar = 1.25 × 2.64 cu m = 3.30 cu m.

[Note: As a thumb rule, dry volume of mortar materials is 30% of brick work] Sum of proportion of cement and sand = 1+6 = 7

Hence, volume of cement = 3.3/7 = 0.47 cu m.

However, cement is available in 50 kg bag whose volume is 0.0347 cu m.

[Mass = 50 kg; Density =1440 kg/m^3; Thus, Volume = 50/1440 = 0.0347 cu m] [Thumb rule: 1 cu m of cement = 30 bags of cement.]

Therefore, number of bags required = 0.47/0.0347 ≈ 13.5 bags.

Volume of sand required = 0.47 × 6 = 2.82 cu m.

(b) Rate Analysis

Assume, the volume of brickwork = 10 cu m.

Particulars	Qnty/Nos.	Rate (Rs.)	Cost (Rs.)
Material Charges			
1. Brick	5000 Nos.	250.00 (/100 nos.)	12500.00
2. Cement	13.5 bags	320.00 per bag	4320.00
3. Sand	2.82 cu m	350 per cu m	987.00
Labour Charges			
1. Head Mason	2 Nos.	450.00 per day	900.00
2. Mason	6 Nos.	350.00 per day	2100.00
3. Mazdoor	16 Nos.	220.00 per day	3520.00
4. Bhisti	08 Nos.	220.0 per day	1760.00
T&P, Sundries, etc.	LS	200.00 LS	200.00
		Total Materials and Labour	**26287.00**
		Add 1.5% water charges	394.30
		Add 10% Contractors profit	2628.70
		Grand Total	**29310**
		Rate per cu m	**Rs. 2931.00**

12 mm thick plaster with cement mortar (1:6)

(a) Estimation of Materials

Assume plastering area = 100 sq m

Hence volume of mortar for 12 mm plaster = 100 m × 0.012 m = 1.2 cum

Add 30% more to the above volume for filling of joints, for making uniform surface well and for wastages

Thus total set volume of mortar including wastages and joint filling etc. = 1.2 + 1.2 × 30/100 = 1.56 cu m.

As, 1.25 cu m of dry volume of mortar materials produces 1.0 cu m set mortar;

Volume of dry materials required for 1.56 cu m of set mortar is
= 1.25 × 1.56 cu m = 1.95 cu m,
Hence, volume of cement = 1.95/7 = 0.28 cu m.
Number of bags required = 0.28/0.0347 ≈ 8 bags.
Volume of sand required = 0.28 × 6 = 1.68 cu m.

(b) Rate Analysis

Assume, the area of plastering = 100 sq. m.

Particulars	Qnty/Nos.	Rate (Rs.)	Cost (Rs.)
Material Charges			
1. Cement	8 bags	320.00 per bag	2560.00
2. Sand	1.68 cu m	350 per cu m	588.00
Labour Charges			
1. Head Mason	2 Nos.	450.00 per day	900.00
2. Mason	6 Nos.	350.00 per day	2100.00
3. Mazdoor	08 Nos.	220.00 per day	1760.00
4. Bhisti	02 Nos.	220.0 per day	440.00
T&P, Sundries, etc.	LS	200.00 LS	130.00
		Total Materials and Labour	**8478.00**
		Add 1.5% water charges	127.17
		Add 10% Contractors profit	847.80
		Grand Total	**9452.97**
		Rate per sq m	**Rs. 94.53**

Cement Concrete (1:2:4) for RC work excluding reinforcement and form work

(a) Estimation of Materials

Assume volume of R.C.C. = 10 cu m (Set volume)

1.54 cu m dry volume of concrete making materials produces 1.0 cu m set concrete

Therefore volume of dry materials required for 10 cu m of set concrete is 15.4 cu m.

Sum of proportion of cement, sand and course aggregate = 1+2+4 = 7
Hence, volume of cement = 15.4/7 = 2.2 cu m.

Number of bags required = 2.2/0.0347 ≈ 64 bags.

Volume of sand required = 2.2 × 2 = 4.4 cu m.

Volume of coarse aggregate required = 2.2 × 4 = 8.8 cu m.

(b) Rate Analysis

Assume, volume of R.C.C. = 10 cu m.

Particulars	Qnty/Nos.	Rate (Rs.)	Cost (Rs.)
Material Charges			
1. Cement	64 bags	320.00 per bag	20480.00
2. Sand	4.4 cu m	350 per cu m	1540.00
3. C. aggregate	8.8 cu m	800 per cu m	7040.00
Labour Charges			
1. Head Mason	Nos. (½)	450.00 per day	225.00
2. Mason	2 Nos.	350.00 per day	700.00
3. Beldar	10 Nos.	220.00 per day	2200.00
4. Mazdoor	10 Nos.	220.00 per day	2200.00
5. Bhisti	05 Nos.	220.0 per day	1100.00
T&P, Sundries, etc.	LS	200.00 LS	200.00
Scaffolding	LS	400.00 LS	400.00
		Total Materials and Labour	**36085.00**
		Add 1.5% water charges	541.28
		Add 10% Contractors profit	3608.50
		Grand Total	**40234.78**
		Rate per sq m	**Rs. 4023.50**

Note: If concrete mixture is employed for mixing of concrete, hiring and running charges may add @ Rs. 100.00 per cu m of concrete; but the labour may be reduced by 2 beldars per 10 cu m of concrete.

Lime Concrete in foundation with 25 mm down brick chips (or jhama chips) with lime surkhi mortar (1:2:5½)

(a) Estimation of Materials

Assume volume of lime concrete = 10 cu m (Set volume)

1.54 cu m dry volume produces 1.0 cu m set concrete

Therefore volume of dry materials required for 10 cu m of set lime concrete is 15.4 cu m. Sum of proportion of cement, sand and course aggregate = 1+2+5½ = 8½

Hence, volume of slaked lime = 15.4/8½ = 1.8 cu m. Volume of surki required = 1.8 × 2 = 3.6 cu m.

Volume of jhama brick chips required = 1.8 × 5½ = 10 cu m.

(b) Rate Analysis

Assume, volume of R.C.C. = 10 cu m.

Particulars	Qnty/Nos.	Rate (Rs.)	Cost (Rs.)
Material Charges			
1. Slaked lime	1.8 cum	600.00 per cum	1080.00
2. Surki	3.6 cu m	250.00 per cu m	900.00
3. Brick chips	10.0 cu m	350.00 per cu m	3500.00
Labour Charges			
1. Head Mason	D Nos.	450.00 per day	225.00
2. Mason	1 Nos.	350.00 per day	350.00
3. Mazdoor	18 Nos.	220.00 per day	3960.00
4. Bhisti	02 Nos.	220.0 per day	440.00
T&P, Sundries, etc.	LS	300.00 LS	150.00
		Total Materials and Labour	**10605.00**
		Add 1.5% water charges	159.08
		Add 10% Contractors profit	1060.50
		Grand Total	**11824.58**
		Rate per sq m	**Rs. 1182.50**

Note: In case of cement concrete in foundation, the labours and T&P will be same as this item. The materials like cement, sand and coarse aggregate can be calculated and accordingly rate analysis can be made.

Providing cold twisted steel reinforcement in R.C.C. slab including bending, binding and placing in position complete

(a) Estimation of Materials

If bar bending schedule is available, then reinforcement quantity may be estimated from the schedule. Alternatively, reinforcement steel for beams and slabs may be taken as @ 1% of volume of concrete and for columns @ 2% of volume of concrete. The weight of 1 cum of steel is 78.5 quintals.

Consider, first 10 m × 10 m of continuous slab of thickness 100 mm.

The volume of reinforced concrete = 10 m × 10 m × 0.1 m = 10 cu m

Reinforcement required by volume = 10×1/100 = 0.1 cu m

Weight of reinforcement required = 0.1 × 78.5 qu. = 7.85 qu.

Increase this amount by 5% for wastages.

Thus the volume of reinforcement required = 7.85 × 5/100 = 8.25 qu.

Black iron wire @ 1kg per quintal = 8.25 kg.

(b) Rate Analysis

Assume, volume of R.C.C. slab = 10 cu m.

Particulars	Qnty/Nos.	Rate (Rs.)	Cost (Rs.)
Material Charges			
1. Reinforcement	8.25 qu.	3800.00 per qu	31350.00
2. Black Iron wire	8.25 kg	45.00 per kg	371.25
Labour Charges			
1. Blacksmith	8.25 Nos.	450.00 per day	3712.50
2. Mazdoor	8.25 Nos.	220.00 per day	1815.00
T&P, Sundries, etc.	LS	300.00 LS	130.00
		Total Materials and Labour	**37378.75**
		Add 1.5% water charges	560.70
		Add 10% Contractors profit	3737.88
		Grand Total	41677.33
		Rate per cu m	Rs. 4167.75

Note: R.C.C. works are paid separately for cement concrete work; for steel reinforcement and for centering and shuttering as per the PWD practices.

25 mm thick cement concrete (1:2:4) damp proof course.

(a) Estimation of Materials

Assume area of DPC is = 100 sq m

The volume of concrete will be = 0.025*100= 2.5 cum.

The quantity of cement, sand and coarse aggregates required for 2.5 cu m concrete are estimated as:

Number of cement bags required = 16½ bags.

Volume of sand required = 1.10 cu m.

Volume of coarse aggregate required = 2.20 cu m.

Quantity of water proofing compound required = 3% by weight of cement = = 3% of 16½×50 kg = 25 kg.

(b) Rate Analysis

Assume, area of DPC = 100 sq m.

Particulars	Qnty/Nos.	Rate (Rs.)	Cost (Rs.)
Material Charges			
1. Cement	16½ bags	320.00 per bag	5280.00
2. Sand	1.1 cu m	350.00 per cu m	385.00
3. C. aggregate	2.2 cu m	800.00 per cu m	1760.00
4. Water proof compound	25 kg	25.00 per kg	625.00
Labour Charges			
1. Head Mason	½ Nos.	450.00 per day	225.00
2. Mason	08 Nos.	350.00 per day	2800.00
3. Mazdoor	08 Nos.	220.00 per day	1760.00
4. Bhisti	01 Nos.	220.0 per day	220.00

(Contd...)

T&P, Sundries, etc.	LS	500.00 LS	100.00
		Total Materials and Labour	**13155.00**
		Add 1.5% water charges	197.33
		Add 10% Contractors profit	1315.50
		Grand Total	**14667.83**
		Rate per sq m	**Rs. 146.70**

Random Rubble Masonry in cement mortar (1:6) in foundation and plinth.

(a) Estimation of Materials

11.7 cu m of undressed stone and 0.80 cu m of through stone (header) is required for 10 cu m of RR masonry. Further, 4.2 cum of dry mortar materials (cement and sand) are required for same volume of RR masonry work.

(b) Rate Analysis

Assume, volume of RR masonry = 10 cu m.

Particulars	Qnty/Nos.	Rate (Rs.)	Cost (Rs.)
Material Charges			
1. Cement	17 bags	320.00 per bag	5440.00
2. Sand	3.6 cu m	350.00 per cu m	1260.00
3. Undressed Stone	11.7 cu m	200.00 per cu m	2340.00
4. Through Stone	0.8 cu m	250.00 per cu m	200.00
Labour Charges			
1. Head Mason	½ Nos.	450.00 per day	225.00
2. Mason	10 Nos.	350.00 per day	3500.00
3. Mazdoor	17 Nos.	220.00 per day	3740.00
4. Bhisti	2 Nos.	220.0 per day	440.00
T&P, Sundries, etc.	LS	200.00 LS	200.00
		Total Materials and Labour	**17345.00**
		Add 1.5% water charges	260.17
		Add 10% Contractors profit	1734.50
		Grand Total	**19339.67**
		Rate per cu m	**Rs. 1934.00**

Rule pointing in cement mortar (1:3) on brickwork on wall.

(a) Estimation of Materials

An empirical quantity of 0.63 cu m (dry) mortar is required for 100 sq. m of Rule and Tuck pointing. In case of Flush pointing 75% of above quantity is required.

(b) Rate Analysis

Assume, area of Rule pointing = 100 sq m.

Particulars	Qnty/Nos.	Rate (Rs.)	Cost (Rs.)
Material Charges			
1. Cement	4.8 bags	320.00 per bag	1536.00
2. Sand	0.48 cu m	350.00 per cu m	168.00
Labour Charges			
1. Head Mason	½ Nos.	450.00 per day	225.00
2. Mason	10 Nos.	350.00 per day	3500.00
3. Mazdoor	09 Nos.	220.00 per day	1980.00
4. Bhisti	1 Nos.	220.0 per day	220.00
T&P, Sundries, etc.	LS	120.00 LS	120.00
Scaffolding	LS	360.00 LS	360.00
		Total Materials and Labour	**8109.00**
		Add 1.5% water charges	121.64
		Add 10% Contractors profit	810.90
		Grand Total	**9041.54**
		Rate per sq m	**Rs. 90.42**

Rate analysis of plastering with cement mortar requires the quantity estimation of materials cement mortar, i.e. quantity of cement, sand and water for various grades of mortar required. Grades of mortar used for plastering is generally same grade as used in the masonry work. There are different grades of mortar that can be used for plastering of masonry structural members such as in CM 1:2, 1:4, 1:6, 1:8 etc. Cement Mortar in ratio 1:6 is generally used for plastering work. For the calculation of quantity of cement and sand in mortar, volume of mortar required for plastering is calculated based on thickness of plastering surface and surface area of the structural member.

Let us take an illustration of a wall to be plastered: Length of wall = 2m, Width of wall = 1.5m. Thickness of plaster to be used = 20mm Plaster to be carried out in two layers of 10 mm each. The quantity of cement mortar required will be: $2 \times 1.5 \times (20/1000) = 0.06$ m^3 of mortar. The rate analysis of mortar need to be done for the calculated quantity required. Note here the rate analysis of cement mortar to know how to calculate quantity of cement and sand in mortar. Once the quantity of cement and sand is calculated, the labour cost required for mixing of mortar is calculated. The coefficients of labour for plastering work are taken from the Rate Analysis by CPWD. The cost of mixing of calculated quantity of mortar remains constant, but the cost of application of mortar varies with number of layers in which the mortar is applied. Thus, the cost of mason depends on area to be plastered and number of layers of plaster to be applied. The cost of other labors varies with the quantity of cement mortar to be mixed and number of layers of plasters to be applied.

The coefficients of mason and labours are as follows per m^2 area per layer of plaster to be applied: Mason - 0.07435 days, Labours - 0.0929 days, Bhishti - 0.0929 days (the one who carries materials and water for mixing). These coefficients are multiplied with the given quantity of cement mortar required for plaster in m^3 and number of layers or plasters. The number of days of mason required = 0.07435 × No. of layers x surface area = 0.07435 × 2 × (2 ×1.5) = 0.44610 days. For Labours and Bhishti = 0.0929 × 2 × (2 × 1.5) = 0.5574 days for each. The labour and bhishti days required for mixing of cement mortar is also calculated as given in rate analysis of cement mortar. For example, labour required = 0.21 × 0.06 = 0.0126 days Bhishti required = 0.0929 × 0.06 = 0.005574 days. Thus total number of days for plastering in two-layers of plaster for 1.5m × 2m × 20mm plaster: Mason = 0.44610 days Labour = 0.5574 + 0.0126 = 0.57 days Bhishti = 0.5574 + 0.005574 = 0.562974 days. The daily wages of masons and labours are multiplied with number of days required by them to get the cost.

Data required for Rate Analysis of RCC

1. Estimation of materials

Material estimation include sand, cement, coarse aggregate and steel for a particular mix design. Let us consider a mix design of 1:1.5:3 for our estimation practice. The dry volume of total materials required is considered as 1.54 times the wet volume of concrete, due to voids present in sand and aggregates in dry stage. Therefore, for our calculation, we will consider the total volume of materials required as 1.54 m^3 for 1 m^3 of wet concrete.

(a) Bags of cement required

Volume of cement required for 1m³ of Concrete $= \frac{1 \times 1.54}{1+1.5+3} = 0.28\ m^3$

Then number of bags of cement (volume of one bag of cement =

0.0347 m³) $= \frac{0.28}{0.0347} = 8.07$ bags of cement.

(b) Volume of Sand required

Volume of sand required $= \frac{1.5 \times 1.54}{1+1.5+3} = 0.42\ m^3$ of sand.

(c) Volume of Coarse Aggregate Required

Volume of Coarse Aggregate $= \frac{3 \times 1.54}{1+1.5+3} = 0.84\ m^3$ of coarse aggregates.

(d) Estimation of Reinforced Steel

Quantity of steel required depends on components of structure, i.e. slabs, beams, columns, foundations, roads etc. To estimate the steel required, there are two methods. First method is, when we have the drawing available, we can calculate the total weight of steel required divided by total volume of concrete for different components. This will give us the weight of reinforcement steel per cubic meter of concrete. Second method is assuming the percentage of reinforcement for different components. Following are the percentage of reinforcement steel generally required per different components. Its values can vary from structure to structure, and can be assumed from past experiences of similar structure.

- For slabs = 1.0% of concrete volume.
- For Beam = 2% concrete volume.
- For column = 2.5% of concrete volume.
- For RCC Roads, 0.6% concrete volume.

Lets take example of RCC Column, where reinforcement required is 2.5% of concrete volume, weight of steel required will be:

$$\frac{2.5 \times 7850}{100 \times 1} = 196.25 \text{ kg.}$$

2. Labour Requirement for 1m^3 of RCC:

Labours required are presented in terms of days required by particular labour to complete its work for the given quantity of concrete. Following are the various labours required: (a) Mason: As per Standard Schedule of Rates and Analysis of Rates, One mason is required for 0.37 days. (b) Labours: One Unskilled labours required for 3.5 days. (c) Water carrier: One water carrier required for 1.39 days. (d) Bar Bender: Bar bender requirement depends on weight of reinforcement. Lets consider one bar bender required for 100 kg of steel as for 1 day. (e) Mixer Operator: One mixer operator required for 0.0714 days. (f) Vibrator Operator: One vibrator operator required for 0.0714 days.

3. Equipment and sundries

Equipment and other charges, such as water charges, miscellaneous items, tools and tackles etc. can be assumed as some percentage of total cost of materials and labours. Let us say it as 7.5%.

4. Contractor's Profit

Contractor's profit depends on place to place, organization to organization and work to work. It varies from 10-20%. Let us assume it as 15% of total cost of materials, labours and equipment. We have calculated the quantity of every item in above 1-3 steps. For rate analysis of RCC, we need to multiply each quantity with their rates to get the amount for every item of work. Rates vary from place to place and time to time. It is advisable to assume local rates or standard rates of the place. The sum total of all the four items above will give the rate or cost for 1m^3 of concrete.

Rate analysis for cement mortar requires estimation of materials for cement mortar, i.e. quantity of cement and sand required for 1m^3 for various proportions, i.e., CM 1:2, 1:4, 1:6, 1:8 etc. So, let us first calculate quantity of cement and sand required for 1m^3 of cement mortar. For 1m^3 of wet cement mortar, 1.3m^3 of cement and sand is required, due to presence of voids in sand in dry state. So, cement and sand quantity is calculated for 1.3 m^3 of cement mortar.

Estimation of Materials for Cement Mortar:

Steps for calculation of quantity of cement and sand for cement mortar: Consider, we want to calculate quantity of cement and sand for CM 1:X, where, 1 is proportion of cement and X is proportion of sand. Then, quantity of cement shall be calculated as: Quantity of cement

$= \frac{1.3 \times 1}{1 + X}$ Above formula gives quantity in volume.

To calculate the quantity of cement in number of bags, we have to divide it by volume of cement of one bags. The volume of one bag of cement is 0.0347 m^3.

Therefore, number of bags of cement = Quantity of sand = For example, for a proportion of CM 1:4,

Quantity of cement in bags = =7.492 bags

Quantity of sand = 1.04 m^3 of sand.

Estimation of Labour and Tools for cement mortar:

For 1m^3 of cement mortar, semi-skilled labour is required for 0.27 days for mixing, unskilled labour for 0.26 for carrying of cement, sand and water. Hire charges for mechanical mixer is taken in lumpsum of can be taken as 0.27 days for mixing 1cum of mortar.

Rate Analysis of Cement Mortar

Rate analysis of cement mortar requires the rate of cement and sand multiplied with their respective quantity. Cost of labour and tools and tackles for the cement mortar is also calculated. Cost of cement, sand, labour and tools depend on local availability and local rates. When, quantity of cement, sand, labour and tools are multiplied by its rates, and a contractor's profit of 10-20% is assumed, the total sum of the amount is taken as the cost of 1m^3 of cement mortar.

Quantity of Cement and Sand Calculation in Mortar

Quantity of cement mortar is required for rate analysis of brickwork and plaster or estimation of masonry work for a building or structure. Cement mortar is used in various proportions, i.e. 1:1, 1:2, 1:3, 1:4, 1:6, 1:8 etc.

Calculation of quantity of cement mortar in brickwork and plaster

For the calculation of cement mortar, let us assume that we use 1m^3 of cement mortar. Procedure for calculation is:

1. Calculate the dry volume of materials required for 1m^3 cement mortar. Considering voids in sands, we assume that materials consists of 60% voids. That is, for 1m^3 of wet cement mortar, 1.6m^3 of materials are required.

2. Now we calculate the volume of materials used in cement mortar based on its proportions. Let's say, the proportion of cement and sand in mortar is 1:X, where X is the volume of sand required.

   Then, the volume of sand required for 1:X proportion of $1m^3$ cement mortar will be $\frac{1.6 \times X}{(1 + X)}$

3. Volume of cement will be calculated as: $\frac{1 \times 1.6}{(1 + X)}$

   Since the volume of 1 bag of cement is $0.0347\ m^3$, so the number of bag of cement will be calculated as:

   $$\frac{\left(\frac{1 \times 1.6}{1 + X}\right)}{0.0347}$$

Illustration: For cement mortar of 1:6, the quantity calculated will be as below:

Sand quantity: $\frac{1.6 \times 6}{(1 + 6)} = 1.371\ m^3$ Quantity of cement (in bags):

Volume of cement $= \frac{1.6}{(1 + 6)} = 0.22857$

There number of bags required $= \frac{0.22857}{0.0347} = 6.58$ bags.

Work Study

Work study is a technique to establish the time which a qualified, skilled worker should take to carry out a given task with reasonable and assured quality in the end product. Of various methods coming under this head, the method of Time Study is the most popular. It involves the following steps:

(a) Firstly, the whole process or the project under study is broken down into elementary operations in commensurate with standard practices and the objects of time study are defined.

(b) Times taken to complete these operations are then carefully noted in a number of observations, repeated a number of times at different locations, at different days and times of the year to take into account climatic seasonal or any other fluctuations in human behaviour.

(c) The results are then analysed to ascertain an acceptable standard time for each of the operations.

(d) The results are then utilised in determining activity dura-tions in the main Network.

Time-measuring devices. To measure times for the different operations, various time-measuring devices like stopwatches, time -recording machines or motion picture camera are used and results are noted in well laid-out charts or table which are simple, user-friendly and easy to decipher.

Minimum Number of Observations required in the Work Study Methods

Statistically it is desirable to obtain a large number of observations (by 'large' it means when the number of observations are more than 30). In that case the dispersion in the observed values will be smaller and smaller, as the sample size is increased. At the same time the estimated values of time observations will be concentrated in the neighborhood of the true value of the whole parameter and the sample will then follow very closely the Statistical Normal distribution Curve.

Method Study

By Method Study a systematic and critical examination is made of any existing process to find out ways and means for an improvement or modification of the system to time, cost, optimise floor area, set of tools and equipment, physical movements and use of materials etc., without compromising quality standards. The Method study thus aims at economising the work content in a process to determine better and improved ways of doing a job to reduce costs.

The technique of the Method Study may be expressed as 'SREDTIM', where

S = Select the work for study.

R = Record all data as available.

E = Examine the data in proper sequence.

D = Develop models, trials or samples for new economical methods.

T = Train the workers in the new technique.

I = Install new method.

M = Maintain new method.

Such technique may be successfully applied to any sphere of project management like *RID* Projects, construction, manufacture, inspection, storage and packing, sales, transporting, maintenance etc. Even for a new project proposed to be implemented the technique of Method Study may be applied to arrive at a better orientation and sequence of activities by reducing the work content of the process and thereby reducing overall costs. It is thus a very effective method of improving in the use of materials, machines, human resources, utilities and infrastructural facilities.

Obsolescence

The importance of adopting a new modern appropriate technology for sustained growth of industry is recognised everywhere now a-days. With the help of new technology in the form of ideas, skill and machines, more goods can be produced in a short time, having a cheaper unit cost and a greater quality standard. For this reason, the manufacturers very often replace old machinery and buy new efficient ones and the existing machines become obsolete and redundant. While replacing a machine because of obsolescence, due consideration should be given to the cost of capital investment, the benefit-cost ratio of acquiring a new machine and also the salvage value. Savings in labour engagement also come into consideration in taking a decision on machinery replacement. As in the case of project profitability analysis, the investment strategy of equipment replacement should be based on the calculations, either by Pay-back Period, Present-value Method or the Rate of Return Method.

Winding up a Project

An existing project may be required to be winded up due to various reasons, a few of which are mentioned below:

1. Technological obsolescence of the existing machinery or the set-up of the production system.
2. Drying up of nearby sources of raw materials may need shifting of the project elsewhere as for example in mining activities, when existing sources are all exhausted.
3. Sudden devastating accidents or faults generated in the system may need the winding up of a project for security reasons as for example, a nuclear plant may have to be dismantled for such reasons.
4. A project may be shifted to another place, where facilities of infrastructure or availability of cheap labour may be alluring.
5. A project may have to be shifted elsewhere due to environmental pollution.
6. A project may be closed due to continued economic recession and loss of market.
7. An industry may be closed due to financial and operational mismanagements or due to lack of working capital.
8. A project may have to be dismantled on expiry of its service life as in the case of river-valley projects or a bridge.
9. A running project, particularly bridge construction projects, continuing for several years in difficult area, may have to be closed during rainy seasons when the site may remain inaccessible or flooded only to be re-opened after the monsoon period.
10. In war-time situations, an industry may have to be winded up at a short notice and shifted elsewhere.

Whatever may be the reason, the winding up operation of a project needs careful advance planning and scheduling. If the reason is technical obsolescence, effort should be given to extract as much salvage value as possible from the discarded machinery and equipment. If the set-up is to be dismantled and shifted elsewhere, it should preferably done in stages, so that no plant or equipment remains idle for long. If the project needs shifting due to reasons of environmental pollution, additional pollution control measures will be necessary at the new location, involving extra costs. In case the project is closed or winded up temporarily due to management failure, effort should be given to revitalise the set-up by re-structuring the organisational set-up and investing additional working capital. Dismantling of a high-tech plant, like a nuclear or power plant, needs an extremely careful scheduling of activities, so that the whole dismantling operation can be done without any safety hazards. It needs the application of technical and scientific expertise to make such a winding-up operation a success.

During the winding-up operation, the inventory procurement should be strictly controlled to avoid accumulation of spares, equip-ment and raw material in stores. In construction sites, particularly, the procurement of inventory should be tapered off gradually, so that raw materials like cement, bricks and aggregates do not remain unused at site, even after completion of all items of work. Retrenchment of labour during winding up a project poses a very serious problem. It often leads to labour and social unrest. The management has to abide by all labour rules and arrange for compensation for retrenchment or engage the workers at some other project, if possible. In any case, any reduction in the labour force needs very tactful handling and humanitarian consideration.

Additional budget allocation needs to be made for the winding-up operation. Fund for winding up, a set-up should be made available in stages depending on the sequence of winding-up operations. It should be borne in mind that the winding up of a project is a project itself and all relevant activities connected with a typical project, like planning, scheduling, implementation, cost control, inventory control, operational control are there to be managed by the project manager.

Project Organisation Design and Human Resource Management

For efficient running of an organisation, the project organisation should be designed properly to develop the human resources alongside the capital resources. Just as a project manager makes optimum use of his material resources, the human resources also need to be managed in an optimised manner to extract maximum benefit out of it. The human resources comprise the shop-floor workers, field staff, sales staff, marketing personnel, construction workers, general office staff, all the supervisory staff, middle-level management personnel and the top management personnel. The management of all these personnel requires a careful planning in advance and involves the following functions:

(a) Recruitment
(b) Training
(c) Motivating
(d) Controlling

Recruitment: An efficient organisation can be set up by an advance manpower planning considering the requirements of manpower required at different work centres of the organisation and the human factors involved in the selection of the team. Phase-wise vacancies likely to occur in the organisation are to be assessed and recruitment planning is to be made accordingly. The recruitment of manpower involves the process of interviewing the prospective candidates, selection of suitable candidates, induction of the new employee in the organisation and his placement. As this whole process takes considerable time sometimes more than three months, the date of notification for recruitment should be fixed accordingly, so that the selected employee is available on job on the specified date. A proper manpower recruitment scheduling should be made for this purpose.

Training: The manpower resources thus inducted in the organisation can be further developed by arranging suitable training programmes in order to augment the skill, aptitude and knowledge of the employee, to enable him to perform his job efficiently and with responsibility.

Motivation: Along with the training programmes, the employees should be motivated in their jobs by setting up suitable pay-scales, arranging fringe benefits, health services, recreational facilities and incentives for good work or production.

Control: Along with the motivation, the employees should be kept under tactful control, so that discipline at work places is maintained and the employees work as per company's regulations and obey the safety rules. Their performances should be periodically assessed and disputes are settled. There should be a proper delegation of powers and responsibilities, so that each employee knows what he has to do and what is his limitations. The initiatives taken by the employees should be encouraged and too much control from above should not be imposed. The employees should not be burdened with infructuous paper work. They should be encouraged in setting up quality circles, maintain the quality control measures and abide by the safety regulations of the organisation.

Human factors are involved in personnel management. Hence all the above functions are to be carried out impartially and a cordial industrial management-worker relationship should be maintained.

Review Questions

1. What are the steps for construction project planning?
2. Explain work breakdown structure in project management.
3. Explain the difference between WBS and work breakdown schedule.
4. What are the things to consider when preparing a WBS?
5. What are the guidance in choosing a method for preparing a WBS?
6. What are the advantages of cost control? List out a WBS cost control techniques.
7. Elaborate the meaning of PDCA.
8. How would you prepare the budget control sheet and cost variance?
9. What do you mean by rate analysis? Write down its objectives, necessity and factors deciding rate of items.
10. Explain the procedure of rate analysis with an illustration.

11. Explain the out-turn method of rate analysis.
12. How would you create a WBS? Explain.
13. What are the characteristics of WBS?
14. Explain the benefits of WBS.
15. Write short notes on :
    - Define the project scope
    - Resource capacity planning
    - Phase design and budgeting
    - Construction financing and cost planning
    - Cost control techniques
    - Direct and Indirect cost
    - Cost control through field supervision
    - Cost control through communication network
    - Break Even Point (BEP)
    - Work Study
    - Method Study
    - SREDTIM
    - Obsolescence
    - Winding up of project

2

Construction Contracts and Specifications

Creating a good construction contract is a skill that every builder should know inside and out, whether he or she is a solo owner-builder or the head of a large construction firm. Although various contractors use different checklists when creating a contract, there are commonalities that should be followed no matter what form the printed contract takes. In most good contracts, the builder starts with the most basic of line items, adapting the contract to the complexity of the project.

Aside from the construction work itself, creating the contract is one of the most important parts of the project. The more detailed the contract the better; even items considered common sense shouldn't be overlooked.

Contract Documents

A so-called "construction contract" is often a combination of individual documents detailing different aspect of the project, or it can be a complex, multi-page document with many sections that offer details on different aspects of the agreement.

There are about 10 different documents or clauses that can be included in a good construction contract. No matter what you are building, having a solid contract represents one of the first steps toward the success of the project.

Contract Types

Different types of construction contracts are used within the building industry, but professionals generally prefer certain ones. Construction contract types are usually defined by the manner in which the disbursement is going to be made, but they may also specify other terms, such as duration, quality, specifications, and other issues of similar importance.

There are four different types of contracts used in the building industry, and how to pick the right one for your purposes you have to learn.

Successful Contract Negotiations

When the time comes for a builder to sit down with union representatives or project stakeholders, he or she needs to be prepared. A good negotiator should have the characteristics and skills that allow for the right deal to be struck—one that achieves set goals while leaving both parties satisfied. Face-to-face negotiations, although rare, can sometimes make things even worse. Certain tricks and methods can ensure that everyone leaves the table feeling that they have a good deal.

Direct Hire or Subcontract

Builders have two options for completing work on a project: direct hire their own employees or subcontract the work to independent contractors. It is not an easy decision.

Subcontracting can offer warranties that can make your work easier while hiring someone directly requires additional efforts as far as oversight and management. But direct hiring often gives the builder more control over costs and quality. How does a contractor decide between the two? Based on the scope of the project and existing relationships with subcontractors, the project manager will make the decision based on what is most cost-effective and timely.

Construction Contract Agreements

A construction contract agreement is a principal document that sets a date and specifies which parties are going to participate in the construction process. Usually, the contract agreement is executed between the owner of the project and the contractor (or supplier) that is providing the requested service. The contract generally contains several sections or clauses (or sometimes appended documents) that define the scope, terms, and conditions of the agreement. The legal writing can be heavy, and both parties often involve a lawyer when dealing with these specific documents.

Differing Site Conditions

Differing site conditions can increase construction costs and can delay breaking ground on the project. A contractor developing contracts needs to know how to handle this possibility and include language that protects against unforeseen circumstances. Normally, differing site conditions surface during the first weeks of the project, potentially affecting the schedule and causing unforeseen delays. Considering the repercussions of delays, it's crucial to document how such impediments affect the general contract.

Construction Contracting Escalation Clause

Escalation clauses are often written into construction contracts. They are more typically included on large construction projects, where the job might require more than one year to complete and where it carries substantial financial backing and risk. For example, the potential for economic changes such as a gas shortage or oil glut may require contract escalation clauses, even on small and medium-size projects. If executed properly, escalation clauses protect the contractor from unpredicted charges.

A construction contract provides a legal binding agreement for both the owner and the builder that says the executed job will receive the specific amount of compensation. It may also say how that compensation will be distributed. There are several types of construction contracts used in the industry, but there are certain types of construction contracts preferred by construction professionals.

Construction contract types are usually defined by the way the disbursement is going to be made. It details other specific terms, like duration, quality, specifications, and several other items. These major contract types can have many variations and can be customized to meet the specific needs of the product or the project.

Lump Sum or Fixed Price Contract Type

This type of contract involves a total fixed price for all construction-related activities. Lump sum contracts can include incentives or benefits for early termination, or can also have penalties, called liquidated damages, for a late termination.

Lump sum contracts are preferred when a clear scope and a defined schedule has been reviewed and agreed upon.

This contract shall be used when the risk needs to be transferred to the builder and the owner wants to avoid change orders for unspecified work. However, a contractor must also include some percentage cost associated with carrying that risk. These costs will be hidden in the fixed price. On a lump sum contract, it is harder to get credit back for work not completed, so consider that when analysing your options.

Cost Plus Contracts

This type of contract involves payment of the actual costs, purchases, or other expenses generated directly from the construction activity. Cost plus contracts must contain specific information about a certain pre-negotiated amount (some percentage of the material and labour cost) covering contractor's overhead and profit. Costs must be detailed and should be classified as direct or indirect costs. There are multiple variations of cost plus contracts and the most common are:

- Cost Plus Fixed Percentage
- Cost Plus Fixed Fee
- Cost Plus with Guaranteed Maximum Price Contract
- Cost Plus with Guaranteed Maximum Price and Bonus Contract

Cost plus contracts are used when the scope has not been clearly defined and it is the owner responsibility to establish some limits on how much the contractor will be billing. When some of the aforementioned options are used, those incentives will serve to protect the owner's interest and avoid being charged for unnecessary changes.

Be aware that cost plus contracts are difficult or harder to track and more supervision will be needed.

Time and Material Contracts When Scope is Not Clear

Time and material contracts are usually preferred if the project scope is not clear, or has not been defined. The owner and the contractor must establish an agreed hourly or daily rate, including additional expenses that could arise in the construction process.

The costs must be classified as direct, indirect, mark-up, and overhead, and should be included in the contract. Sometimes the owner might want to establish a cap or specific project duration to the contractor that must be met, in order to have the owner's risk minimized. These contracts are useful for small scopes or when you can make a realistic guess on how long it will take to complete the scope.

Unit Pricing Contracts

Unit pricing contracts are another type of contract commonly used by builders and in federal agencies. Unit prices can also be set during the bidding process as the owner requests specific quantities and pricing for a pre-determined amount of unitized items.

By providing unit prices, the owner can easily verify that they're being charged with uninflated prices for goods or services being acquired. Unit price can easily be adjusted up and/or down during scope changes, making it easier for the owner and the builder to reach agreements during change orders.

There are three main types of construction contract, identified according to the mechanism for calculating the sum due to be paid by the employer: lump sum contracts, re-measurement contracts and cost-reimbursable contracts. The different types vary primarily with regard to who takes the risks involved, which party has to pay for the cost over runs, and which party can keep the savings if the project costs are less than the estimated costs.

Other types of contract and descriptions of contractual purpose include:

- Commercial contract
- Domestic building contract
- Percentage rate contract
- Item rate contract or Unit price contract
- Lump sum and scheduled contract
- Cost plus fixed fee contract
- Cost plus percentage of cost contract
- Subcontract agreement
- Special contracts

Special Contracts

Special contracts are further classified into five types:

- Turn key contract

Owner exerts governance over the contractor. Contractor exerts governance over internal units and sub suppliers.

- Negotiated contract
- Package contract
- Continuing contract
- Running contract

Specifications

To maintain quality and to build a durable structure, the construction manager should have a thorough idea about the specifications that should be adopted. There are many standard specifications and codes of practices prepared by the Bureau of Indian Standards (BIS), Indian Roads Congress (IRC) and Railways covering materials, workmanship, quality control, safety, testing, investigation etc. related to different types of structures. These standards provide important and valuable guidelines to the planner, designer, contractor, supplier, owner, project manager, in fact, everybody concerned and connected in some way or other to the construction work. Over the years, these codes have been and are being constantly revised, updated and modified to suit the needs and changes of the style and nature of construction activities. In addition there are manuals available from different manufacturers giving specifications and instructions of using their equipment.

Specifications describe the nature and the class of the work; materials to be used in the work, workmanship etc. and is very important for the execution of the work. The cost of a work depends much on the specifications. Specifications should be clear.

Specifications are the instructions, meant to supplement the drawings and provide technical requirement of the work. The drawings with the specifications will give the complete requirement of the structure.

Aim of Specifications:

1. It checks the strength of materials for a work involved in a project.
2. The cost of a unit quantity of work is governed by its specification.
3. It is required for court cases.
4. Specifications gives detailed information regarding quantity and quality of the material which enables a contractor to estimate the cost correctly.
5. Specifications give detailed information regarding the equipments, tools and plants to complete the project which enables us to procure them before hand in time.
6. Cost of work depends mainly on the specifications. If specifications are changed, the cost of work is also changed. Specifications give correct idea about the estimated cost so that funds can be arranged accordingly.
7. It also serves as a guide.

Principles of Specification writing:

(a) Description of materials

(i) The quantity and quality of materials required should be given.

(ii) The proportion of mixing or treatment of materials before use should be clearly mentioned.

(b) ***Workmanship***
   (i) Tue method of mixing to the proportion
   (ii) Method of laying
   (iii) Method of preparation of surface
   (iv) Compaction, finishing and curing etc. should be clearly mentioned in the specifications.

(c) ***Equipments:*** The equipments, tools and plants should be described clearly to carry out the work. The method of operation also should be stated in the specifications.

(d) ***Expressions:*** The following points should be kept in mind while preparing specifications:
   (i) The sentence should be brief, simple and concise without any specifications.
   (ii) The writing style and tense should be kept same throughout.
   (iii) There should not be any grammatical mistake.
   (iv) The omission or misplacement of comma, fullstop etc. should not change the meaning of the sentence.
   (v) There should not be any ambiguity in the sentence.

(e) Practical limitations of materials, workmanship etc. should be kept in mind while writing the specifications.

(f) The specifications should be fair to both the parties and it should not favour any party.

(g) In case of any foreseen difficulty or hazards, it should be specified.

(h) Specifications are legal documents and they should be prepared keeping constitution in mind.

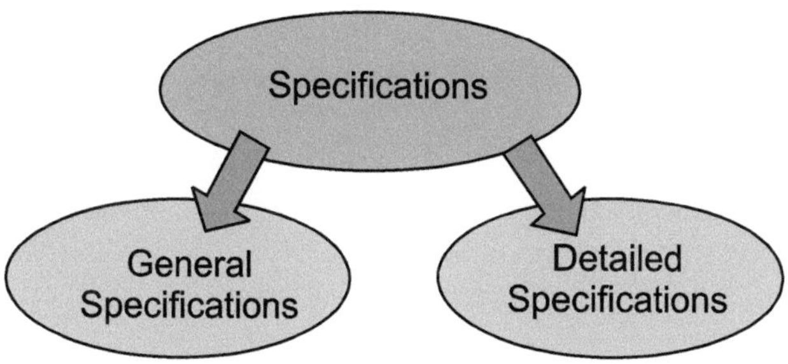

Fig. 2.1: Types of Specifications

Purpose of giving Specifications

- The cost of a unit quantity of work is governed by its specifications.
- Specification of a work is required to describe the quality and quantity of different materials required for a construction work and is one of the essential contract documents.
- This also specifies the workmanship and the method of doing the work. Thus, specification of a work serves as a guide to a supervising staff of a contractor as well as to the owner to execute the work to their satisfaction.

- A work is carried out according to its specification and the contractor is paid for the same. Any change in specification changes the tendered rate.
- As the rate of work is based on the specification, a contractor can calculate the rates of various items of works in tender with his procurement rates of materials and labour. Thus, tender rate without specification of works is baseless, incomplete and invalid.
- Specification is necessary to specify the equipment tools and plants to be engaged for a work and thus enables to procure them beforehand.
- The necessity of specification is to verify and check the strength of materials for a work involved in a project.

Types of Specifications

Following are the three most common types of building construction specifications used in commercial projects:

Performance Specifications address the operational requirements of an installation. The focus is on the project outcome, indicating how the final project must be able to function. Here, architects provide direction to the general contractor about what is needed and the general contractor must determine the best path to achieve the desired outcome. While this approach gives the general contractor flexibility in his construction approach, it limits how much authority an architect maintains once construction is underway.

Prescriptive Specifications contain detailed descriptions of what specific materials must be used as well as the installation instructions. This type of specification usually involves three key components:

1. *General provisions:* requirements surrounding codes and standards
2. *Required products:* the type of products required based on performance and structural stipulations
3. *Execution procedures:* how to do the install and measure its effectiveness

Prescriptive specification put a greater burden on the architect (as opposed to the general contractor) to ensure proper installation.

Proprietary Specifications demand that only one specific product be used for a given installation. It is commonly utilized if the portion of a project requires a certain performance that only one product can achieve. This could ultimately drive up project costs.

Along with these three types of construction specifications, specifications can also be classified as "open" or "closed," a division often determined by the project's stage in design development.

An open specification is when an architect does not name a specific supplier or product and allows for substitutions to be made by the contractor. It does not limit competition, but rather is dictated by a set of standards that more than one manufacturer can meet, allowing for many alternatives to be submitted for approval. Performance specifications are often considered to be open.

A closed specification lists specific products, systems and manufacturers, with no alternatives or mechanisms to apply a substitution. Closed specifications are most often seen when matching a specification to an existing building, or when an exact duplication is important. By default, closed specifications are proprietary. However, they can be made "open" by not referring to a singular brand or providing requirements applicable only to a specific product. This is commonly done by adding "or equal" after the listed brand.

It is critical for building/design product manufacturers to understand the various classifications and types of specification in construction, as well as in which specification scenarios their products will most frequently be considered.

Detailed Specifications

The detailed specifications form a part of a contract document. They specify the qualities, quantities and proportions of materials and the method of preparation and execution for a particular item of works in a project. The detailed specifications of the different items of the work are prepared separately and they describe what the work should be and how they shall be executed. While writing the detailed specifications, the same order sequence as the work is to be carried out is to be maintained.

Illustration-Brickwork Detailed Specification

{Step By Step Process of Brickwork Construction by Vicky, 2019}

Brickwork specification specifies the nature and the class of the work, materials to be used in the work, workmanship etc. It is very important for the execution of the work. The cost of the work depends much on the specification of work. So let us see the brief specification of Brickwork and how it is carried out step by step.

Materials required for the construction of Brickwork and its process

Bricks

All the Bricks shall be made from clean and good soil. The size of the brick should be 19*19*9 cm. The bricks should be well-burnt and should have proper reddish colour. The bricks should have plane rectangular faces with parallel sizes and right angle edges. The brick should not break when dropped from height of minimum 1 meter. When the brick is immersed in water, it should not absorb water more than 20% of its dry weight. The brick should have minimum crushing strength of 105 kg/cm^2.

Sand

It shall be natural sand and should be hard, strong and durable. The sand should be purely obtained from river bed and it should be clean and free from shells, alkali, salt, dust, silt etc. If the sand is not clean and not as per I.S specifications and contains more than 5% of clay, dust or silt then it should be properly washed off before using it for brickwork construction.

Cement

It shall be fresh portland cement with standard specifications. The cement should have minimum compression strength of 175 kg/cm^2. The initial setting time of cement should not be less than 30 minutes and final setting time should not be more than 10 hours.

Water

Water should be free from alkaline, salt and other acid matter.

Proportion

Proportion of the cement sand mortar may be 1:3 to 1:6 or as specified. It means that one part of cement and three parts of sand or one part of cement and six parts of sand.

Mixing of Mortar for Construction of Brickwork

In Brickwork construction the mixing of cement sand mortar is generally done by 2 methods:

- Hand Mixing
- Machine Mixing

But generally on site and mainly for small works of brickwork, hand mixing method is mostly preferred. In this method sand and cement are mixed thoroughly in dry state on mortar platform or iron sheet tray. After mixing properly, water is added to the material and then again it is uniformly mixed. The quantity of the prepared cement mortar can be used within 30 minutes of its preparation. Stale mortar shall never be used for brickwork construction. The quantity of water in the mortar at the time of mixing should not exceed 25 litres per bag of cement.

Laying of Bricks

Bricks shall be laid in English bond, unless if mentioned. The frog of the brick should be laid upward except in case of top course where the frog is laid downward. Every course shall be truly horizontal and wall shall always be in plumb. No damaged bricks shall be used. The brickwork constructed in one day should not exceed the height greater than 1 meter. The joints in brickwork are racked in 12 mm and it should not exceed the thickness of 6 mm.

Curing

Brickwork shall be well watered continuously for 10-14 days. After the end of days work the top of wall shall be flooded with water.

Scaffolding

Necessary and suitable scaffolding should be provided for safety during the wall construction of brickwork. It should be built strong enough to withstand the load of labours and tools coming upon it.

Measurement

The brickwork shall be measured in cu.m (cubic meter) and thickness of wall shall be taken as multiple of half brick 10 cm, 1 brick 20 cm, 1½ brick 30 cm and so on. The rate shall be for complete brickwork including scaffolding and tolls and plants.

Specification for Cement Concrete

The cement concrete shall be prepared by mixing graded stone or brick aggregate of nominal size as specified with fine aggregate and cement in specified proportions with required quantity of water. Toe grading and quantity of aggregate shall be such as to give maximum compressive strength.

Proportioning: It shall be done by volume. Boxes of suitable size generally 35×25×40 cm. shall be used for the measurement of sand and aggregate. The unit of measurement for cement shall be a bag of 50 kg and this shall be taken as 0.0347cum.

Mixing: Brick aggregate shall be well soaked with water for a minimum period of 2 hours and stone aggregate shall be washed with water to remove dirt, dust or any other foreign materials. Toe mixing shall be done in mechanical mixer. Mixing by band shall be employed only in special case with the prior permission.

Machine mixing: Measured quantity of dry coarse aggregate shall be first placed in the hopper. This shall be followed with measured quantity of fine aggregate and then cement. In case damp sand is used, add half the quantity of coarse aggregate followed by cement and sand. Finally add balance quantity of the coarse aggregate. The total quantity of water for mixing shall be introduced before 25% of the mixing time bas elapsed and shall be regulated to achieve the specific water cement ratio. When the mixer is closed down for the day or at any time exceeding 20 minutes, the drum should be washed with water.

Mixing Time: The material shall be mixed for a period of not less than 2 minutes and until a uniform colour and consistency are obtained.

Hand mixing: The mixing shall be done on a smooth, water tight, clean platform of suitable size in the following manner:

(i) Measured quantity of sand shall be spread evenly.

(ii) The cement shall be dumped on the sand and distributed evenly.

(iii) The sand and cement shall be mixed with spade, turning the mixture over and over again, until it is of even colour throughout and free from streaks.

(iv) The sand cement mixture shall be spread out and measured quantity of coarse aggregate shall be spread on its top. Alternatively the measured quantity of coarse aggregate shall be spread out and the sand, cement mixture shall be spread on its top.

(v) This shall be mixed at least three times by shovelling and turning over by twist from centre to side, then back to the centre and again to the sides.

(vi) Three quarters of the total quantity of water required shall be added while the material is turned in towards the centre with spades. The remaining water shall be added by a water can fitted with rose bead, slowly turning the whole mixture over and over again until a uniform colour and consistency is obtained throughout the pile.

(vii) The mixing platform shall be washed at the end of the day's work.

Consistency: The quantity of water to be used for rich mix of 50 kg cement, to give the required consistency shall not be more than 3- litres for 1: 3:6 mix, 30 litres for 1:2:4 mix, 27 litres for 1:1.5: 3 mix and 25 litres for 1:1:2 mix.

Laying: The entire concrete used in the work shall be laid gently (not thrown) in layers not exceeding 15 cm, shall be thoroughly vibrated by means of mechanical vibrators till a dense concrete is obtained. Hand compaction shall be done with the help of tamping rods so that concrete is thoroughly compacted and completely worked into toe comers of the formwork. Compaction shall be completed before the initial setting starts i.e. within 30 minutes of addition of water in the dry mixture.

During cold weather, concreting shall not be done when the temperature falls below 4.5 °C. The laid concrete shall he protected against frost by suitable covering. During hot weather, precautions shall be taken to see that temperature of wet concrete does not exceed 38 °C.

Where it is found to be necessary to deposit any concrete underwater, the method, material, equipment and mix shall first to be approved. The concrete hall be deposited under water by one of the approved methods such as Treamic method, Drop bottom bucket, bags etc.

Curing: After 1-2 hours of laying the concrete, it shall be protected with moist gunny bags, sand or any other material approved, against drying. 24 hours after laying the concrete, its surface shall be cured by flooding with water of minimum 25 mm depth, or by covering with wet absorbent materials.

The curing is done for at least 14 days. In special cases the curing period can be increased.

In case of foundation concrete the masonary work may be started after 48 hours of it's laying, but the curing of cement concrete shall be continued along with the masonry work for a minimum period of 14 days.

Form work: If centering and shuttering are required to be done for this work, it shall be done in accordance with the specifications for form work and shall be paid for separately unless otherwise specified.

Rate: The rate is inclusive of the cost of labour and materials involved in all the operations described above except form Form Work" in case of cast-in-situ work.

Review Questions

1. Explain the meaning of contract documents and contract types.
2. Give classification of special contract.
3. Define specification and explain its aim.
4. Describe the principles of specification writing.
5. What is the purpose of giving specification?
6. Give a detailed specification of brickwork.
7. Explain the specification for cement concrete.
8. Write short notes on:
   - Differing site conditions
   - Construction contracting escalation clause
   - Cost plus contracts
   - Time and material contract when scope is not clear
   - Unit price contract
   - Performance specification
   - Prescriptive specification
   - Proprietary specification
   - Detailed specification

3

Project Scheduling and Network Technique

Project Planning and Project Scheduling

Project planning is the process of identifying all the activities necessary to successfully complete the project. Project scheduling is the process of determining the sequential order of the planned activities, assigning realistic durations to each activity, and determining the start and finish dates for each activity. Thus, project planning is a prerequisite to project scheduling because there is no way to determine the sequence or start and finish dates of activities until they are identified.

However, the terms of project planning and scheduling are often used synonymously because planning and scheduling are performed interactively. For example, a specific list of activities may be planned and scheduled for a project. Then, after the schedule is reviewed, it may be decided that additional activities should be added or some activities should be rearranged in order to obtain the best schedule of events for the project.

Planning is more difficult to accomplish than scheduling. The real test of the project is his or her ability to identify all the work required to complete the project. The preceding chapters of this book focused on identifying work activities and grouping those activities into meaningful categories. For example, the process of developing a well-defined work breakdown structure (WBS) as discussed in chapter-1, must be performed to com plete a project. Garold D. Oberlender,. (2014).

After the activities are identified, it is relatively easy to determine the schedule for a project. Many methods and tools have been developed for scheduling. The computer is universally used to perform the calculations for a project schedule.

Time Management

- **WHAT** is the time limit for the project?
- **WHAT** products have long delivery and turnaround times?

- **WHAT** are the critical tasks to get the project finished on time?
- **HOW** can it be ensured that work is progressing in such a way that the project will finish on time?
- **HOW** can planned activities (and which ones) be adjusted to meet the agreed date of completion?
- **HOW** can the uncertainties in time estimates be addressed?
- **WHEN** are decisions needed, request documents sent out, and work or products ordered?
- **WHEN** must certain activities be finished in order to coordinate with other project interests and interfaces?
- **WHEN** must production or installation not take place due to bad weather conditions?

General

The schedule is an important part of planning and involves most of the knowledge areas. The time schedule is one of the main instruments to control a project. The schedule is an activity plan that shows when to start doing a task, the task's duration, and when the activity must be completed.

The purpose of time management is to:

- Ensure that the client gets what is agreed to on time.
- Select methods of production within the planned production times.
- Know when products with different delivery times must be ordered so that they will be delivered in time.
- Through resource management, less and over time, reduce the cost of temporary peak loads.
- Know if and how to plan activities (on the critical path) if the schedule is not followed or know how much time activity outside the critical path can be allowed to slip without changing the day.
- Know when to send documents to authorities, boards, clients, etc., to obtain approvals and decisions in time so as not to cause a delay of completion.

During the design and production phases, meetings to monitor how the work is proceeding are held. Is it possible to "plod" on or must the date of completion be honored?

Today, work is primarily controlled with four different schedule types:

- The **milestone plan** shows the big picture and critical dates (milestones) that must be met-for example, approved construction documents, start of excavation, water-tight building, final inspection, start of moving in.
- The **Gantt chart** graphically shows the period within which an activity will be done. This chart is an excellent tool for reconciliation and information.
- The **network diagram** shows the structure/flow chart for the project.

Activities are shown as blocks and the logical links and restrictions as arrows. The network diagram is the PM's tool for scheduling optimization, resource allocation, replanning, etc. It provides answers to whether an activity delay will result in the delay of the date of completion or not.

- The flow-line diagram for site work shows the periods when many activities are going on in the same areas. The diagram helps to avoid "work jams" in the same workspace.
- Other types of schedules are the critical chain, spreadsheet, and project calendar, but these are not covered in this book.
- In the market there is much project planning software, such as MS Project and Primavera. A person is not a planner if he or she can cope with these programs, even if the good-looking plans may induce someone to believe that he or she has project control. It is necessary to know and understand how the programs are working-not to suboptimize. When this is understood, the programs are excellent tools.
- Hofsteader's law states:
- *"It always takes longer than you think, even when you take* Hofsteader's *law into account."* Another law reads: *"A project takes as long as you give it."*

These expressions may be playful, but "there is no smoke without fire."

Watch out if there is suddenly more time for a project than is needed. Remember the project triangle. Neither side can be changed without any of the other sides being affected. More time is usually burdened with indi-rect costs and perhaps lower productivity among employees.

Definitions of Scheduling Terms

Effort: labor units required to complete an activity

Working time: time period (hours, days, weeks ...) needed if normal working hours and time for sleep are considered

Milestone: significant event or delivery; has no duration

External milestone: milestone governed by dates not controlled by the project team (board meetings, contract dates, permits)

Internal milestone: milestone decided on by the project team to be able to control the development of the project (start excavation, start erection of superstructure, start installation of special equipment) *Toll gate:* an external or internal milestone that requires approval before the work can continue (approved basic design)

Activity: element of work that is required by the project; can be subdivided into tasks; includes time and costs.

Relationship/restriction: shows in which order activities can be done; foundations must be done before erection of superstructure.

Successive planning: detailed planning for the first period and outline planning for the rest: production cannot be planned before basic design is ready.

Compelling logic: an imperative sequence of activities: foundation before erection of superstructure.

Preferred logic: sequence of activities based on tradition, but alterna-tives exist.

External relationships: relationships between project relationships and activities outside the project.

Lag: a forced amount of time that does not consume any resources: con-crete curing before striking formwork.

Lead: opposite of lag; an activity can be started before the prior one is complete: painting/laying floor (laying floor on the second floor can start before all painting in the building is finished).

Total float (sometimes called slack or just float): amount of time a start of an activity may be delayed without delaying project end date.

Free float: amount of time a start of an activity may be delayed without delaying the subsequent activity.

Critical path: the series of activities that determines the earliest completion date; a delay on the critical path means a delay of the completion date; there can be many critical paths in a network diagram.

Project Scheduling

Introduction

There are three objectives in the Project Scheduling System:

- To identify a scheme for achieving prescribed goals through the development of a well-defined and realistic plan.
- To provide a visual means of conveying the information contained in this plan to all levels of management.
- To facilitate timely updating and monitoring procedures as necessary.

The Schedule is a document which links the scope of work, budget, resources, etc. in a time phased manner. Without a schedule it would not be possible to establish at any point of time how much work should have been completed or how much cost should have been expended. It is the starting point for any Project Controls effort.

- Schedule determines critical paths, which derives the project completion date or any other milestones. This helps the Project Manager to ensure that various departments (Design, Engineering, Procurement and Construction) deal all the matters related to the activities on critical path as a top priority.
- With good task linking and *logic* flow the *schedule* helps to determine the criticality of the activities in the program. It particularly helps in dealing with the scenarios such as value engineering, change orders, tendering an alternative supplier, decision awaited from the Client, suspension of any part of the works, etc. where the team wants to know how long a decision can be *delayed* without effecting the project completion date.
- Schedule is a live document usually updated on a weekly basis to determine how much was planned and what has actually been achieved.

It helps to determine the area(s) where the team is failing to meet the planned dates thus the Project Manager can take the necessary timely measures by adding resources, expedite procurement, etc.

- Look-ahead schedule helps the team to establish short-term goals from the master schedule.

Revised, Recovery schedule, accelerated schedule, "What If" scenarios, etc. are all possible only when the project has a baseline schedule.

Schedule is actually a live document usually updated on a weekly/monthly basis to determine how much was planned and what have actually been achieved. Without having a Schedule in place it is impossible to ascertain project's health and team performance.

Different Types of Schedules

For management of the Project, the following Schedules are developed:

- Master Schedule
- Project Schedules
- Revised Schedules
- Recovery Schedules
- Special Schedules

Master Schedule

At the outset of the project, the Project Manager prepares a Master Schedule that integrates the Design, Procurement and Construction phases of the Project. This shall be further broken down to the various packages that constitute the Project.

Once the scheduling objectives for both the short- and long-term have been agreed upon, the Project completion date is determined and "locked in". The Master Schedule thus developed shall be the 'Baseline' for comparison with updated actual data and will be the basis for developing more detailed schedules for individual trade packages.

The Master Schedule is updated periodically with actual data to compare against the baseline. Revisions to the Master Schedule shall be made in case of changes in design, design schedule, package strategy, etc.

As the design and procurement phases progress, on the basis of the Master Schedule, the Project Manager develops schedules with reasonable detail to demonstrate the feasibility and strategy for executing the project within the constraints of the approved Master Schedule. The objectives of such an exercise shall be to:

- Identify the sequence of works for execution.
- Determine the inter-relationships (FS, SS, SF, FF) between activities forming the Schedule.
- Analyze and Establish interface constraints between different trades.
- Form a framework within which the execution of Project should proceed.

This enables to produce indicative schedules for individual trade packages and establish interim Milestone Dates within those schedules.

Once the 'Schedule of Works' of any trade package Contractor is finalized, the detailed schedule of that trade package is removed from the PM's Schedule. However, the Project Manager will update the progress of each trade package on his Master Schedule, which will be in a summary form, based on update information provided by the Contractor.

Project Schedules

- Review the Project Scope of Work
- Review the Project Execution Plan
- Review the WBS developed
- Establish the level of reporting and coding structure in P3/P6 (Primavera Project Planner)/Any other planning software
- Enter all activities into P3/P6/Any other planning software
- Obtain a copy of the Project Estimate
- Enter the estimated man-hours against the activities into P3/P6 of any other planning software
- Discuss with the discipline engineers, the crew size for each activity, to establish the activity duration
- Discuss and establish with Discipline Engineer, the logic links between activities
- Discuss and establish logic links between disciplines
- Run a time analysis (first pass) to check end date against Contract Completion date
- Make adjustments to durations and logic to maintain Contract Completion Date
- Print the detailed schedule, manpower histograms and planned progress "S" curves for comments
- Code activities to enable the various levels of summary schedules to be produced
- Incorporate all comments
- Issue schedule for approval by the Project Manager and issue to Client
- Incorporate any comments received from the Client
- Re-issue Project Schedule for issue to Client

At this stage, the mutually agreed Project Schedule is also called as "Baseline Schedule".

This schedule is frozen as a target schedule in P3/P6/ Any other planning software and a copy of the same is used for subsequent updates.

The Client or Client Representative shall require all participants, including the Design Consultant and the Construction Contractors to submit their Project schedules or detailed schedules of their relevant activities while at the same time abide by the several milestones and constraints set by the Client.

To ensure reasonable compatibility and reduce the problems of collecting, organizing and reporting the information in a regular, coordinated manner, the format and structure of all schedule information and reports submitted by any party to the Project is specified by the Client or Client Representative and is a part of the contract documents.

Revised Schedules

If, at any time:

Actual progress is too slow to complete within the Time for Completion or progress has fallen (or will fall) behind the current program and/or changes in scope work, the Client may instruct the Contractor to submit a revised program and supporting report describing the revised methods.

Recovery Schedules

In the event that the project completion has been delayed, a recovery schedule must be prepared and issued to the Client, for review and approval.

Prior to the issue of the recovery schedule, certain questions must be asked:

- Who is responsible for the delay?
- Can the delay be covered by a Change Order?
- What is the cost impact of the recovery schedule?
- Are there Liquidated Damages involved?
- Review cost of recovery against LDs (Liquid Damages).
- Can the schedule be recovered?

After asking all the above questions, a recovery plan must be established; this plan will be reflected in the recovery schedule. Development of the recovery plan is as follows:

- Review the critical path activities and discuss with the Discipline Engineer's methods to reduce durations in line with the recovery execution plan.
- Review the critical path to check for logic errors or potential changes in the logic to recover the overall project duration.
- Review progress reports to highlight unproductive working methods.
- Review manning levels. Are they realistic and within the budget?
- Highlight activities that will be undertaken by working extended hours or shift system.
- Incorporate all comments and policies established in the recovery execution plan.
- Issue the recovery schedule for review and comments.
- Issue approved schedule to the Client.
- Incorporate Client's comments.
- Issue recovery schedule for Client approval.

Special Schedules

Special Schedule can be one of the following:

- Sub-contract schedule
- Shut-down schedule

- Detailed vendor erection schedule
- Change Order fragnet
- Delay analysis
- Intermediate recovery schedule

The procedure for development of these special schedules is basically the same for producing any schedule, with the exception that more detail is usually required; and therefore, more involvement by the Discipline Engineer or specialist.

In the case of the Change Order fragnet, all disciplines will be involved:

- Create a sub-project within the main project schedule that will contain the special schedule.
- Obtain all the information required to prepare the schedule.
- Prepare a scope of work for the schedule, issue to all parties for comments.
- Prepare the schedule independent of the main schedule, to ensure all activities are captured.
- Issue the schedule to all parties for review and comments.

"If you have no plan, you can't possibly have control."

Need of the Project Control

In earlier times, as well as in present days, many projects are implemented by the project manager by applying his intuition and experience. Sometimes he applies arbitrary decisions, which may or may not be the correct ones to tackle the particular problem situations. If the project is small and the activities are well defined and known to the project manager, he may be able to execute the project in an efficient manner. But in the present days, many of the projects are very big, complex and involves co-ordination of thousands of persons, hundreds of participating agencies; the time of completion is limited in the present competitive markets; many of such projects are of multidisciplinery nature and there are innumerable activities spreading over considerable span of time and space-all these are no simple affair for any project manager to tackle by applying arbitrary and intuitive decisions, however experienced he may be in his own field of knowledge. To overcome these problems he needs to control the activities, to know the status of the project at any time and to find out ways and means to locate the problem areas and solutions, if the project gets delayed and involves cost over-run or the limits of tolerance in the quality is exceeded. Among the methods most commonly used now-a-days are the application of Bar Charts, Tables, Reports and the PERT/CPM method of network Analysis. Among the tools, the project manager takes the help of the modern communication network systems and the help of computers. In addition to these he also applies sometimes the statistical methods of solutions or adopts the techniques of Resource Levelling, Resource Allocation and optimisation to meet problem situations when there are constraints in resources.

Desired Results of Planning

Project planning is the heart of good project management because it provides the central communication that coordinates the work of all parties. Planning also establishes the benchmark for the project control system to track the quantity, cost, and timing of work required to successfully complete the project. Although the most common desired result of planning is to finish the project on time, there are other benefits that can be derived from good project planning (see Table 3.1)

Planning is the first step to project scheduling. Planning is a process- and not a discrete activity. As changes occur, additional planning is required to incorporate the changes into the schedule. There are many situations or events that can arise that can impact a project schedule. Examples are changes in personnel, problems with permits, change in a major piece of equipment, or design problems. Good planning detects changes and adjusts the schedule in the most efficient manner.

A common complaint of many design engineers is they cannot pro duce their work because of interruptions and delays. The cause of this problem is usually a lack of planning, and in some instances no planning at all. Planning should clearly identify the work that is required by each individual and the interface of work between individuals. It should also include a reasonable amount of time for the exchange of information between project participants, including the delay time for reviews and approvals. Garold D. Oberlender,. (2014).

Table 3.1: Desired Results of Project Planning and Scheduling

1.	Finish the project on time
2.	Continuous (uninterrupted) flow of work (no delays)
3.	Reduced amount of rework (least amount of changes)
4.	Minimize confusion and misunderstandings
5.	Increased knowledge of status of project by everyone
6.	Meaningful and timely reports to management
7.	You run the project instead of the project running you
8.	Knowledge of scheduled times of key parts of the project
9.	Knowledge of distribution of costs of the project
10.	Accountability of people, defined
11.	Clear understanding of who does what, when, and how much
12.	Integration of all work to ensure a quality project for the owner

Project planning and scheduling can serve as an effective means of preventing problems. It can prevent delays in work, a major cause of late project completion and cost overrun, which often leads to legal disputes. It can also prevent low worker morale and decline in productivity that is caused by lack of direction.

Principles of Planning and Scheduling

There must be an explicit operational plan to guide the entire project. The plan must include and link the three components of the project: scope, budget, and schedule. Too often, planning is focused only on schedule without regard to the important components of scope and budget.

To develop an integrated total project plan, the project must be broken down into well-defined units of work that can be measured and managed. This process starts with the WBS. Once this is completed, the project team members who have the ex pertise to perform the work can be selected. Team members have the ability to clearly define the magnitude of detail work that is required. They also have the abil ity to define the time and cost that will be required to produce the work. With this information a complete project plan can be developed. Garold D. Oberlender,. (2014).

The project plan and schedule must clearly define individual responsibilities, schedules, budgets, and anticipated problems. The project manager should prepare formal agreements with appropriate parties whenever there is a change in the project. There should be equal concern given to schedule and budget, and the two must be linked. Planning, scheduling, and controlling begin at the inception of the project and are continuous throughout the life of the project until completion. Table 3.2 lists key principles for planning and scheduling.

Table 3.2: Key Principles for Planning and Scheduling

1.	Begin planning before starting work, rather than after starting work
2.	Involve people who will actually do the work in the planning and scheduling process
3.	Include all aspects of the project: scope, budget, schedule, and quality
4.	Build flexibility into the plan, include allowance for changes and time for reviews and approvals
5.	Remember the schedule is the plan for doing the work, and it will never be precisely correct
6.	Keep the plan simple, eliminate irrelevant details that prevent the plan from being readable
7.	Communicate the plan to all parties; any plan is worthless unless it is known

Responsibilities of Parties

The principal parties of owner, designer, and contractor all have a responsible role in project planning and scheduling. It is erroneous to assume this role is the responsibility of any one party. Each must develop a schedule for his or her required work and that schedule must be communicated and coordinated with the other two parties, because the work of each affects the work of the others.

The owner establishes the project completion date, which governs the scheduling of work for both the designer and contractor. The owner should also set priorities for the components that make up the project. For example, if the project consists of three buildings, the relative importance of the buildings should be identified. This assists the designer in the process of organizing his or her work and developing the design schedule to produce drawings that are most important to the owner. It also assists in the development of the specifications and contract documents that communicate priorities to the construction contractor.

The design organization must develop a design schedule that meets the owner's schedule. This schedule should include a prioritization of work in accordance with the owner's needs and should be developed with extensive input from all designers who will have principal roles in the design process. Too often, a design schedule is produced by the principal designer, or the project manager of the design organization, without the involvement of those who will actually do the work.

The construction contractor must develop a schedule for all construction activities in accordance with the contract documents. It should include procurement and delivery of materials to the job, coordination of labor and equipment on the job, and interface the work of all subcontractors. The objective of the construction schedule should be to effectively manage the work to produce the best-quality project for the owner. The purpose of construction scheduling should not be to settle disputes related to project work, but to manage the project in the most manner.

For some projects, it may be desirable for one party to maintain the schedule and the other parties to participate in monitoring it. Ultimately each one of the parties will be responsible for his or her portion of the schedule. Maintaining one common schedule as a cooperative effort between parties can reduce problems associated with maintaining three separate schedules.

Bar Chart

Definition and Introduction

A bar chart is a graphical representation of project activities that are shown in time-scaled bar lines with no links shown between the bars (activities) defines it as "a graphic display of schedule-related information. In the typical bar chart, schedule activities or work breakdown structure components are listed down the left side of the chart, dates are shown across the top, and activity durations are shown as date-placed horizontal bars." The bar chart was originally developed by Henry L. Gantt, an American mechan- ical engineer, in 1917 and is alternatively called a Gantt chart. It quickly became popular—especially in the construction industry—because of its ability to graphically represent a project's activities in a clear, simple, and time-scaled manner. Saleh A. Mubarak,. (2019).

Before a bar chart can be constructed for a project, the project must be broken into smaller, usually homogeneous components, each of which is called an activity or a task. None of the methods is a uniquely "correct" way to break down a project into activities, nor can we describe other ways as incorrect. However, the scheduler should take a balanced approach and break the project down into a reasonable number of activities that are easily measured and controlled without being overly detailed.

An activity, or a task, may be as large as laying the foundation of a building, as small as erecting the formwork of one footing, or anywhere in between. The duration of each activity must be estimated. Bars are then drawn to show each activity: the duration and the starting and ending points. As mentioned previously, links between activities are not usually shown.

On a bar chart, the bar may not indicate continuous work from the start of the activity until its end. For example, the activity Getting a Building Permit may be represented by a 2-month-long bar. However, most of this time is a waiting period. Likewise, a Concrete Foundation summary activity may include several days of waiting for the concrete to cure. Noncontinuous (dashed) bars are sometimes used to distinguish between real work (solid lines) and inactive periods (gaps between solid lines).

Bar charts have become a vehicle for representing many pieces of a project's information. Many variations of bar charts have evolved; some simply show the start and the end of each activity (Figures 3.1 and 3.2), some are loaded with resource or budget numbers (Figures 3.3 and 3.4), and others compare the as-planned schedule with the as-built schedule (Figure 3.5).

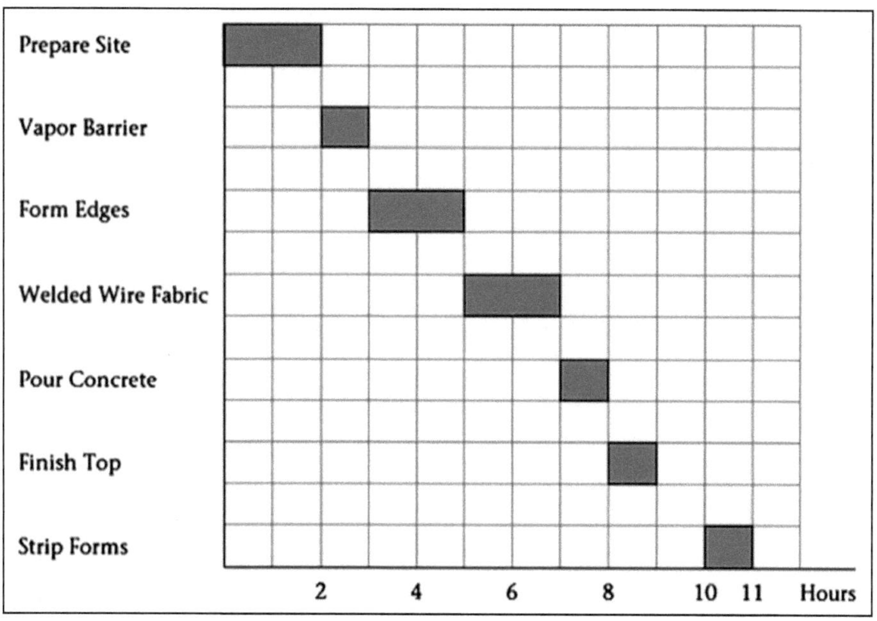

Fig. 3.1: Bar chart for placing a simple slab on a grade

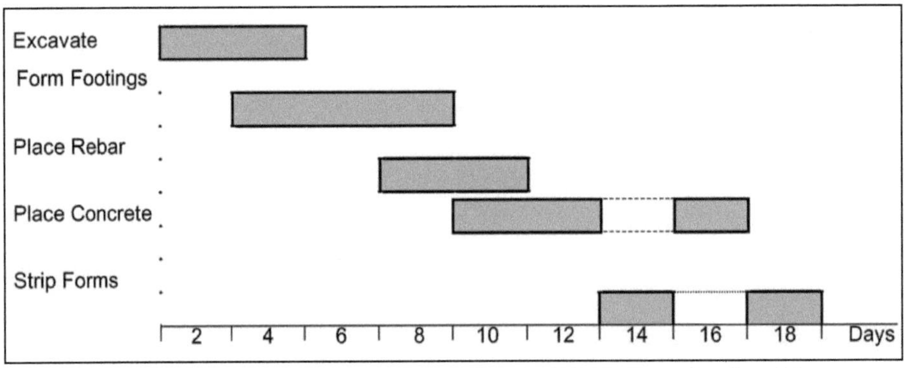

Fig. 3.2: Alternative bar chart for placing a simple slab on a grade placed in two parts

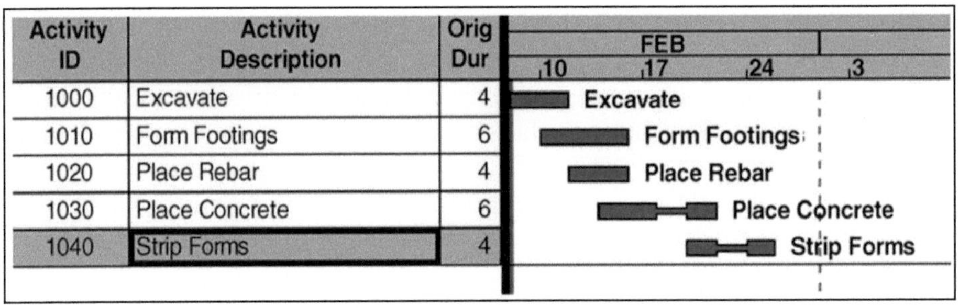

Fig. 3.3: Bar chart for placing a simple slab on a grade, with activities "necked" during nonwork periods

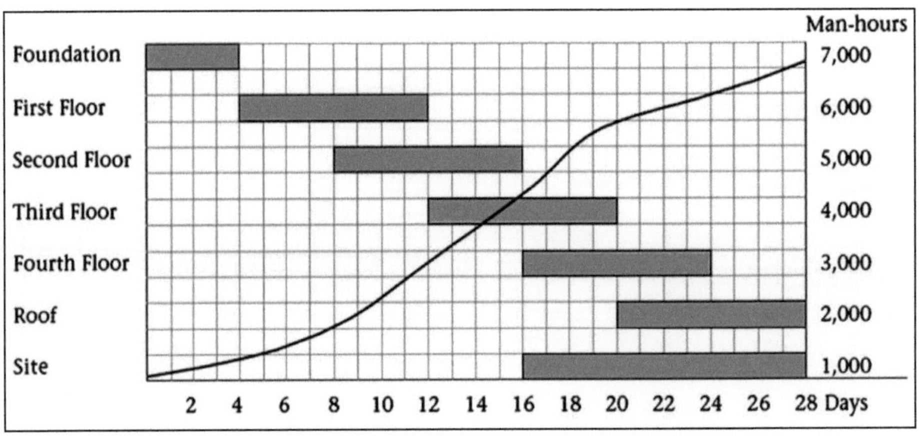

Fig. 3.4: Bar chart—loaded with man-hours—for construction of a two-story building

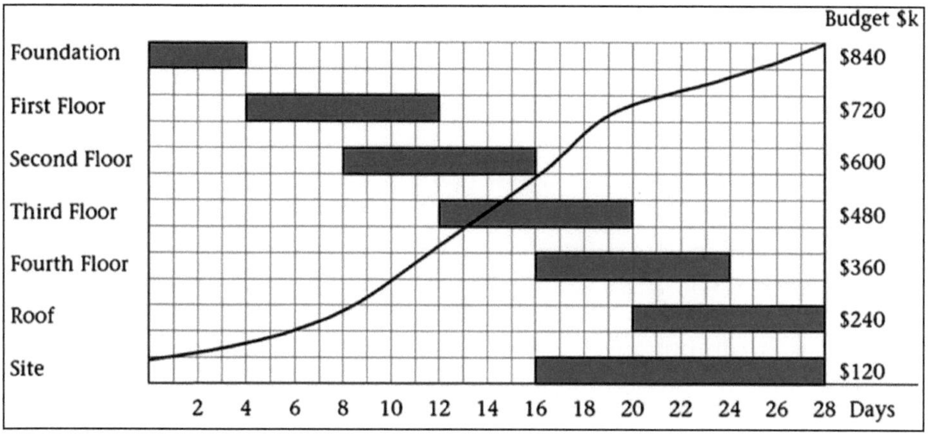

Fig. 3.5: Bar chart—loaded with the budget—for construction of a two-story building

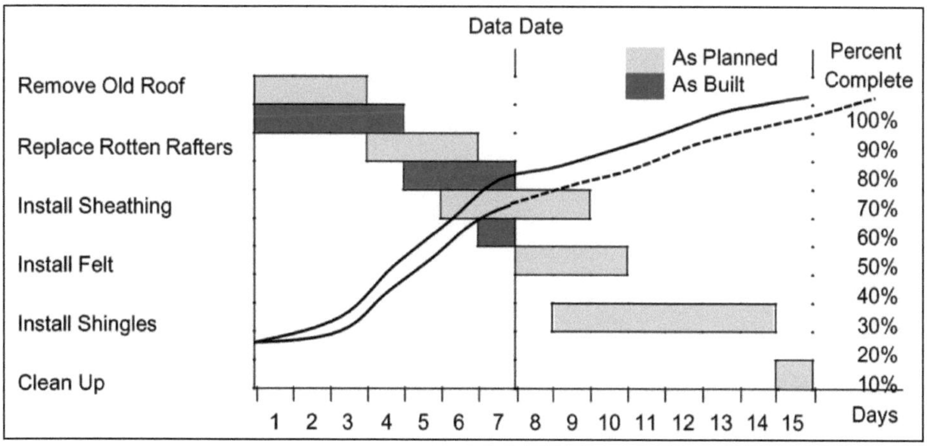

Fig. 3.6: Bar chart for replacement of an old roof, showing a comparison between the percent complete of the planned (as-planned) activities (thin line) and that of the actual (as-built) activities (thick line). The dashed line is an extrapolation of the actual curve

Advantages of Bar Charts

Bar charts have gained wide acceptance and popularity mainly because of their simplicity and ease of preparation and understanding. No "theory" or complicated calculations are involved. Anyone can understand them. They can be prepared anywhere with just a pencil and paper. So, although bar charts can carry—or be loaded with—other information, the user must be careful not to overload them and, thus, eliminate their main advantage: simplicity. Saleh A. Mubarak,. (2019).

Unlike networks, bar charts are time-scaled; that is, the length of a bar representing a certain activity is proportional to the duration of that activity. Just by looking at the chart, you can get an idea of the duration of each activity and the entire project.

Another advantage of bar charts is that they particularly appeal to persons who do not have a technical background. For example, some clients and upper-level managers may better understand the plan for carrying out a construction project by looking at a bar chart than by looking at a schematic of a logic network.

A final advantage of bar charts is the ability to roll up or roll down a schedule, either on the basis of a specific activity code (e.g., area, phase, responsibility, or floor) or the project's work breakdown structure (WBS).

With the advent of logic networks and the critical path method (CPM) and the evolution of powerful computers and software, bar charts have not perished or lost importance. Instead, they have evolved from a scheduling technique to a reporting tool that has made them more valuable and popular.

Disadvantages of Bar Charts

The main disadvantage of bar charts is their lack of logical representation (relationships); for example, why an activity started on a certain date (bar charts don't reveal the answer). The reason can be a logical relationship, a resource constraint, or a subjective decision by the project manager. Although some software programmers try to depict logical relationships on bar charts, the result is not always clear. The logic lines get tangled, and unlike networks, bar charts do not allow the length of the bars to be subjectively changed or the bars to be moved around to make the chart look or read better. Saleh A. Mubarak,. (2019).

Another limitation, rather than a disadvantage, of bar charts is the size and complexity of projects. Bar charts may not be practical for projects with a large number of activities, unless you use them in two ways:

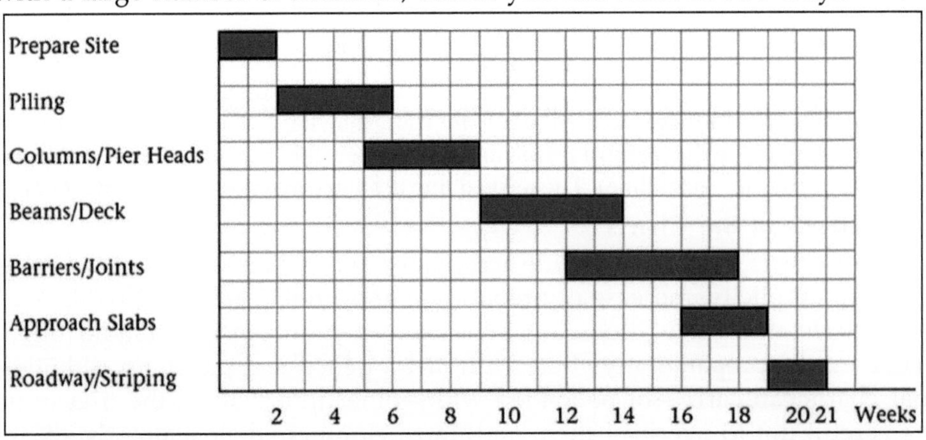

Fig. 3.7: Bar chart representing the summary activities for constructing a bridge

Difference Between PERT and CPM

Any project is involves scheduling, controlling, and planning several interrelated activities with the use of limited resources, namely money, manpower, material, and time.

The project may be complex, large, and small such as shopping complex, nuclear power plant, highways, and housing, etc. The main aim of the project management is to complete the task within the time limit.

They are following two methods used for planning, scheduling, and controlling the activities.

1. Critical Path Method
2. Programme Evaluation and Review Technique

Above both techniques are used for project management, but there is a significant difference between PERT and CPM.

Meaning of CPM

Full from of CPM is Critical path method.

The critical path method is a project managing tool. It is a one–type network diagram that shows the different activities in a step-by-step algorithm, methodology, and technique for planning the projects. CPM was developed by DuPont company. Technically, the critical path shows the longest path of the network. It has the least amount of float.

In project management, a float is defined as the amount of time that a task can be delayed without affecting delay to subsequent tasks and project completion tasks.

In the critical path method activities are shown in a network diagram, as a network of precedence relationships using activity on node network. CPM is followed the deterministic approach, it is calculated single estimation time for activity.

Meaning of PERT

Full from of PERT is Program Evaluation and Review Technique.

A PERT chart is a project management tool that shows a graphical representation of different activities in the timeline. The Program Evaluation Review Technique breaks down the whole activities in individual tasks of a project for analysis purposes.

PERT charts are considered to the Gantt bar chart, because they are identified task dependencies, but they are often more difficult to interpret the activity and event, which was developed in 1958.

It helps a project manager, to analyze a project's various tasks and estimate the amount of time required to complete each task in the project activities. Using this subjective information, the project manager can easily estimate/calculate the minimum amount of time required to complete the whole project activity.

This information also helps in a budget and to determine all types of resources needed to complete the project. PERT is followed the Probability approach. It calculates the three-time estimation for activity.

The Basic Difference Between PERT and CPM

The following are the key difference between PERT and CPM techniques.

PERT (Program Evaluation Review Technique)	CPM (Critical Path Method)
It is Event oriented technique	It is Activity oriented technique
PERT manages unpredictable activities	CPM manages the predictable activities
It is focused on time control	It is focused on cost optimization
It is suitable for non-repetitive projects like building construction, road, infrastructure, maintenance work, etc.	It is suitable for repetitive projects like building construction, road, infrastructure, maintenance work, etc.
It was developed in 1958	It was developed in 1957
It is a three-time estimate	It is single time estimates
It is a probability model	It is a deterministic model
It applies to the crashing concept	It does not apply to the crashing concept

PERT and CPM Comparison

Advantage of CPM

- CPM indicates the activities that can run parallel to each other. It is easily found out the most critical elements of the project.
- CPM gives a practical base visualization that helps in determining how to reach the aim of the project within the time limit.
- CPM is effective in new project management. It can strengthen a team's perception if it is applied properly.
- CPM provides a demonstration of dependent activities. CPM helps in the scheduling of individual activities. It shows the various project activities and their outcomes as a network diagram.
- CPM gives a fair and parallel procedure of documenting of project. It helps in determining the slack time of the activities.
- A clear–cut approach to communicating project plans, schedules, time, and it also developed cost performance.
- CPM is mostly used in the construction industry. It helps in time optimization by determining the project duration.

Disadvantages of CPM

- In CPM, it is difficult to estimate the completion time of the activities.
- The critical path is not always clear in the critical path method. when succeeding and preceding activities are more.
- For larger projects, critical path method networks can be complicated too.
- It also does not manage the scheduling of all types of resource allocation.
- In the critical path method, a critical path needs to be calculated very accurately. Because that affects the completion time of the particular project.

Advantage of PERT

- PERT (Program Evaluation and Review Technique) is suitable for large types of projects, also, scheduling of different activities is done easily.
- Each subordinate manager has to plan the events for which he or she is responsible. It's the main focus on critical elements of the project, that may need correction.
- It makes a forward–looking concept for the project manager.
- The main aim of the PERT reports and pressure for action at the particular spot and particular organization level at the right time.

Disadvantages of PERT

- Another biggest disadvantage has been it gives time importance but not on costs.
- It is more suitable for large and complex projects and not suitable for small and repetitive works.
- Also, the time required to complete each activity is highly subjective, so the programmer evaluation review technique is a high chance of an incorrect critical path.

Component of CPM and PERT Network

In CPM and PERT networks involves following two major components:

1. Activity
2. Event

Activity: An activity shows action and utilization of resources such as money, time, and material required to complete the project.

- Activity is represented by an arrow.
- In the network two types of activities are there, Preceding and Succeeding activities.
- **Preceding Activity:** Activities perform before a given event are called preceding activities
- **Succeeding Activity:** Activities perform after a given event are called succeeding activities.

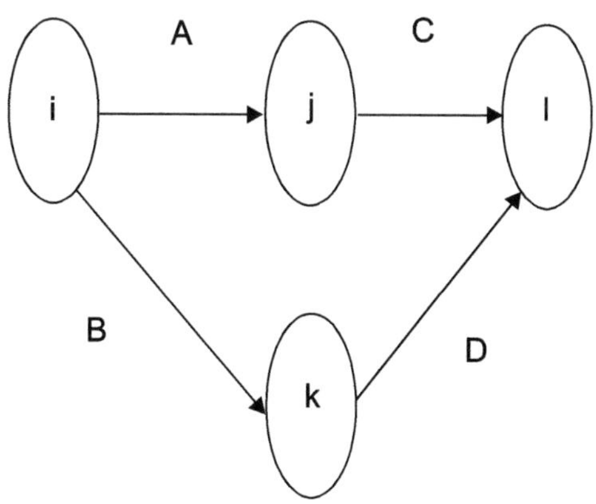

Preceding and Succeeding Activity

In figure activities A and B preceding activities C and D respectively. Where activities A and B are succeeding activities for Event J and K.

Event: An event will always occur at the beginning of the activity and end of the activity.

- In network numbering the event applying Fulkerson' RULE.The event has no resources.
- Event is denoted by a node. A node may be shaped in a circle, square.
- In the network, various types of events are there, namely Head event, tail event, merge and burst event.
- In the above figure event I is Head event and J is Tail event for activity A, also L event is merged event and I is burst event.

Dummy activity: An imaginary activity that does not consume any time and resources is known as a Dummy activity.

- **Dummy activities** are denoted by the Dotted line.
- The main function of dummy activity is representing a connection between events to maintain the sequence logically.

Critical Path Analysis

The critical path for any network is the longest path of the whole project network, all the activities will be complete then complete the whole project. If the project is to be completed in the shortest time, all activities on the critical path must be started as soon as possible. These activities are called critical activities.

Any delay in the critical activity will increase the project duration. Any type of delay considers here such as bad weather, accident, and some technical reason. Project duration directly affects the budget of the project, activities do not lie on the critical path is known as non–critical activities.

These types of non-critical activities may have some slack. The slack is the amount of time by which the start of an activity may be delayed without affecting the entire completion time of the project. But critical activities have no slack. To reduce the overall project time it would require more resources like man, material, machinery at extra cost to reduce the time taken by the critical activities to complete.

Example 3.1: How to Calculate Project Completion Time

A project schedule has the following data

Sr. No.	Activity	Name	Time	Sr. No.	Activity	Name	Time
1.	1-2	A	4	7.	5-6	G	4
2.	1-3	B	1	8.	5-7	H	8
3.	2-4	C	1	9.	6-8	I	1
4.	3-4	D	1	10.	7-8	J	2
5.	3-5	E	6	11.	8-10	K	5
6.	4-9	F	5	12.	9-10	L	7

- Construct PERT network.
- Compute TE and TL for each activity.
- Find the critical path.

From the given data, the activities network are constructed as shown in the given figure 3.8 below.

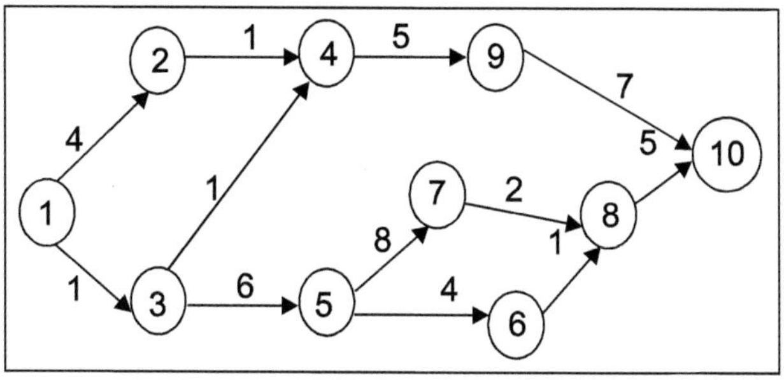

Fig. 3.8

To determine the critical path, compute the earliest time TE and latest time TL for each of the projects. The calculation of TE and TL are as follows:

To calculate TE for all activities

- TE 1 = 0
- TE 2 = TE1 + t1,2 = 0+4 = 4
- TE3 = TE1 + t1, 3 = 0+1 = 1
- TE4 = max (TE2 +t2, 4 and TE3 +t3,4)
- TE4 = max (4 + 1 and 1+ 1)
- TE4 = 5 Days
- TE5 = TE3 + te3, 6 = 1+6 = 7
- TE6 = TE5 + te6 = 7+4 = 11
- TE7 = 15
- TE8 = 17
- TE9 = 10

To calculate TL for all activities

- TL10 = TE10 =22
- TL9 = TE10 – t9,10 = 22 – 7 = 15
- TL8 = TE10 – 18,10 = 22 – 5 = 17
- TL7= TE8 – t7,8 = 17 – 2= 15
- TL5 = 7
- TL4 = 10
- TL3 = 1
- TL2 = 9
- TL 1= 0

Activity	Name	Normal Time	Earliest Start Time	Earliest Finish Time	Latest Start Time	Latest Finish Time	Total Float
1-2	A	4	0	4	5	9	5
1-3	B	1	0	1	0	1	0
2-4	C	1	4	5	9	10	15
3-4	D	1	1	2	9	10	8
3-5	E	6	1	7	1	7	0
4-9	F	5	5	10	10	15	5
5-6	G	4	7	11	12	16	5
5-7	H	8	7	15	7	15	0
6-8	I	1	11	12	16	17	5
7-8	J	2	15	17	15	17	0
8-10	K	5	17	22	19	22	0
9-10	L	7	10	17	15	22	5

Techniques for Planning and Scheduling

The technique used for project scheduling will vary depending upon the project's size, complexity, duration, personnel, and owner requirements. The project manager must choose a scheduling technique that is simple to use and is easily interpreted by all project participants. There are two general methods that are commonly used: the bar chart (sometimes called the Gantt chart) and the Critical Path Method (some- times called CPM or network analysis system).

The bar chart, developed by Henry L. Gantt during World War I, is a graphical time-scale of the schedule. It is easy to interpret; but it is difficult to update, does not show interdependences of activities, and does not integrate costs or resources with the schedule. It is an effective technique for overall project scheduling, but has limited application for detailed construction work because the many interrelationships of activities, which are required for construction work, are not defined. Many project managers prefer the bar chart for scheduling engineering design work because of its simplicity, ease of use, and because it does not require extensive interrelationships of activities. However, it can require significant time for updating since the inter relationships of activities are not defined. A change in one activity on the bar chart will not automatically change subsequent activities. Also, the bar chart does not integrate costs with the schedule, nor does it provide resources, such as hours, that are important for management of design. Garold D. Oberlender,. (2014).

Some designers argue that they cannot define the interrelationships between the activities that make up a design schedule. They use this argument to support the use of a bar chart. They will also argue that resources change constantly on a design project, resulting in a schedule that is too difficult to maintain. Either of these situations may occur at times on some projects. However, if these situations exist on every project, it is likely that the projects are not well planned, managed, or controlled.

The Critical Path Method (CPM) was developed in 1956 by the DuPont Company, with Rernington Rand as consultants, as a deterministic approach to scheduling. The CPM method is commonly used in the engineering and construction industry. A similar method, Program Evaluation and Review Technique (PERT), was developed in 1958 by the U.S. Navy, with Booz, Allen & Hamilton Management consultants, as a probabilistic approach to scheduling. It is more commonly used by the manufacturing industry; however, it can be used for risk assessment of highly uncertain projects. Both methods are often referred to as a network analysis system. The CPM provides interrelationships of activities and scheduling of costs and resources. It also is an effective technique for overall project scheduling and detailed scheduling of construction. However, it does have some limitations when applied to detailed engineering design work during the early stages of a project because it requires an extensive description of the interrelationshipsof activities.

Although the CPM technique requires more effort than a bar chart, it provides more detailed information that is required for effective project management. Using a network schedule to plan a project forces the project team to break a project down into identifiable tasks and to relate the tasks to each other in a logical sequence in much greater detail than a bar chart. This up-front planning and scheduling helps the project team to identify conflicts in resources before they occur. The project manager must use his or her own judgement and select the method of scheduling that best defines the work to be done and that communicates project requirementsto all participants.

Network Analysis Systems

A network analysis system (NAS) provides a comprehensive method for project planning, scheduling, and controlling. NAS is a general title for the technique of defining and coordinating work by a graphical diagram that shows work activities and the interdependences of activities. Many books and articles have been written that describe the procedures and applications of this technique. It is not the purpose of this book to present the details of network methods because so much material has already been written. The following paragraphs and figures present the basic fundamentals of NAS to guide the project manager in the development of the project plan and schedule. The basic definitions shown in figure 3.1 are presented to clarify the following paragraphs because there are variations in terminology used in network analyses. Garold D. Oberlender,. (2014).

Activity - The performance of a task required to complete the project, such as, design of foundations, review of design, procure steel contracts, or form concrete columns. An activity requires time, cost, or both time and cost.

Network - Adiagram to represent the relationship of activities to complete the project. The network may be drawn as either an "arrow diagram" or a "precedence diagram."

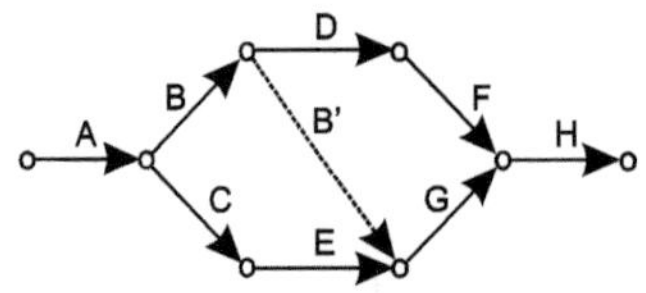

Arrow Diagram

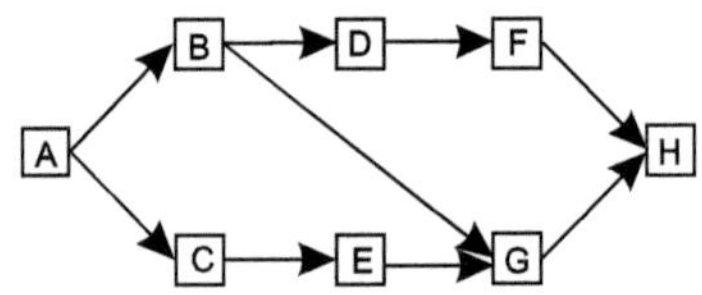

Precedence Diagram

Duration (D) - The estimated time required to perform an activity. The time should include all resources that are assigned to the activity.

Early Start (ES) - The earliest time an activity can be started.

Early Finish (EF)- The earliest time an activity can be finished and is equal to the early start plus the duration.

$$EF = ES + D$$

Late Finish (LF) - The latest time an activity can be finished.

Late Start (LS) - The latest time an activity can be started without delaying the completion date of the project.

$$LS = LF - D$$

Total Float (TF) - The amount of time an activity may be delayed without delayingthe comple- tion date of the project.

$$TF = LF - EF LS - ES$$

Free Float (FF) - The amount of time an activity may be delayed without delaying the early start time of the immediately following activity. = - EF,, where the subscriptirepresents the preceding activity and the subscript represents the following activity.

Critical Path - Aseries of interconnected activities through the network diagram, with each activity having zero, free and total float time. The critical path determines the minimum time to complete the project.

Dummy Activity - An activity (represented by a dotted line on the arrow network diagram) that indicates that any activity following the dummy cannot be started until the activity or activities preceding the dummy are completed. The dummy does not require any time.

Basic Definitions for CPM

For project management the CPM is the most commonly used NAS method. The concept is simple, the computations only require basic arithmetic; and a large number of computer programs are available to automate the work required of CPM scheduling. The most difficult task in the use of CPM is identifying and interfacing the numerous activities that are required to complete a project, that is, development of the CPM network diagram. If a well-defined is developed the task of developing a CPM diagram is greatly simplified.

There are two basic methods of drawing CPM diagrams: the arrow diagram (sometimes called activity on arrow) and the precedence diagram (sometimes called activity on node). Although both methods achieve the same results, most project managers prefer the precedence method because it does not require the use of dummy activities. The precedence method can also provide the start-to-start, to-finish, start-to-finish, and finish-to-start relationship of activities, which can significantly reduce the number of activities that are required in a network diagram. However, many individuals prefer to not use these relationships because of potential confusion in the network scheduling.

Figure 3.9 is a simple precedence diagram that is presented to illustrate the time computations for analysis of a project schedule by the CPM. Each activity is described by a single letter. The number at the top of the activity is the assigned activity number, and the number at the bottom of each activity represents the duration in working days. A legend is shown in the lower left hand corner to define the start and finish days. All calculations for starts and finishes are based on end-of-day. Garold D. Oberlender,. (2014).

Time Computations for Simple Precedence Diagram

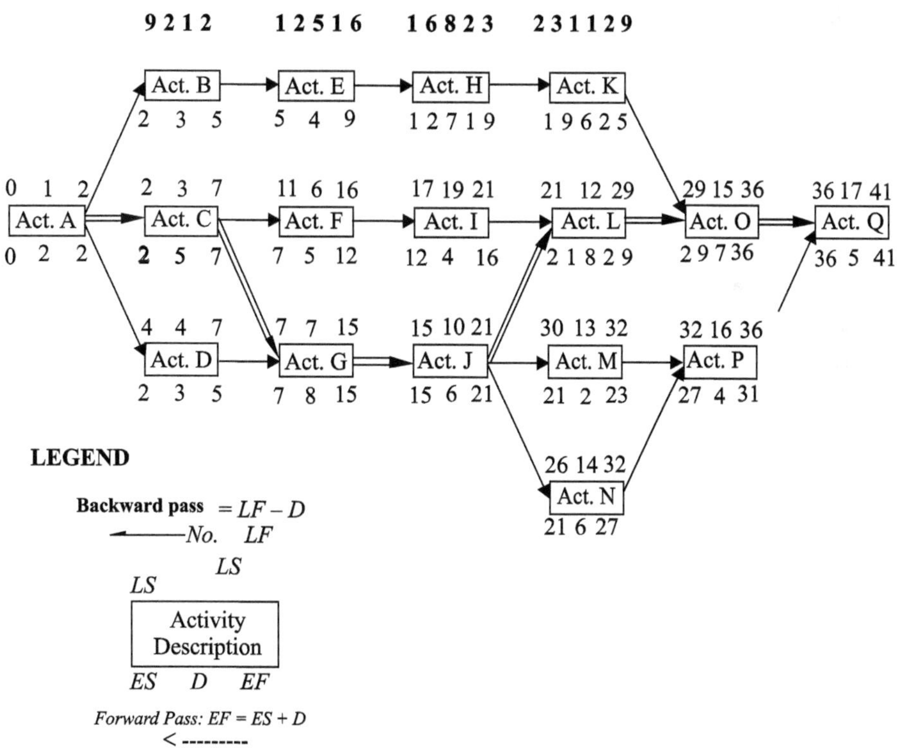

After the CPM diagram has been prepared, the duration of each activity can be assigned and the forward-pass calculations performed to calculate the early start and early finish of each activity. The largest early finish of all preceding activities defines the early start of all following activities. For example, activity H cannot be started until activities E and F are both completed. Since the largest early finish of the two preceding activities is 12, the early start for activity H is 12. The pass calculations are performed on all activities from the first activity A to the last activity Q. The early finish of the last activity defines the project completion, which is 41 days for this particular project. This project duration is a calculated value based upon the duration and interdependences of all activities in the project.

A backward pass can be performed to calculate the late start and late finish of each activity. The smallest late start of all following activities defines the late finish of all preceding activities. For example activities H and I cannot both be started until activity F is completed. Since the smallest early start of the two following activities is 16, the late finish for F is 16. The backward-pass calculations are performed on all activities from the last activity Q to the first activity A.

The difference between starts and finishes determines the amount of free and to tal float. For example, the total float for activity M is 9 days, the difference between its late start (30) and early start (21). The free float for activity M is 4 days, the dif ference between its early finish (23) and the early start (27) of the immediately fol lowing activity P.

The critical path, as noted by the double line on figure 3.9, is defined by the series of interconnected activities that have zero total float. Since these activities have no float time available, any delay in their completion will delay the completion date of the project. Therefore, they are called critical activities.

Table 3.3 lists the basic steps to guide the process of developing a CPM diagram for project planning and scheduling. It is not always possible to complete a step without some readjustments. For example, the CPM diagram of step 2 may need readjusting after evaluation of the time and resources of steps 4 and 5. Some activities that were originally planned in a series may need to occur in parallel to meet a time requirement. Each project manager and his or her team must work together to develop a project plan and schedule that achieve the required project completion date with the resources that are available.

Development of CPM Diagram from the WBS

Table 3.3 provided the list of basic steps that can be used to guide the process of developing a network analysis system for project planning and scheduling. The development of the WBS is an important first step that is often neglected. Attempting to draw the CPM diagram without a WBS usually leads to numerous revisions to the diagram. Garold D. Oberlender,. (2014).

Table 3.3: Steps in Planning and Scheduling

Steps
1. Develop a work breakdown structure (WBS) that identifies work items (activities) (a) Consider activities that require time (b) Consider activities that require cost (c) Consider activities that you need to arrange (d) Consider activities that you want to monitor
2. Prepare a drawing (network diagram) that shows each activity in the order it must be performed to complete the project (a) Consider which activities immediately precede each activity (b) Consider which activities immediately follow each activity (c) The interrelationship of activities is a combination of how the work must be done (constraints) and how you want the work to be done
3. Determine the time, cost, and resources required to complete each activity (a) Review work packages of the WBS (b) Obtain input from project team members
4. Compute the schedule to determine start, finish, and float times (a) Perform a forward pass to determine early starts and finishes (b) Perform a backward pass to determine late starts and finishes (c) Determine the differences between start and finish times to determine float time and critical activities
5. Analyze costs and resources for the project (a) Compute the cost per day for each activity and for the entire project (b) Compute the labor-hours per day and/or other resources that are required to complete the project
6. Communicate the results of the plan and schedule (a) Display time schedule for activities (b) Display cost schedule for activities (c) Display schedule for other resources

Resource Allocations for Design

Efficient utilization of resources is critical to successful management. The primary resource during design is the work-hours of the design team. The project manager depends on the design team to create design alternatives, produce drawings, and write specifications for the proposed project. To properly coordinate all aspects of the design effort, the project manager must ensure the correct expertise is available when needed. Generally, design team members are assigned to the project by their respective home departments. Since each designer is often involved in several projects at the same time, the project manager must develop a resource allocation plan for each project. The plan should then be distributed to each design team member's home department to ensure that each resource will be available when needed.

The project manager can resource load the project plan to include the number of work-hours required for each design discipline. The resource plan is similar to the cost distribution analysis presented earlier in this chapter, except work-hours are used in place of dollars of cost. Thus, the resource plan is simply a histogram of work-hours versus time for each design discipline. The project manager should provide the resource plan for each project to the design team's manager. The design managers can then integrate the resource plans of all active projects into their department's demand for technical expertise. This is necessary to ensure adequate resources are available for the projects when required.

Resource Allocations for Construction

During the construction phase, the primary resources are labor, materials, and equipment. The correct quantity and quality of material must be ordered and delivered to the job-site at the right time to ensure efficiency of labor. Equipment to be installed in the project often requires a long lead time from the fabricator. Thus, the project plan should include material and equipment required by the construction work force.

Labor represents a major cost of construction. The labor force on the job operates construction equipment and installs the materials. A resource allocation plan is required to ensure high efficiency and productivity during construction. The project manager can resource load the project plan to include the number of work-hours required for each craft of construction labor. The resource plan is a histogram of work-hours versus time.

The construction plan shows the desired sequence of work. However, to be workable, the plan must also show the distribution of resources, such as the required labor for each craft on the job. The demand for labor should be uniformly distributed for each craft on the project, to prevent irregularities. The resource plan can be used as a tool to ensure a relatively uniform distribution of labor on the project.

Program Evaluation and Review Technique (PERT)

In the Critical Path Method of scheduling projects, the duration of each activity is usually defined with a reasonable degree of certainty. For most projects the type and amount of work is known, which enables the project manager to establish the approximate duration for each work activity. For example, the time to produce PERT uses a weighted average of the three times to find the overall project duration. This average is called the expected t_e time, and is found by the following simple equation:

$$t_e = \frac{a + 4m + b}{6}$$

where t_e is the expected time of an activity.

The above equation is used to calculate a single duration for each activity in the PERT network diagram. Then a forward pass can be performed to calculate the early start and early finish for each activity in the PERT network and a total project duration, similar to the CPM scheduling method.

A backward pass can be performed to calculate the last start and late finish for each activity. Total float, free float, and the critical path can be identified. At this stage of a PERT analysis it is identical to the CPM analysis.

The three time estimates of PERT can be used to measure the degree of uncertainty involved in the activity. The measure of the spread of the distributionis called the *standard deviation* and is denoted by the Greek letter a. To determine the probability of the project completing earlier or later than expected using PERT, the variance of each activity along the critical path must be calculated. The variance of an activity is the square of the standard deviation of the activity and can be calculated using the following equation:

$$v = \sigma^2 = \left(\frac{b-a}{6}\right)^2$$

where v is the variance of an activity.

Since the duration (t_e) of each of the activities in the arrow network diagram is uncertain, the time of occurrence of each event is also subject to uncertainty. The expected time of an event is denoted as Although the distribution of uncertainty of individual activities may not necessarily be symmetrical, the distribution for event times is assumed symmetrical because there are numerous activities in the chain ahead of the event. For example, the T_E for the final event in a project has all the critical path activities in the chain that lead up to the final event in the project.

The measure of uncertainty of the final event in a PERT diagram is the standard deviation of the expected time, denoted as $\sigma_{TE.}$ The a, is the square root of the sum of the activities ahead of the event. Thus, the a, for the last event in a PERT diagram is the square root of the sum of the variance of all activities along the critical path. The a, is calculated with the following equation. Garold D. Oberlender,. (2014).

$$\sigma_{TE} = \sqrt{v_{1-2} + v_{2-3} + v_{3-4} + \ldots + v_{i\text{-}j}}$$

where σ_{TE} is the standard deviation of the expected time.

The final calculation necessary to perform PERT calculations is the deviation, which is denoted by the symbol z in the following equation:

$$z = \frac{T_s - T_E}{\sigma_{TE}}$$

where z is the deviation.

Table 3.4: PERT Probability Table

Value of deviation z	Probability of completion
-3.0	0.00
-2.5	0.01
-2.0	0.03
-1.5	0.07
-1.4	0.08
-1.3	0.09
-1.2	0.11
-1.1	0.14
-1.0	0.16
-0.9	0.18
-0.8	0.21
-0.7	0.24
-0.6	0.27
-0.5	0.31
-0.4	0.35
-0.3	0.38
-0.2	0.42
-0.1	0.46
-0.0	0.50
0.1	0.54
0.2	0.58
0.3	0.62
0.4	0.66
0.5	0.69
0.6	0.73
0.7	0.76
0.8	0.79
0.9	0.82
1.0	0.84
1.1	0.86
1.2	0.88
1.3	0.90
1.4	0.92
1.5	0.93
2.0	0.98
2.5	0.99
30.	1.00

Table 3.4 provides the value of z for various probabilities of certainty. The term T_E is the expected time of the event, which is calculated from the PERT diagram, and is the scheduled time. Garold D. Oberlender,. (2014).

Example 3.2

The following PERT diagram shows the optimistic time (a), the most likely time (t_m) and the pessimistic time (b) at the bottom of each activity. The number shown in parentheses below each activity is the expected time (t_e) for the activity, calculated from the equation, t_e = (a + 4m + b)/6. All times denote working days in the project.

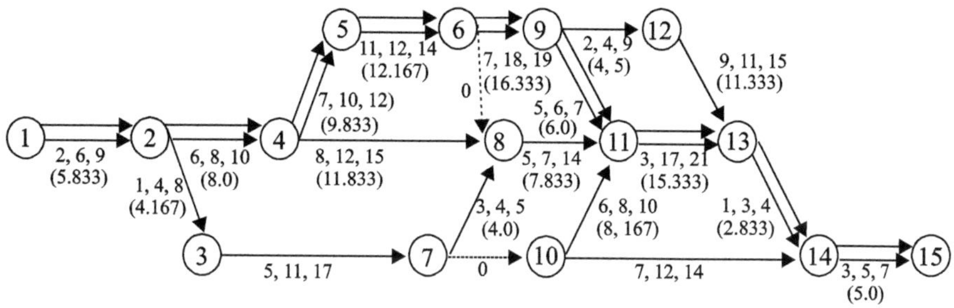

The optimistic duration of the project can be calculated by performing a forward pass to calculate the start and finish times using the optimistic duration (smallest duration) of each activity. For the data in this PERT diagram, the optimistic project duration is 48 days, which follows through events 1, 2, 4, 5, 6, 9, 12, 13, 14, and 15.

The pessimistic duration of the project can be calculated by performing a forward pass to calculate the start and finish times using the pessimistic (largest duration) of each activity. For the data in this PERT diagram, the pessimistic project duration is 103 days, which follows through events 1, 2, 4, 5, 6, 9, 11, 13, 14, and 15.

The project duration (T_E) for event 15 can be calculated by performing a forward pass to calculate the start and finish times using the expected duration (t_e) of each activity. For the data in this PERT diagram, the expected time of the last event number 15 T_E is = 81.3 days, which follows through events 1, 2, 4, 5, 6, 9, 11, 13, 14, and 15.

In summary, for this PERT diagram the optimistic project duration is 48 days, the pessimistic project duration is 103 days, and the expected project duration is 81.3 days. The probability of completing the project ahead or behind day 81.3 will depend on the a, of event 15. The value of a, can be calculated as the square root of the sum of the variances of all activities along the critical path. Based on the expected time of each activity, the critical path for this PERT network follows events 1, 2, 4, 5, 6, 9, 11, 13, 14, and 15. The T_E for the final event, number 15, is the square root of the variances of these events, which can be calculated as follows:

$$\sigma_{TE} = \sqrt{V_{1-2} + V_{2-4} + V_{4-5} + V_{5-6} + V_{6-9} + V_{9-11} + V_{11-13} + V_{13-14} + V_{14-15}}$$

$$= \sqrt{\left(\frac{9-2}{6}\right)^2 + \left(\frac{10-6}{6}\right)^2 + \left(\frac{12-7}{6}\right)^2 + \left(\frac{14-11}{6}\right)^2 + \left(\frac{19-7}{6}\right)^2 + \left(\frac{7-5}{6}\right)^2 + \left(\frac{21-3}{6}\right)^2 + \left(\frac{4-1}{6}\right)^2 + \left(\frac{7-3}{6}\right)^2}$$

$$= \sqrt{1.3661 + 0.4444 + 0.6944 + 0.2500 + 4.0000 + 0.1111 + 9.0000 + 0.2500 + 0.4444}$$

$$= \sqrt{16.5554}$$

$$= 4.07$$

All the calculations necessary to perform an analysis of the project finishing ahead or behind the expected project duration of 81.3 days can be determined from the deviation equation. For example, it may be desired to determine the probability of the project finishing later than expected, later that 81.3 days. The probability of completing the project by the eighty-fourth day can be calculated as follows:

$$z = \frac{T_s - T_E}{\sigma_{TE}}$$

$$= \frac{84.0 - 81.3}{4.07}$$

$$= 0.663$$

From table 3.4, based on a deviation z of 0.663, the probability is 0.75. Thus, there is a 75% probability that the project will be completed by the fourth day.

Other probability scenarios can be calculated. For example, the probability of the project finishing earlier than expected, such as in 78 days, can be calculated as follows:

$$z = \frac{T_s - T_E}{\sigma_{TE}}$$

$$= \frac{78.0 - 81.3}{4.07}$$

$$= -0.811$$

From table 3.4, based on a deviation z of -0.811, the probability is 0.21. Thus there is only a 21% probability that the project will be completed by the eighth day.

It may be desired to calculate the number of days to complete the project based on a 70% probability. This scenario can be calculated as follows. For a 70% probability the value is 0.5333:

$$z = \frac{T_s - T_E}{\sigma_{TE}}$$

$$0.533 = \frac{T_s - 81.3}{4.07}$$

$$T_s = 83.5 \text{ days}$$

Thus, there is a 70% probability that the project will be completed by 83.5 days.

The following is a graphical plot of the probability versus the project duration. This curve provides the full range of probabilities for assessing the risks of finishing the project ahead of or behind the expected duration of 81.3 days.

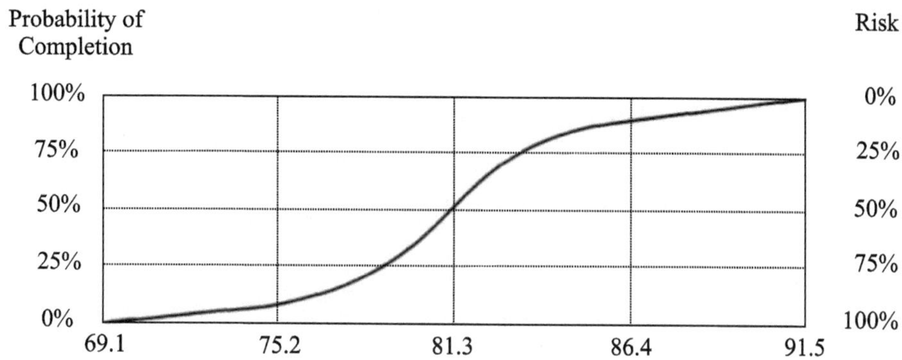

This analysis is based on a, a for event 15 based on the critical path activities. A full analysis is more complex if all near-critical paths are analyzed. For example, it is possible to have another path through the PERT diagram that is near critical, which would yield a shorter for event 15 and a higher value for ,a due to larger variances in the activities in the non-critical path. Garold D. Oberlender,. (2014).

Success of Predecessor Relationships

The CPM diagram is a graphical representation that shows the sequencing of activities in the project. The purpose of the CPM diagram is to model the logical flow of work, showing each activity and one or more activities that follow each activity. In the activity relationships, the preceding activity is called the predecessor activity and the following activity is called the successor activity. In a pure CPM network diagram a finish-to-start (F-S) relationship of each activity is assumed. That is, the preceding activity must be completed before starting the following activity. All the CPM diagrams and calculations presented earlier in this book are based on an F-S relationship of activities to depict a pure CPM logic network diagram.

Most commercially available computer software programs for CPM provide activ- ity relationships other than F-S. These additional activity relationships include start- to-start (S-S), start-to-finish (S-F), and finish-to-finish (F-F). The S-S relationship means the successor activity can start at the same time or later than the predecessor activity. The F-F relationship means the successor activity can finish at the same time as or later than the predecessor activity. The S-F relationship means the predecessor activity must start before the successor activity can finish. There is no practical application of the S-F relationship in the engineering and construction industry. It is presented here just to show all possible successor/predecessor relationships.

Delays in the relationship of one activity to another is defined by lag. Lag is the amount of time that an activity follows or is delayed from the start or finish of its predecessor. For example, an F-S relationship with a two-day lag means the start of the successor activity cannot occur until two days after the finish of the predecessor activity. Lag can be assigned to any activity relationship, including F-S, F-F, S-S, and S-F. In addition, the lag can also be a negative number. Lead is the opposite of lag, which is the amount of time that an activity precedes the start or finish of its successor. Thus, the use of successor/predecessor relationships provides many options for modeling the network logic to show constraints between activities and the sequential flow of work in the project. Garold D. Oberlender,. (2014).

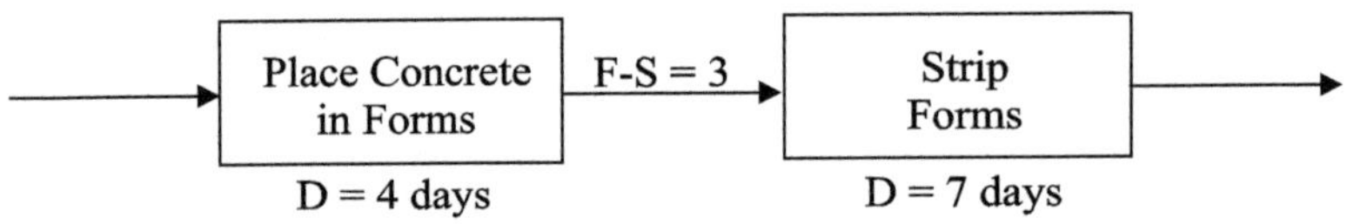

a. Modern with Successor/Predecessor with F-S = 3 days of lag

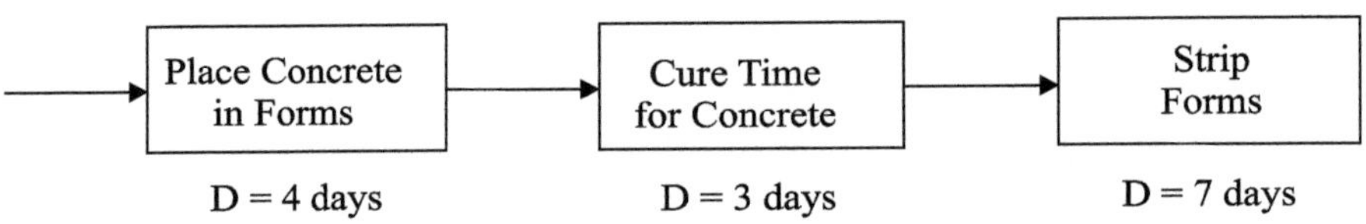

b. Modern without Successor/Predecessor Relationship

Fig. 3.9: Comparison of CPM Diagrams with and without a Finish-to-Start (F-S) Relationship

A model of the F-S with lag relationship is illustrated in figure. An alternative of this model is shown in figure which shows the same results without using the F-S with lag relationship. Thus, adding one intermediate activity between the successor and predecessor activities clearly produces the same results as the F-S relationship.

The S-S relationship is used to overlap activities, allowing the successor activity to start before the predecessor activity is completed. To tie the finish of the predecessor activity to the successor activity, the F-F relationship is normally used together with the S-S relationship. Figure 3.10a illustrates S-S and F-F with lag relationships between two activities. Figure 3.10b is a model of the activities with- out the S-S and F-F relationships. Thus, for any CPM diagram the S-S and F-F activity relationships can be eliminated by adding additional activities. Although additional activities may appear as a disadvantage, the additional activities provide a clearer understanding of the sequence of work.

As previously stated, there is no practical application of the S-F relationship in the engineering and construction industry. Some computer scheduling software packages do not allow the use of the S-F relationship.

Problems using Successor/Predecessor Relationships

Successor/predecessor relationships between activities allow overlapping of concurrent activities, thereby reducing the number of activities in CPM precedence diagrams. However, misunderstandings, confusion, and serious problems have been created by assigning successor/ predecessor relationships to activities in the project schedule. Too often a good CPM schedule at the beginning of the project becomes a schedule that is incoherent and unmanageable due to relationships that are assigned to activities, particularly during schedule updates. Garold D. Oberlender,. (2014).

In a pure CPM diagram, activity relationships are not used. During the forward pass, the path that yields the largest early finish of the preceding activities becomes the early start of the following activities. During the backward pass, the path that yields the smallest late start of the following activitiesis the late finish of the preceding activities. Thus, in a pure CPM diagram there are no questions regarding the start and finish times of activities.

When successor/predecessor activity relationships are introduced into a CPM diagram, the start, finish, and float times may be governed by the successor and predecessor relationships, which makes the calculations more complex. Using activity relationships with positive or negative lag provides numerous opportunities to model the network logic, but it can also create numerous opportunities for creating confusion, and misunderstandings in the project schedule.

Figure 3.10 illustrates problems that can occur in calculating the start and finish dates when modeling multiple activity relationships. Each successor/predecessor path must be evaluated to determine start and finish times. A forward pass for the activities in figure 3.10 shows Activity B can start as early as two days or as late as seven days after the start of Activity A, depending on the path chosen. Also, depending on the successor/ predecessor path chosen, the early finish for Activity B is 9, 12, 14, or 19 days. Similarly, a backward pass will result in multiple late starts and late finishes for these activities. All these multiple calculations can cause confusion and a lack of confidence in the project schedule. Garold D. Oberlender,. (2014).

Figure 3.12 is a CPM diagram with successor/predecessor relationships of construction work related to foundation construction work. Pure CPM diagram, without successor/predecessor activity relationships, for the same construction work. These figures are presented to illustrate the complexity of successor/predecessor calculations. As shown in figure 3.12, the early start of 10 days for Activity 30 is calculated by subtracting its duration from the latest early finish time. The largest early finish of Activity 30 is 16 days, which is governed by the F-F relationship between Activities 20 and 30.

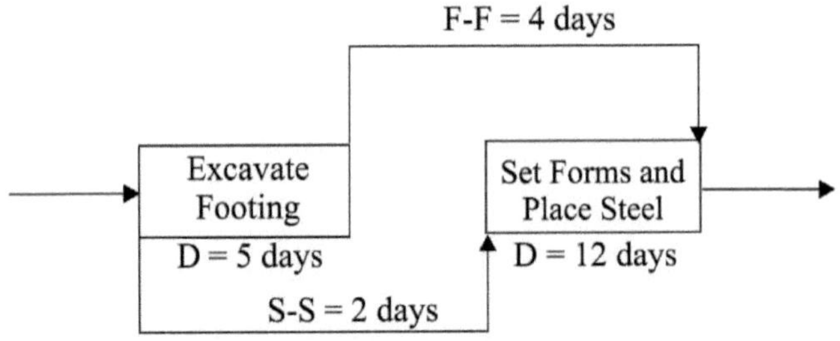

a. Model with Successor/Predecessor S-S = 2 days and F-F = 4 days of lag

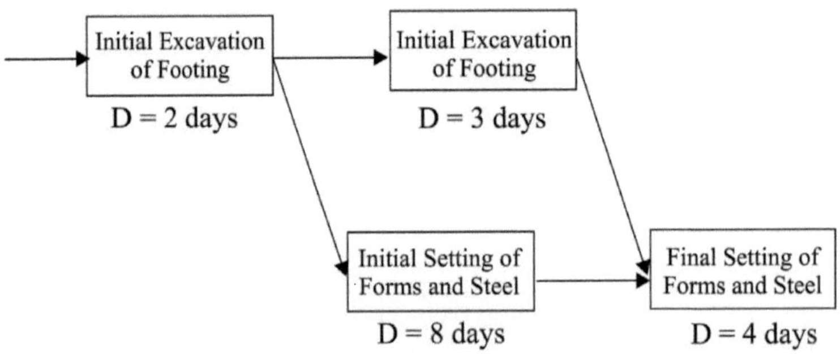

b. Model without Successor/Predecessor Relationships

Fig. 3.10: Comparison of CPM Diagrams with and without a Start-to-Start (S-S) Successor/Predecessor Relationship

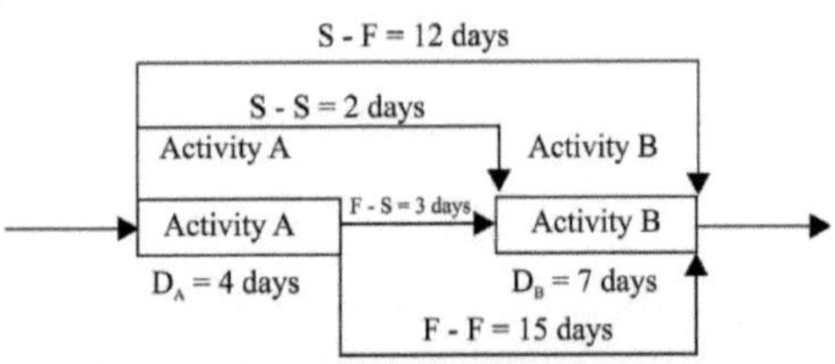

Fig. 3.11: Example of Multiple Successor/Predecessor Activity Relationship

When multiple S-S and F-F relationships are used to relate the predecessor activity to its successor activities, the calculations of total and free float are very complex and require a close examination of the relationships. Using multiple relationships, it is possible to show activities that are critical, when they are really not critical.

A review of figure 3.12 indicates that all the activities are critical, although that is not the case. Figure 3.12 is a pure CPM diagram for the same work shown in figure 3.11. Figure 3.12 eliminates the S-S and F-F relationships by splitting activities to show the actual sequencing of work. As shown in figure 3.12 much of the work is actually non-critical, compared to that of figure 3.11. For example, the network of figure 3.12 shows all form works as critical, but most of the excavation work and placement of concrete is non-critical. A pure CPM diagram as illustrated in figure 3.12 provides an accurate description of the work without raising questions regarding start/finish times, float, or critical path activities. Thus, a pure CPM diagram is recommended, rather than modeling with successor/ predecessor activity relationships.

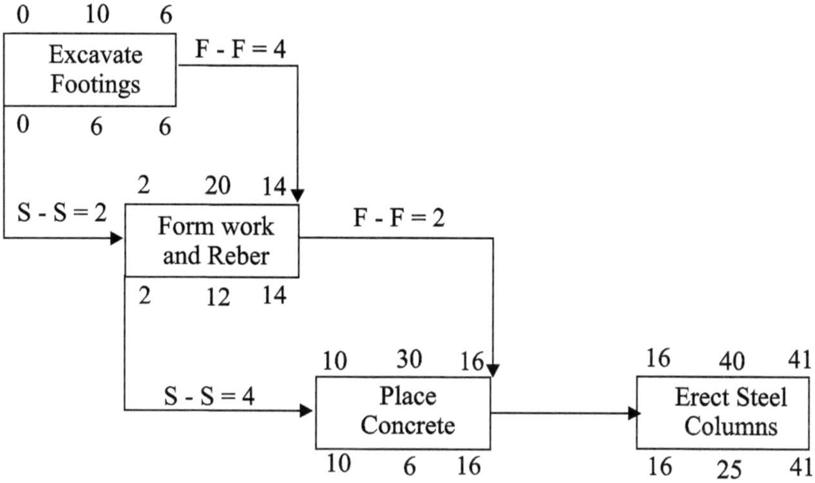

Fig. 3.12: Successor/Predecessor Activity Relationships. (Note all activities appear critical.)

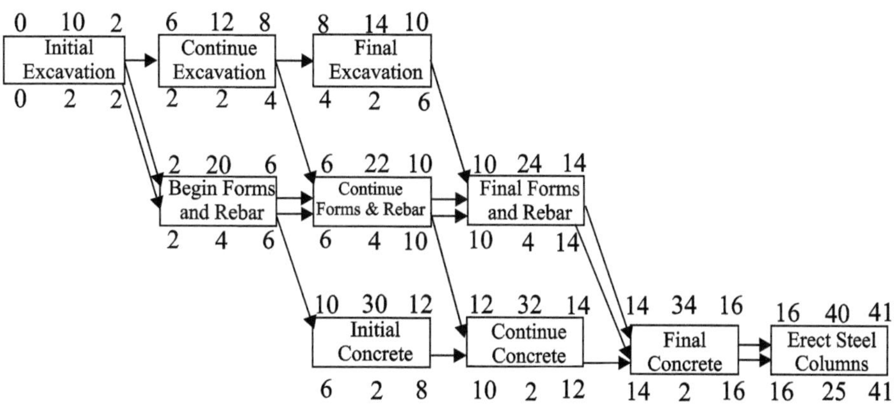

Fig. 3.13: Pure CPM Diagram without Activity Relationships. (Double arrows denote path of critical activities: 10, and 40)

Unfortunately, there is no industry standard to define how to finalize the early/late starts and finishes using successor/predecessors relationships with lags.

Each computer software developer defines the algorithm used by his or her software to determine start and finish dates using successor/

predecessor activity relationships. Since different software packages handle the calculations differently, it is possible to get different results in the schedule, depending on the software used, which creates problems in understanding and interpreting project schedules. Therefore, take extreme caution when using successor/predecessor activity relationships because serious problems can arise. To prevent these problems, a pure CPM diagram can be developed without successor/predecessor relationships by simply adding additional activities. Although additional activities may appear as a disadvantage, adding additional activities provides a clear understanding of the sequence of work, thus preventing confusion and misunderstandings of the project schedule.

The following examples are presented to illustrate calculations based on each of the successor/predecessor activity relationships, F-S, S-S, F-F, and S-F. These examples show how each path must be carefully investigated to determine the start, finish, and float times. When the calculations are performed by computer software, it is possible to get project schedules that sometimes do not make logical sense. Garold D. Oberlender,. (2014).

Thus, it is important to understand the calculations and be able to check the results of computer outputs to ensure the results are presented as intended by the person who prepared the schedule as well as by the persons who will be interpreting and using the schedule.

Example 3.3: Finish-to-Start Activity Relationships

Calculate the start, finish, and float for each activity in the following CPM precedence network diagram. Activity C and E have an F-S relationship with a 5-day negative lag. The 5-day negative lag is equivalent to a 5-day lead, which means the finish of Activity C must occur no later than 5 days after the start of Activity E, given by the relationship: $EF_c \leq E_{SE} + 5$. The use of negative lags is not rec- ommended because they lead to much confusion with respect to the meaning of the relationship and the schedule as discussed in the following paragraphs. The following diagram and accompanying table show the early and late start, early and late finish, and the total free float for each activity.

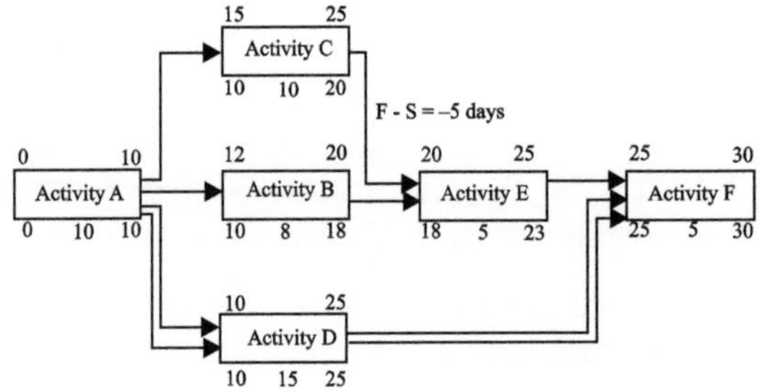

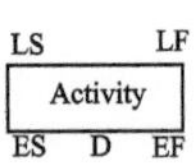

For Activity C, the calculations for total and free float are:

$TF_C = LF_C - EF_C$
$= 25 - 20$
$= 5$ days

$FF_C = ES_E - EF_C - Lag$
$= 18 - 20 - (-5)$
$= 3$ days

Activity	Duration	Early start	Early finish	Late start	Late finish	Total float	Free float
A*	10	0	10	0	10	0	0
B	8	10	18	12	20	2	0
C	10	10	20	15	25	5	3
D*	15	10	25	10	25	0	0
E	5	18	23	20	25	2	2
F*	5	25	30	25	30	0	0

* Denotes critical activities.

Based on the forward and backward pass calculations, the critical path is through Activities A, D, and F. Note the impact on the schedule and possible confusion due to the F-S relationship with five days of negative lag between Activity C and E. The is 20 days (governed by the equation $EF_c = ES_c + D$ 10 + 10 = 20). The EF_c is not governed by the equation $EF_c = \leq ES_E + 5 = 18 + 5 = 23$ days, because the ES_E is governed by 18 from the EF_B. Also, a review of the diagram shows the preceding activity $EF_c = 20$, which is greater than the following activity $ES_E = 18$, which is illogical. A computer schedule that shows the ES of a following activity to be earlier than the EF of a preceding activity will make no sense to workers in the field and will cause them to lose confidence in the schedule.

Similar confusion exists for the free float of Activity C. The FF_c is governed by the equation = $FFc = ES_E - EF_c - Lag = 18 - 20 - (-5) = 3$ days. However, a review of the diagram shows the $EF_c = 20$, which is greater than the $ES_E = 18$. Thus, the use of F-S relationships with negative lag is not recommended.

Example 3.4: Finish-to-Finish Activity Relationships

Calculate the start, finish, and float for each activity in the following CPM prece- dence network diagram. Activity C and E have an F-F relationship with a lag of 7 days. The following diagram and accompanying table show the early and late start, early and late finish, and the total and free float for each activity.

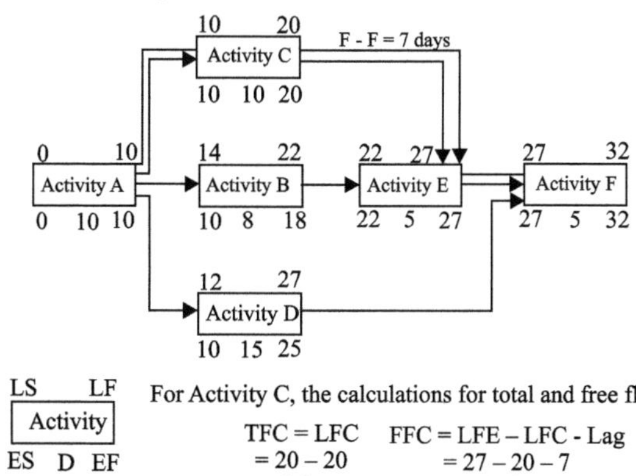

For Activity C, the calculations for total and free float art

TFC = LFC
= 20 – 20
= 0 days

FFC = LFE – LFC - Lag
= 27 – 20 – 7
= 0 days

Activity	Duration	Early start	Early finish	Late start	Late finish	Total float	Free float
A*	10	0	10	0	10	0	0
B	8	10	48	12	20	2	0
C	10	10	20	15	25	5	3
D*	15	10	25	10	25	0	0
E	5	18	23	20	25	2	2
F	5	25	30	25	30	0	0

* Denotes critical activities.

Based on the forward and backward pass calculations, the critical path is through Activities A, C, E, and F. Note that the free float of Activity C is governed by the F-F relationship.

Example 3.5: Start-to-Start Activity Relationships

Calculate the start, finish, and float for each activity in the following CPM prece- dence network diagram. Activity C and E have an S-S relationship with a lag of 10 days. The following diagram and accompanying table show the early and late start, early and late finish, and the total and free float for each activity. Garold D. Oberlender,. (2014).

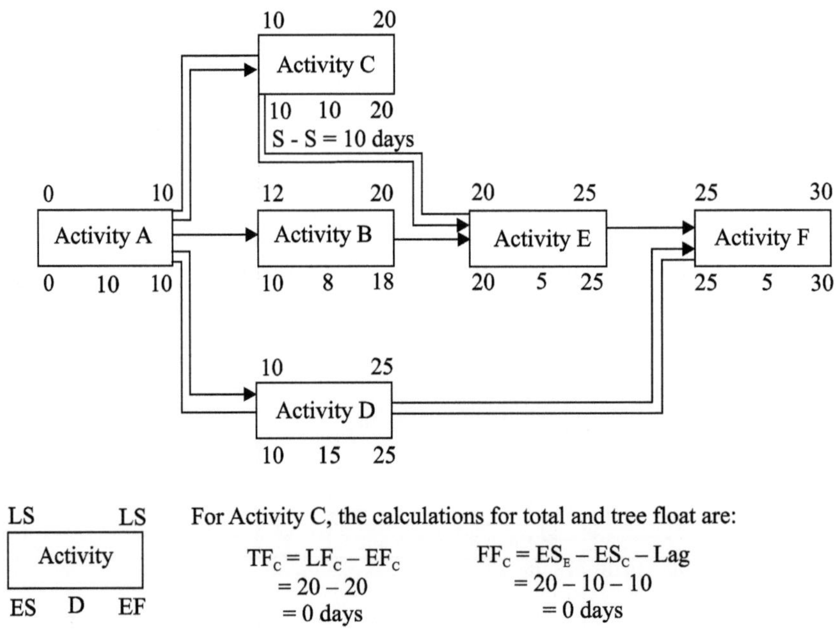

For Activity C, the calculations for total and tree float are:

$$TF_C = LF_C - EF_C = 20 - 20 = 0 \text{ days}$$

$$FF_C = ES_E - ES_C - Lag = 20 - 10 - 10 = 0 \text{ days}$$

Activity	Duration	Early start	Early finish	Late start	Late finish	Total float	Free float
A*	10	0	10	0	10	0	0
B	8	10	18	14	22	4	4
C*	10	10	20	10	20	0	0
D	15	10	25	12	27	2	2
E*	5	22	27	22	27	0	0
F*	5	27	32	27	32	0	0

* Denotes critical activities

Based on the forward and backward pass calculations, the critical path is through Activities A, C, D, E, and Note the free float of Activity C is governed by the S-S relationship between Activity C and E.

Example 3.6: Start-to-Finish Activity Relationships

Although the start-to-finish relationship is seldom used, this example is pre- sented to illustrate the calculations for an S-F relationship with a lag of 7 days between Activity C and E. The following diagram and accompanying table show the early and late start, early and late finish, and the total and free float for each activity. Garold D. Oberlender,. (2014).

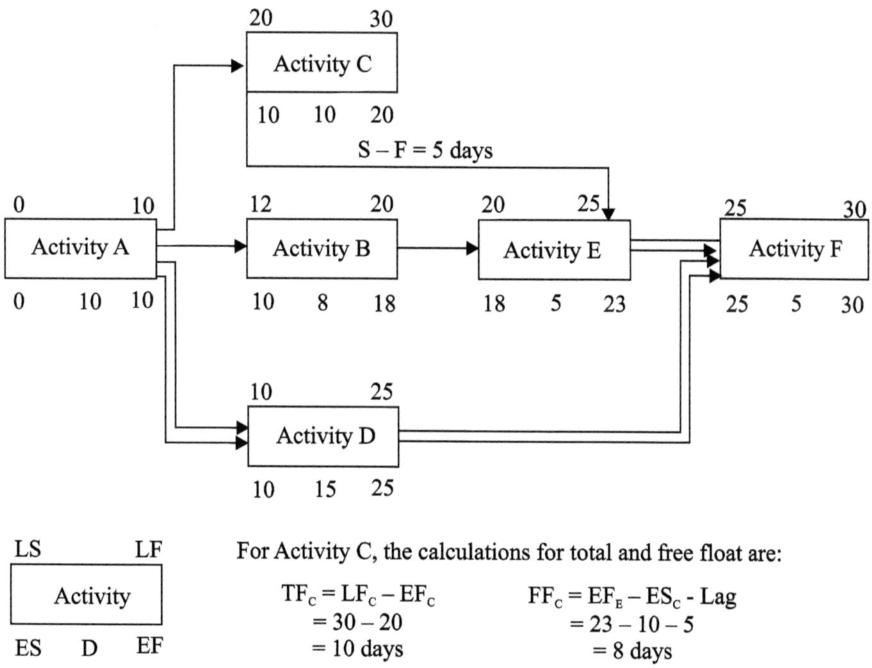

For Activity C, the calculations for total and free float are:

$$TF_C = LF_C - EF_C = 30 - 20 = 10 \text{ days}$$

$$FF_C = EF_E - ES_C - Lag = 23 - 10 - 5 = 8 \text{ days}$$

Activity	Duration	Early start	Early finish	Late start	Late finish	Total float	Free float
A*	10	0	10	0	10	0	0
B	8	10	18	12	20	2	2
C*	10	10	20	10	20	0	0
D*	15	10	25	10	25	0	0
E*	5	20	25	20	25	0	0
F*	5	25	30	25	30	0	0

* Denotes critical activities

Based on the forward and backward pass calculations, the critical path is through Activities A, D, and F.

Review Questions

1. What do you understand by Project Planning & Project Scheduling?
2. Explain different types of schedules.
3. What is the need of Project Control?
4. What do you understand by desired result of planning?
5. Explain the meaning of Bar chart, its advantages and disadvantages.
6. Explain the meaning of CPM, PERT and basic difference between CPM & PERT.
7. Explain the concept of CPM & PERT Network.
8. How CPM diagram is developed from WBS?
9. Explain the problem using successor/predecessor relationship.
10. Write short notes on:
    - Advantages and disadvantages of CPM& PERT
    - Critical Path Analysis
    - Network Analysis system
    - Resource Allocation for design
    - Resource Allocation for Construction
    - Successor/predecessor relationship
    - Master Schedule
    - Recovery Schedule
    - Revised Schedule

4

Construction Resource Allocation

Resource allocation is the process of assigning and scheduling available resources in the most effective and economical way possible. Projects will always need resources but they can often be scarce. The task, therefore, lies with the project manager to determine the proper timing and allocation of those resources within the project schedule.

Resource allocation or loading is nothing but the assigning required resource to each activity in amount and timing. On another word, resources in construction express 4 main categories contracted as 4Ms (Manpower, money, material and machinery). Assigning the required resources to each activity in amount and timing is known as resource allocation or loading. Resource leveling is the process of allocating resources among various activities such that peak requirement of resource is reduced with a constraint on time. To put it simply, resource leveling is basically minimalizing the variations resource use during the project execution. That means, if the resources are unlimited or infinite, then smoothing or leveling is performed to see the fluctuation. If the resources are restricted or scarce resource is predicted, then allocate it. Resource leveling is required in the project to avoid difficulties with major differences in resource use in the construction works. The resource allocation and leveling have attracted many researcher's attraction in the last decades.

Projects in the construction sector require cautious coordination of various moving parts. Resource allocation, which means managing people and equipment between different projects and locations is not an easy task. When talking of construction resource allocation, there are multiple processes and sub-processes under it. You need to have clear insights into your organizational objectives and capacity for effective resource planning and management. When there are loopholes in the process, it can have rippling effects on the entire workflow. For example, if the challenges in the construction resource allocation are not resolved, it can lead to lack of productivity, wastage of materials, and idle equipment. Thus, resource allocation or management issues cause a loss of time, effort, and money on projects. Hence, it's vital to align the right resources to suitable projects at the right time. Sounds tough?

Challenges in Resource Allocation and Management in Construction

Huge Number of Projects

When there are many residential and non-residential projects, contractors end up taking on more work than they can handle. Imagine managing multiple people, projects, and several deadlines. Won't it create resource allocation challenges? Besides, an overburdened workforce cannot be very productive and can be more detrimental in the long run due to low-quality work.

Changing Project Scope

Though your projects might be meticulously planned, you may still face unexpected deviations, crises, and complexity of work. May be the number or type of resources required also changes. The construction sector, in particular, can be quite unpredictable. There could be delays in material delivery, changes in client's requests, or sudden weather changes that might stop the work. All this can cause your project plan to go for a toss. The contractors can easily monitor the status of the project and check if any deadline could get missed. The tool also helps in estimating the resources or time required to complete new projects. This Cloud-based resource planning and management tool update all changes in real-time. This helps identify viable alternatives if there is any change in the project scope and needs more or fewer resources.

Shortage of Skilled Resources

Though the demand for construction services is huge, there is scarcity of skilled resources or labor. It is tough to manage complex and large projects with limited workforce as that over burdens existing people and may lead to burnout. Scheduling your resources in such scenarios becomes very complex as you need to ensure that your people are utilized effectively and in a balanced way.

Late Delivery of Materials

Work is delayed at many construction sites as they do not have the required material reaching them on time. Some sites do not have adequate storage for materials. Therefore, careful resource planning and management should be done to ensure that materials are being delivered at the right time. For example, if materials are delivered too early, then storing them securely will be a challenge. But if the materials are delivered very late, then the projects can be delayed, or you may need twice the number of resources to complete the work as per the schedule.

Outdated Workflow

Many construction companies rely on paperwork or spreadsheets for resource allocation and management. Though that's a very low-cost alternative, it is highly prone for errors and manual work is very slow and inefficient too. If you do resource allocation and planning manually on documents, you can easily miss out important resources or miscalculate the schedule. So, how can you achieve accurate and effective construction resource allocation?

Effective Resource Allocation in Construction

Resource allocation in the construction industry plays a critical role in project management. Staying ahead of, and prioritizing, your team member allocations helps to build workforce planning strategies to ensure organizational goals are being met and projects are progressing efficiently.

Effective resource allocation will mean understanding the scope of projects and availability of team members within the organization. This will often include input from the Project Manager in terms of project scope, task dependencies, required resources, and resource dependencies. The Project Managers will then identify available team members that meet the skills and experience requirements demanded by the project.

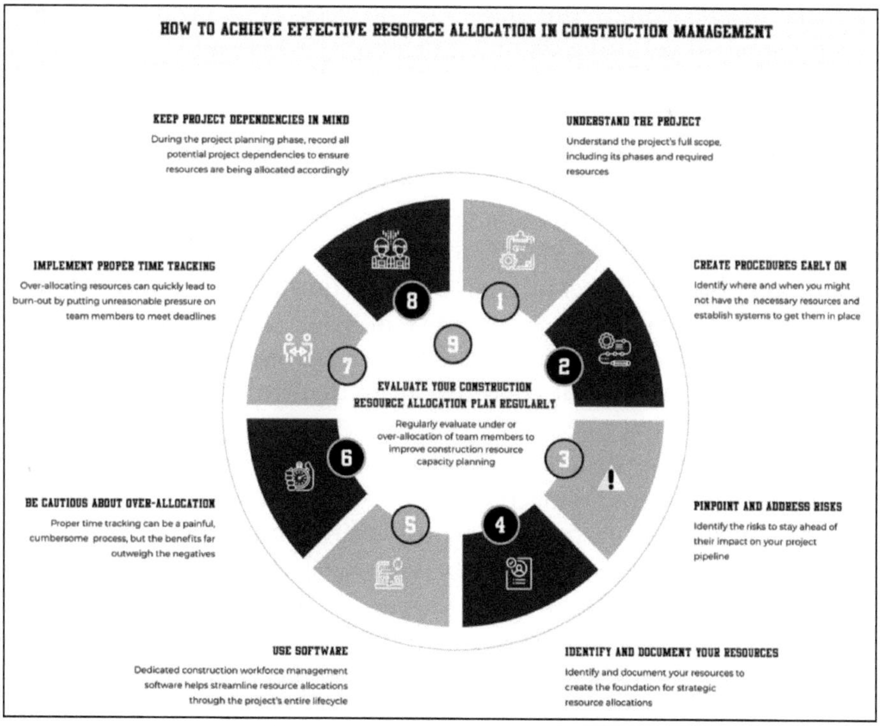

Tips to achieve effective resource allocation in construction management

Understand the Project Thoroughly

Understanding the full scope of the project, individual project phases, and the required resources will help to create strategic resource allocations that positively impact each project. Developing an accurate project roadmap will help provide an overview of what needs to happen and when. It can help to identify your performance metrics, help the Project Manager determine whether the project is on schedule, and help to allocate resources accordingly.

Developing a schedule for the required resources is one of the most important steps in managing a construction project. With a thorough understanding of the project, the Project Manager is capable of identifying all of the moving parts and required resources.

It also helps to develop a clear modification process. Having an in-depth understanding of the project helps to reduce the response time of change orders and create contingency plans for their resource allocations.

Create a Procedure to Allocate Construction Resources Early on

When should you allocate resources to a project? The earlier the better. Projects are won and lost in the pre-construction planning phase. Resource allocations should be no different. Waiting too long to make adjustments will make it increasingly difficult to get projects back on track.

Take some time during planning to identify where and when you might not have the available resources required, project risks, or project dependencies.

Setting up a procedure to allocate construction resources early helps to identify potential red flags and figure out different ways to respond to them beforehand. This helps to handle any allocation issues efficiently when they inevitably happen.

Pinpoint and Address Risks in your Construction Resource Allocation Plan

In a perfect World, every member of your team is readily available and perfectly qualified to take on whatever project role you need them for. Sounds pretty nice – but we're all aware that isn't a realistic expectation in the real World.

Risks in your construction resource allocation plan can vary:

- Unskilled workers taking on project roles that require skilled work
- New team members not being fully trained on safety procedures
- Project schedules slipping due to unrealistic expectations around workforce availability
- Errors due to lack of proper oversight.

It may be difficult to address every potential risk in your resource allocation plan. However, taking the time to identify the risks can help to stay ahead of their impact on your project pipeline.

Identify and Document your Resources

Identifying and documenting your resources helps to create the foundation for strategic resource allocations. Construction resource management solutions will allow you to identify the people you need to track and customize the workforce data you plan to capture and document. This should include things like:

- Project history
- Relevant experience
- Safety training and certifications
- Performance evaluations
- Availability

Properly identifying and documenting your resources will also help to identify skills gaps and inform recruitment based on project demand. It will also help to ensure you are allocating the best-fit resources for the task at hand.

Use Software Designed for Resource Allocation in Construction

Construction is a stand alone industry given the dynamic nature of each project. While there are plenty of project and workforce management software solutions available to general contractors, most (if not all) aren't optimized for the construction industry. Meaning they quickly wind up becoming a temporary "band aid" solution.

Managing construction projects and allocating resources productively can be extremely complex since every individual project has so many moving parts. Construction workforce management software that was designed to simplify the complex nature of a construction project. They can help to drive operational efficiencies and streamline resource allocations through the entire lifecycle of the project.

Tools like Bridgit Bench were built exclusively for the construction industry. They allow general contractors to create custom fields for construction resource management, and do simple workforce allocation filtering for smart allocation suggestions.

Be Cautious about Over-Allocation, too

Over-allocation of your workforce happens when more project assignments are assigned than your team members can reasonably handle. It's inevitable that organizations will have a single team member assigned to multiple projects. However, the more projects a team member is assigned to, the higher the risk of over-allocation. It's important to monitor your team members that are assigned to multiple projects to ensure they aren't being spread too thin.

Over-allocating resources can quickly lead to burn-out by putting unreasonable pressure on team members to meet deadlines. Construction workforce management tools can help to identify over-allocated resources and create balanced workloads across the project pipeline. If over-allocation is unavoidable, be sure to clearly communicate high priority projects. This can help to lift some of the pressure and also helps identify which lower priority tasks or projects can be assigned to an available team member.

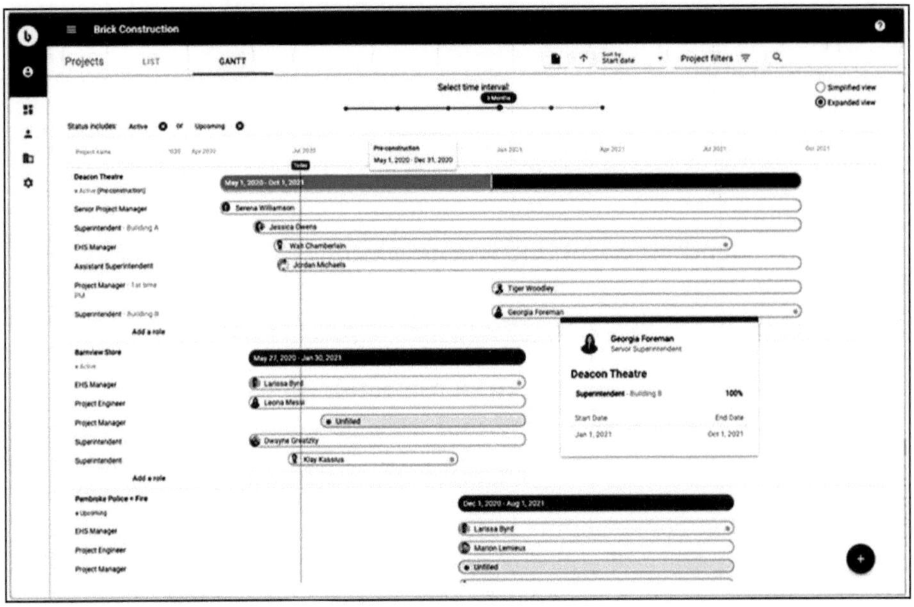

Implement Proper Time Tracking for Resources

Proper time tracking can be a painful, cumbersome process, but the benefits far outweigh the negatives. Time tracking allows general contractors to:

- Keep your team on track – With proper time tracking, team members can:
  - Easily reference deadlines for specific tasks
  - Better understand how they are performing against deadlines
  - See what tasks are lined up for them ahead of time.
- Track contractors – Keeping an accurate record of work being done ensures you aren't going over-budget with resources, and track performance to help allocate the right work to the right people.
- Prioritize projects – Easily make adjustments to high priority project allocations based on project progress.
- Performance reviews – Organizations can base evaluations on real data and compare directly to other team members.
- Avoid under and over-allocations – Have a holistic view of your resource utilization to understand where you're spread too thin and where you can take on more work.

Keep Project Dependencies in Mind

Construction projects can come with a lot of internal, external, and mandatory dependencies. It's important to keep these dependencies in mind when allocating resources to prevent idle hands waiting for tasks to be completed and costing the organization unnecessary costs. During the project planning phase, efforts should be made to identify and record all potential project dependencies to ensure resources are being allocated accordingly.

Be sure to continuously monitor dependencies and regularly discuss how different linked tasks are progressing and if adjustments to the workforce plan are needed. Project dependencies are unavoidable, but better communication with the project team and relevant stakeholders will help to ensure that projects are kept on track.

Evaluate your Construction Resource Allocation Plan Regularly

Operations teams will often meet on a regular basis to discuss and create a strategic workforce plan. These resource planning meetings are a good opportunity to reflect on and evaluate how effective your resource allocation plan is. Reports can be generated to give a holistic view of individual projects and how they're progressing. This can help to balance your resource allocations on a regular basis to maintain your organization's utilization rate.

Regularly evaluating your resource allocations also helps to identify periods of time when team members may be under or over-allocated, and help with construction resource capacity planning. Forecasting resource availability will also help to inform recruitment and layoff decisions to ensure the supply of workers will consistently meet the demand of work.

Challenges made Easier by Effective Construction Resource Allocation

Changes in Project Scope

Even the most carefully planned projects will encounter times when the expected complexity of work, and the resources required to complete the work, will change. Construction can be unpredictable, whether it be time-loss incidents, material delivery delays, changes in client demands, or whether that can bring work to a stop. Having an effective resource allocation process in place can help organizations pivot when needed and easily adjust to project changes.

Construction resource management tools have streamlined project allocations that allow general contractors to track when project phases and sub phases are expected to be completed. They also help to reference historical projects of similar scope to better estimate the resources and time needed to complete new projects in the pipeline.

Construction resource management tools that are cloud-based help to ensure all changes are updated in real time. They also help to identify adequate alternatives should the change in project scope require more or fewer resources.

Conflict between Various Projects

There are times when a project can be delayed by the lack of availability of skilled workers. It can often be assumed that the workforce required to build the project will be ready, willing and waiting whenever you need them. In reality, the availability of a general contractor's workforce is influenced by supply and demand.

Effective construction resource allocation helps when forecasting field and office labor requirements. This helps prevent a lack of supply or demand from creeping up on you and causing delays.

Resource allocation in construction management can also help to identify potential allocation conflicts before they become problematic. Overlapping project schedules can often lead to high-performing team members being allocated beyond their capacity, causing a stressful work environment.

On the other hand, having too many team members available for work when demand is low can be very costly to an organization. Having a holistic view of your workforce allocations can prevent potential conflicts before they begin to cost you money or team members.

Availability Issue Regarding Resources

There are many different factors that can impact a team member's availability to take on a new project role such as:

- Injuries
- Training
- Vacation
- Maternal or paternal leave

Having an accurate record of availability and being able to forecast your resource availability is challenging for even well-crafted spreadsheet systems. Construction resource allocation software helps track periods of unavailability for your individual team members. They provide an accurate, up to date record of availability to inform strategic resource allocations.

Review Questions

1. What are the ways to allocate resources?
2. What is the importance of resource allocation?
3. What does resource allocation mean?
4. What is an example of resource allocation?

5

Construction Disputes, Claims and Settlements

Although it may be the general contractor (GC)'s intent to avoid it at all costs, it may happen that litigation is forced upon them. A project manager who is doing a proper job of documentation and who is somewhat familiar with legal terminology, previous court decisions, and past industry practices will be better prepared to deal with that inevitable dispute.

The practice of law is best left to the lawyers, but as the name "contractor" implies, we deal with contracts (legal papers), and therefore need to become familiar with the legal implications and responsibilities that go hand-in-hand with the administration of all of these documents.

Global Perspective on Construction Disputes

Arcadis, a global design, engineering, and management consulting firm based in the Netherlands, for the past 5 years tracked the value of construction disputes Worldwide as well as their duration.

Arcadis found that, for the second year in a row, the most common source of construction disputes in North America was errors and omissions.

Differing conditions ranked second and failure to understand or comply with the contact requirements, be it the failure of owner, GC, or subcontractor, ranked third. It also found that the three most common methods of dispute resolution were (1) party-to-party negotiations, (2) mediation, and (3) arbitration.

Causes which Triggers Claims and Disputes

The principal reasons for misunderstandings leading to disputes and claims are as follows:

- Plans and specifications containing errors, omissions, and ambiguities, or which lack the proper degree of coordination.
- Incomplete or inaccurate responses or nonresponses to questions— or resolutions of problems—presented by one party in the contract to another party in the contract.

- The inadequate administration of responsibilities by the owner, architect/engineer, contractor, subcontractors, or vendors.
- An unwillingness or inability to comply with the intent of the contract or to adhere to industry standards in the performance of work.
- Site conditions that differ materially from those described in the contract documents.
- Unforeseen subsurface conditions.
- Shift work and an owner's failure to recognize the added costs when requesting such work, which are generally made to accelerate the project or make up for a delay, whether compensable or excusable.
- Loss of productivity—both cause and effect.
- The uncovering of existing building conditions which differ materially from those indicated in the contract drawings—situations that occur primarily during rehabilitation or renovation work.
- Extra work or change-order work.
- Breaches of contract by either party in the contract.
- Disruptions, delays, or acceleration to the work that creates any deviation from the initial baseline schedule.
- Inadequate financial strength on the part of the owner, contractor, or subcontractor.

The key to dispute resolution is promptness in addressing the issue(s) and conducting negotiations with the goal of early settlement. Experience has shown that the longer a dispute lingers, the more entrenched each party's position becomes, and the more difficult it is to resolve. Rapid resolution can be achieved if there is complete, accurate, and indisputable documentation regarding the events surrounding the disagreement.

Only when efforts at resolution fails, there a need to consult a claims consultant and/or the company attorney, and again, if this is done at an early stage, the strengths and weaknesses in both party's claim may become apparent and the decision to pursue or dismiss the claim made somewhat easier.

The Bid Proposal Process and the Potential for Disputes

Disputes often arise even before a contract is awarded. In fact, a dispute can occur even before a sealed bid proposal is submitted at a formal bid opening. In public sector work, and frequently in private industry, a formal bid procedure is established: bids are submitted on a pre-issued bid proposal form, the form is completed and signed by an officer of the construction company, sealed in an envelope, and presented to the owner's representative at a pre-determined place, time, and date. If bid bonds are required, a certified check or letter of credit may also be acceptable.

The bid bond offers assurance to the owner that if the contractor's bid is accepted, the owner will be protected if that low bidder is unable or unwilling to accept a contract when offered. Other requirements may also accompany the bid proposal and, per the bid instructions, any deviations may be cause for rejection.

In the public sector, strict compliance with all aspects of the bid proposal must be adhered to, otherwise formal protests may be lodged by other bidders protesting acceptance of a bid that fails to meet all the required criteria.

In the private sector, compliance is not so strict, and the owner may waive any or all requirements in their selection of an acceptable bidder if it appears to be in their best interest. In a way, so can public officials.

Impact of Late Bids

What's described next is very familiar to project managers as they dash from their office with an incomplete bid form, several pens, and a cell phone on their way to the bid opening. The scenario generally is as follows: The project manager arrives at the bid opening, looks for a spot away from the crowd and calls the office for final instructions and late-breaking price adjustments. With only a few minutes to bid closing, they find themselves looking at a blank bid form that must be filled in with a page or two of alternates, allowances, unit prices, and who knows what. Time is running out and only a minute or two is left, but many blanks remain to be filled in before that mad dash to the bid office. A cold sweat begins to form on the project manager's brow: "What will happen if I make a mistake on this form or enter the wrong number in the wrong place—or worse yet, if the battery in my cell phone dies before I'm finished consulting with the office?"

Don't despair! Complete the form as quickly and as accurately as possible and proceed, post haste, to the office where the bids are to be submitted.

Turn the bid into the clerk and have it date/time stamped. Just because a bid is received after the exact specified time on a public work project does not mean it will automatically be disqualified.

In private bid situations, the owner is free to waive any prebid qualifications, but in public bid openings, the courts have ruled that the public bidding requirements are there for the public's benefit. If a bid is received a few minutes late and there is no evidence of fraud, collusion, or intent to deceive, all is not lost. If the local authority refuses to accept the late bid, an immediate protest must be filed and can be voiced in the presence of the official refusing to accept the bid, as well as in the presence of witnesses.

In the meantime, the project manager should remain during the entire bid opening process and keep detailed notes of the other bidders' proposals, noting their competitive bids and any exceptions that they may have taken to the bidding instructions. As quickly as possible, upon return to the office, a written protest should be filed citing the circumstances involved in the late submission (all other portions of the bid having been met with strict compliance) and stating that the bid was delivered at, say, 2:05 pm instead of 2:00 pm as required.

If your late bid happens to be the low bid, but could be rejected because it was submitted 5 minutes after the deadline, it may well be worthwhile to lodge a formal, legal protest and challenge. It is possible that the public officials—or the court, if it gets that far (which it probably won't)—will rule that it is in the public interest to accept the low bid even though the late submission must be considered a violation of the bidding procedure. In fact, if other "minor" deviations are noted during the bid opening process and another contractor requests that your bid be disqualified, or the public official opening the bids indicates your bid may be disqualified, continue to take notes relating to all other bids being opened, then return to the office and discuss lodging a protest.

When a "non-conforming" bid is significantly lower than the second bid, the public agency may well decide to interview your firm to ascertain that all other aspects of the bidding procedure have been met, that your company fully understands the scope of the work, has included costs for the full scope of work, is a reputable builder, and is ready, willing, and able to enter into a contract. The agency may then rule that it is in the best interests of the public to accept this bid even though there was a "minor" discrepancy in the bidding process. Emphasis must be placed on the word "minor."

The Low Bidder

Unless the bid documents state otherwise, the low bidder on a lump-sum contract is the "apparent" low bidder, subject to a review of compliance with all other provisions of the bidding instructions.

When a bid contains not only a lump-sum price but several alternates, as additions or deletions to the base bid, the determination of who is actually the "low" bidder becomes less clear. In the absence of any language to the contrary, usually the owner may select the alternates that will be accepted.

Therefore, the low bidder may not be apparent until all accepted alternates are included with the lump-sum bid. It is also possible that the "apparent low bidder" wishes they were not the lowest bidder, which brings up another point.

Withdrawing a Bid

In the rush to assemble a competitive bid, it is not unheard of to make a mistake, either in the extension of unit prices, the omission of a few cost items, or to just plain make a mathematical error. If these items are minor and if the GC has been designated "apparent low bidder," they will usually accept the contract and chalk up the mistakes to experience—afterward having a few words with the person that put the bid together.

Other Potentials for Disputes Oral contracts

Oral contracts are often recognized as creating the same obligation between two parties as a written contract. A case tried in a North Carolina court was based upon an architect performing work for an owner without benefit of a written contract. The owner, a developer, requested the architect to complete some preliminary design work on a condominium project. The architect, according to the owner, had prepared such a complex contract for the design work that it was rejected. The owner indicated that he would be willing to sign a simpler contract if presented. The architect's fees, at that time, were not a subject of disagreement, and the owner directed the architect to proceed with the design work. Months later, the owner advised the architect that another designer had been hired to work on the project, but by that time the original architect had completed a major portion of the work and so submitted an invoice for it. The owner refused to pay the invoice and the architect filed suit, claiming breach of contract.

Disputes regarding Contract Interpretation

Ambiguities in the contract documents are usually resolved by reasonable parties taking a reasonable approach. When reasonableness does not prevail and one party pursues a hard line, what to do? There are no tried and true procedures for contract interpretation and each case seems to stand on its own. What documents have priority over others? Do the plans or specifications take precedence? Will the specifications take precedence over full-scale drawings? Some contract requirements specifically establish an order of the precedence such as: contract requirements—first precedence; schedules—second; drawings—third; and so forth.

Errors and Omissions Concerns

Another provision that architects often include in the contract requires the contractor to advise the architect of a discrepancy, error, or omission before submitting a change order relating to that error or omission. This provision may appear in the bid documents or most certainly in the contract for construction.

Dealing with Problem Drawings

With barely enough time to completely scrutinize the drawings and specifications before and during the hectic competitive bidding process, some things will undoubtedly slip through the cracks. Last-minute calls from sub-contractors, and suppliers e-mailing their bids along with estimate adjustments by the boss as the clock ticks away, leave little time to concentrate on potential drawing errors and omissions.

Doesn't the contractor have a right to assume that the plans and specifications submitted by the owner's architect/engineer are reasonably complete for bidding purposes? If there are any major discrepancies, shouldn't they be able to rely on everyone's good faith to find an equitable solution to correct major errors?

Subsurface, changed, and differing conditions

A significant number of disputes involve the site work phase of a project.

Even with numerous test borings and other geotechnical site investigations, conditions uncovered during excavation may be at variance with the conditions assumed by the information in the geotechnical survey.

Test borings accurately display the subsurface soil strata in the exact location where they have been taken, but another boring drilled just yards away may reflect totally different subsurface conditions. In fact, a standard clause in most geotechnical reports recognizes this condition and state.

Regardless of the thoroughness of a geotechnical engineering exploration, there is always the possibility that conditions will vary from those encountered in the test borings, or that conditions are not as anticipated by the designers.

About "differing conditions"?

This is probably the first question that the architect/owner will ask and for which the contractor must have an answer. A differing conditions clause in a contract/specification serves two purposes: it requires the contractor to identify those conditions on the site which it deems materially different from those ordinarily encountered; it alerts the owner to the need to promptly investigate those conditions and determine whether an equitable adjustment should be made to the contract price and time. There are a number of conditions that exist that can be used to qualify for differing conditions:

1. Is there more or less of something component in the soil or rock, or is the quality of soil or rock different? Do quantities or qualities of the following vary?
   - Soil
   - Boulders
   - Gravel
   - Sand
   - Silt
   - Clay
   - Vegetation
   - Permeability
   - Behavior
   - Rock
   - Top of rock
   - Quantity of rock
   - Strength, hardness of rock
   - Boreability
   - Deformability
   - Permeability
2. Is there a difference between what was anticipated (based upon the information contained in the contract documents) and what was actually encountered? What was reasonably expected "quantitatively," and what was reasonably expected qualitatively?

3. What is the difference in performance expected?
   - Is material strength higher so that it would cause slower drilling, trenching, and tunneling?
   - Is compressibility lower than anticipated so that it would cause more soil modification to achieve design strength?
   - Is permeability lower and requiring higher dewatering requirements?

All of the above may lead to slower advance rates, higher tool costs, over excavation, or under-excavation.

Differing Site Conditions (DSC)

The Federal Highway Administration developed a dual test system to define the nature of a differing soil condition claim. A type 1 claim involves a situation where the site surface or subsurface conditions deviate substantially from those represented in the bid package. A type II claim refers to a situation where the site surface or subsurface conditions are substantially different from those that might be "reasonably" expected. These are terms that are often used in describing conditions on private projects and can be found in various court decisions where public project claims are adjudicated.

How can the risks associated with differing site conditions be reduced?

1. If there are no differing conditions clauses in the contract documents, very strongly suggest that adding such a clause is in everyone's best interest. It is better to negotiate if the occasion arises than to proceed on to a court case.
2. Be very knowledgeable about the benefits and limitations of the contract language regarding differing conditions and the extent and limitations of the information furnished by the owner regarding subsurface conditions.
3. Conduct a thorough investigation of the site prior to submitting the bid.
4. Promptly notify the owner of any real or suspected differing conditions, giving the owner ample time to determine how best to handle this situation, review their options, and hopefully, resolve the matter without delaying the project and adding more potential costs to the project.
5. Differing conditions can manifest themselves in any number of ways, adding to the costs for foundations and installation of site utilities, and so there may be more at issue than replacement of unsuitable soils, blasting of rock, and/or schedule delays.

Some Possible Contractual Problems

It has to be remembered that the parties assigned to a construction contract are not competitors among themselves, but associates who have different functions to perform to achieve the common goal for accomplishing the prescribed end-product. Despite this fact, differences of opinion leading to conflicts do arise, since in the final analysis, each party has to protect its interest and financial gain. Development of disputes leads to the creation of claims and then to seeking of settlements either by direct discussion or arbitration or court action, depending upon the degree of disagreement and enormity of claim.

Most of the happenings that could occur during the currency of a contract cannot be foreseen. Consequently, no universal formula can be postulated to foreclose their effects in advance. However, their gravity can be diluted by following ethical practices and by performing business in an un-emotional and above-board manner. Interpretation of contract clauses, differences of opinion regarding their application, effects of unforeseen subsoil and climatic conditions, riots and strikes, etc. can create contractual problems. These need to be solved equitably and expeditiously.

The client may rightfully demand completion of work by the contractor despite the latter's protestations and petitions. The client may threaten to stop the work, take over the contractor's plant and equipment, and also the materials and stores, and have the remaining work carried out either departmentally or through another contractor at the risk and cost of the original contractor.

A vast majority of contractual problems arise from lacunae in and misinterpretation of the clauses, pertaining to the following 13 subjects:

(i) Changes in contract work
(ii) Differing or unusual site conditions actually encountered
(iii) Suspension of work
(iv) Variation in quantities
(v) Damage due to natural disasters s and force-majeure
(vi) Re-inspection and acceptance
(vii) Termination for the convenience of the client
(viii) Possession prior to completion
(ix) Escalation of price due to inflation
(x) Acceleration of work progress
(xi) Ripple effect
(xii) Currency fluctuation effect
(xiii) Ambiguity in specifications and drawing,

There can be many other causes too, but most problems seem to emanate from the subjects indicated above. A very brief description of these is given below in order to make the constructor and his management aware of their possibility so that they may try to tighten up their contract conditions III their bid proposal and be on guard for the fall-out during execution.

Changes in Contract-Work (Extra Work)

This basically means the client permitting an increase or forcing a decrease in the agreed contract cost, owing to his unilaterally desiring changes in the otherwise agreed contract work. Owing to rather restrictive wording for working out an equitable price adjustment under this clause, the contractor should read "this clause most carefully before submitting his bid. During execution, it is of utmost importance for the contractor to first notify the client in time as to his intent to fit: the resulting claim, and then prepare and submit this claim in time, giving all the supporting data. However, the constructor may not want to volunteer to perform such extra work unless the authorisation for the same is contractually validated by the client.

Differing or Unusual Site Conditions Actually Encountered

This basically means the client permitting an increase in the agreed contract cost, owing to the actually encountered site conditions being either different or unusual, compared to those stated in the original contract document. The basic purpose of this 'changed site conditions' clause is to prevent the contractors from adding high contingency margins as a protection against the possibility of differing or unusual site conditions (such as, type of sub-strata-loose or dense sand, hard clay, conglomerates, soft weathered rock, lenses of hard clay in sand-boulder strata, hard rock. intensity of subsurface water percolation and the possibility of its connection with a surrounding water source which could affect dewatering operation, etc.). The idea behind such a clause is to take some of the gamble out of the bid price. But then during the actual execution the client's engineer, who might have authored the erroneous site conditions in the contract document in the first place, is likely to try his hardest not to acknowledge the truth about the changed site condition. This is where the dispute starts. When the actual substrata is clearly different from that indicated in the original contract document (bore hole data and strata profiles), then the document is clearly in error and the contractor can rightfully claim he was misled by the client's error. Even where the substrata and the subsoil water conditions are not different, but quite clearly of unusual nature which was not revealed by the bore hole data given in the original contract document, the contractor has a case for claim since he could not have carried out all the necessary subsurface investigations prior to bid submission however much the contract document might have asked him to do so in advance. This kind of clause is purely diversionary on the part of the author of the contract document. It only exposes the lack of homework on his part, and has no backbone in the court of law. The contractor must never forget to give timely notice of the changed site conditions (either 'different' or 'unusual') to the client (who can then investigate immediately) and convey the intent of filing the consequent claim as well as file the claim promptly, in a workman-like manner.

Suspension of Work

Through this clause the client reserves the right to suspend the contractor's work or any part thereof, but pay an equitable compensation to the contractor for doing so. Generally this compensation is limited only to the constrictor's standby costs and extended overheads, determination of which may not be easy-particularly if verification of its estimates is subject to client's audit. Where suspension of work is ordered because of poor performance of the contractor which can be linked directly to the client's error in specifications for design or construction, the compensatory amount payable to the contractor can become considerably more. Again, if a work is suspended because of inordinate delay in the contractor's construction (the delay not being attributed wholly to fault on his behalf), the compensatory amount payable to the contractor can be significant.

Variation in Quantities

In an item-rate contract the client may want a price adjustment owing to variation (increase or decrease) between actual and initially estimated quantities, due to various reasons (e.g. desired change in extent of work, change in design assumptions, change in site conditions, etc.). Generally if there is a variation in quantities by up to 10 to 15¾ the contract document-may not allow any compensatory increase in unit prices of the work-items. The contractor should clearly specify in his bid as to by how much his quoted unit prices will stand increased in respect of the balance work, should the estimated quantities shown in the bill of quantities in the original contract document increase by more than the specified percentage. Likewise he may specify an upward revision in his unit prices for the whole work, should the quantities reduce to less than 75% of the initially estimated in the bill of quantities. He can also suggest a staggered scale of increase in unit prices corresponding to a staggered scale of change in quantities. Such a change in unit prices is plausible since the contractor can rightly attribute it to change in his hid strategy, revision in the deployment of manpower and machinery, and his standby and extended overhead costs. This clause and the mechanics of price adjustment to suit it should be covered even in a lump sum contract.

When there is a substantial under run in quantities (actual quantities less than the estimated quantities), consideration should be given to claim this under the 'changed condition' clause (differing or unusual conditions clause). If the 'changed condition' is established, then the restrictive 10 or 15% stipulation (or as specified) does not have to be operated. Further, when a change-order causes a substantial reduction in quantities the contractor may like to claim under the 'suspension of work' clause. All this has to be carefully examined by the contractor's contract manager and work surveyor.

Damage due to Natural Disasters and 'Force-Majeure'

Natural disasters cover unprecedented or unusual floods, rains, heat, cold, wind, earthquake, fire, and the like. These, together with occurrence of riots, civil commotion, insurrection, and all other such happenings are outside the control of the contractor and not attributable to his fault. The contract document usually specifies as to which of these risks the contractor should cover against financial loss through an insurance policy in his own interest. Except for financial loss due to such risks, the contractor can rightfully claim compensation from the client for financial loss due to the remaining risks, and for the time loss due to all of them, but this must be so specified in his bid conditions.

Force-majeure usually includes hostilities, restraint of rulers or peoples, revolution, civil commotion, strikes, epidemic, unusual accidents, fire, flood, wind, earthquake, explosion, Sabotage, blockade, embargo, failure of transportation facilities. or any law, proclamation, regulation or ordinance, demand or requirement of any government or governmental agency having or claiming to have jurisdiction over the work, or with respect to materials purchased for the work or over the parties to the contract or any natural calamity or other act of government, or any cause, whether of the same or different nature, existing or future, which is beyond the control of and without the fault or negligence of the parties to the contract.

The delays due to all such occurrences and the financial loss from events other than those covered by insurance are excusable under the contract law. However. the contractor must immediately after the event put in his claims with the client as well as the insurance company prefacing the request with the words ' ... because of unforeseeable causes beyond the control and without the fault or negligence of the Contractor'.

Re-Inspection and Acceptance

Sometimes the client's inspector or any other authorised agent alleges that a particular work, already completed, is defective, and forces the contractor to tear-open or uncover the same for re-inspection. The contractor does as directed and if the work is discovered to be acceptable, then the contractor should promptly claim for all his costs in tearing open (uncovering), exami-nation and reconstruction as also for the loss of time, notwithstanding the demoralisation. If however the allegation proves right, then the contractor cannot make the claim and instead all the involved expenses are to his account with no compensation for loss of time.

Termination of Work for Convenience of the Client

Should the client, purely for his own convenience (may be due to revision of his plans for his own reasons, or whatever), decide to terminate the contract (which is different than suspension of work consequent to contractor's faulty or delayed construction, etc.), then settlement bas to be made with the contractor on cost basis, duly compensating him for not only his standby costs, extended and already stretched overhead costs, all dismantling and transportation for redeployment, terminal hand-shake pay-packets to those required to be so treated, but also for the loss of his anticipated profit on the component of work terminated. However, if the contract document has a suitable clause to suit this contingency of 'client wanting to terminate the work for his convenience', then the contractor may not be able to claim for the loss of his anticipated profit on the component of work terminated. In the absence of such a clause, if the client has to pay for this profit, he may well audit the contractor's earlier accounts and financial statements to uncover the profit actually shown up to then and perhaps pay only pro-rata for the anticipated profit on the terminated work. The contract document has to be studied by the contractor very carefully before finalising his bid conditions.

Possession Prior to Completion

Sometimes the client may, under this clause, take possession of the partially completed facility and begin to use it prior to the contractor having completed and formally banded over the entire work. In such an event, the contractor should promptly bring to the client's notice that he stands relieved of the responsibility for loss or damage to the work other than that resulting from the contractor's fault or negligence. If such prior possession or use by the client delays the progress of the work or causes additional expense to the contractor, an equitable adjustment in the contract price and in the comple-tion time should be claimed promptly.

Escalation of Price due to Inflation

Where the work has to be executed in an inflation-prone area and the labour wages and material prices are expected to suffer an upward revision incident to inflation, a balanced and realistic document should build into the contract the necessary escalation payment clause on the basis of any standard recognised empirical formula commonly in vogue these days. This way the client is bound to receive fairly competitive bids (other things being equal), eliminating the variable contingency margin that the bidders may build into their quotations fearing escalation of labour wages and material prices over the duration of the contract period.

If the contract document is silent on this aspect, the contractor should include a reasonable clause in his bid proposal to cover the effect of escalation, clearly explaining through short numerical illustrative examples how the labour and material cost escalations can be worked out over a specified period of time in respect of the balance work remaining from the date the cost index numbers for minimum wages and material prices arc revised upwards. He has to commit what percentages the labour and the material components his bid price is based on. (Figures usually mentioned are 30 and 70 respectively).

Whether the escalation clause is included in the contract document or is proposed by the contractor himself, it is absolutely essential to clearly define the base index numbers and their dates from which the escalations are to be computed. It is also essential to stipulate as to how much is the assumed increase in the labour and material cost components, for every (say) 1% increase in their base index values. (The base index value for labour wages could be taken as the minimum labour wage on the base date and the cost of labour escalation computed based on percentage increase in this wage. as promulgated from time to time.)

In order to avoid the problems of escalation calculations in relatively smaller projects, sometimes a certain lump sum amount is agreed to be paid by the client to the contractor (in a certain manner over a certain period) and the cost escalation calculation is altogether avoided. This may be desirable in some cases where the amount could even be played with and used as a final lever in order to clinch the contract but is not recommended for major contracts.

Acceleration of Work Progress

It has to be recognised that all said and done, organising and managing modern-day construction is a very complicated exercise involving long-term planning and scheduling of operations, optimised deployment of manpower and machinery, and delicate balancing of resources, loans and funds. Unless the contractor is delayed in the construction work due to his own fault, the client will have to pay extra compensation in terms of money should he force the contractor to change and accelerate his performance for an earlier completion of the project. This is different from the 'bonus-penalty· provision where a certain completion period is agreed to and the contractor then qualifies for the bonus for an earlier completion (because that enables the client an earlier use of the facility) or suffers the penalty for over-running the agreed date (because that forces the client to suffer liquidation damages). Against this, if the client refuses excusable extension of contract time (sought for valid reasons and for causes beyond the contractor's control), he in effect is accelerating the contractor's performance, for which he may have to pay a heavy price. In modern-day construction, sometimes it can be costly for the client not to permit valid extension of the period of construction.

Ripple Effect

If a contractor can prove that his previously planned work schedule had been disrupted because he had to meet the client's dictum of accelerating a part of the whole activity, then, should the balance work be delayed (as a ripple effect), the contractor can claim extension of time for overall completion of the whole work. Also, in attempting to accelerate or hasten the completion of a part of the work (as dictated by the client) if the contractor had to put the concerned labour on extra shifts or pay them overtime to beat the clock, the labour in the balance project work will also demand extra time, as a ripple effect. The contractor can claim for the above monetary compensations, thrust on him as a ripple effect. Sometimes when a work is divided into phases that have to be constructed in sequence one after the other and there is delay in any one of them except the last phase, then all subsequent phases will be automatically delayed as a ripple effect. Again, the contractor has the right for extension of period of construction if the delay was not attributable to his performance. In each case the construction engineer will very promptly make out the case for the time extension and monetary compensation (ripple effect) and duly submit his claims in time.

Currency Fluctuation Effect

While working in a foreign country and having imported a sizeable amount of hard currency for meeting various mobilisation charges, initial expenditures and purchase of equipment, etc., the contractor may wish to seek protection against a serious downward fluctuation in the currency of that forcing country within whose territory he is working. This matter has to be built into the bid proposal and negotiated in depth, as it can have serious repercussions on the overall resource planning of the contractor.

Ambiguity in Specifications and Drawings

Where the contract specification is ambiguous and can have more than one interpretation, precedent legal procedure has proved that it has to be interpreted against the client who authored it or used it as his designated specification. The contractor will carefully study such specifications and use them to his cost-advantage by actually following sound engineering practice.

It must be clearly remembered that the specific provisions govern over the general provisions, the typed provisions govern over the printed provisions, and what has been specified in writing will govern over what has been shown on the drawings. Conflicts are easy to develop in the absence of mature experience on either side.

Equitable Adjustment

This is one phrase used easily by all but agreed to by few. It is not t easy to work out what an equitable price adjustment-satisfactory to both parties would be. Some believe that neither party should profit from a windfall or suffer owing to action by the other. Others believe that an equitable adjustment is one when both parties leave the table shaking hands but a little bit unhappy. It has to be fair to both parties. In the final analysis, there is no substitute for playing the game with ethics and honesty, with the two parties working as associates rather than competitors. They have different roles to play in reaching the same goal. But since contracts without disputes are impossible, each party must do its homework thoroughly and participate with more objectivity and practicality rather than emotion, to minimise the disputes.

Defective Specifications

Defective specifications are not a subject area of the contract like a differing site condition or notice requirement. However, there is an important area of the law that considers the impact of defective specifications under implied warranties. The theory of implied warranty can be used to resolve disputes originating in the specifications or the plans; the term defective specification will refer to both. The contract contemplates defects in the plans and specifications and requires the contractor to notify the designer when errors, inconsistencies, or omissions are discovered.

Defective specifications occur most frequently when the contractor is provided a method specification.

A method specification implies that the information or method is sufficient to achieve the desired result. Because many clauses are mixtures, it is imperative to identify what caused the failure. For example, was the failure caused by a poor concrete specification or poor workmanship? Another consideration in isolating the cause of the failure is to identify who had control over the aspect of performance that failed.

When the contractor has a performance specification, the contractor controls all aspects of the work. If a method specification was used, it must be determined that the contractor satisfactorily followed the specifications and did not deviate from the work. If the specification is shown to be commercially impractical, the contractor may not be able to recover if it can be shown that the contractor assumed the risk of impossibility. Defective specifications are a complex area of the law, and competent legal advice is needed to evaluate all of the possibilities.

Misrepresentation

Misrepresentation is often used in subsurface or differing site condition claims, when the contract does not have a differing site conditions clause. In the absence of a differing site conditions clause, the owner assigns the risk for unknown subsurface conditions to the contractor [Jervis and Levin, 1988]. To prove misrepresentation, the contractor must demonstrate that he or she was justified in relying on the information, the conditions were materially different from conditions indicated in the contract documents, the owner erroneously concealed information that was material to the contractor's performance, and the contractor had an increase in cost due to the conditions encountered. More commonly, a differing site condition clause is included in the contract.

Differing Site Conditions

One of the more common areas of dispute involves differing site conditions. However, it is also an area in which many disputes escalate due to misunderstandings of the roles of the soil report, disclaimers, and site visit requirements. The differing site condition clause theoretically reduces the cost of construction, because the contractors do not have to include contingency funds to cover the cost of hidden or latent subsurface conditions. The Government differing site conditions (DSC) clause, or a slightly modified version, is used in most construction contracts. The clause is divided into two parts, commonly called Type I and Type II conditions. A Type I condition allows additional cost recovery if the conditions differ materially from those indicated in the contract documents. A Type II condition allows the contractor additional cost recovery if the actual conditions differ from what could have been reasonably expected for the work contemplated in the contract. Courts have ruled that when the wording is similar to the Government clause, federal precedent will be used to decide the dispute.

Type I Conditions

A Type I condition occurs when site conditions differ materially from those indicated in the contract documents. With a DSC clause, the standard of proof is an indication or suggestion that may be established through association and inference. Contract indications are normally found in the plans and specifications and may be found in borings, profiles, design details, contract clauses, and sometimes in the soil report. Information about borings, included in the contract documents, is a particularly valuable source because they are commonly held to be the most reliable reflection of the subsurface conditions.

While the role of the soil report is not consistent, the courts are often willing to go beyond the contract document boundaries to examine the soil report when a DSC clause is present. This situation arises when the soil report is referred to in the contract documents but not made part of the contract documents.

Groundwater is a common problem condition in DSC disputes, particularly where the water table is not indicated in the drawings. Failure to indicate the groundwater level has been interpreted as an indication that the water table exists below the level of the borings or that it is low enough not to affect the anticipated site activities.

The contractor must demonstrate, in a DSC dispute, that he or she was misled by the information. To show that he or she was misled, the contractor must show where his or her bid incorporated the incorrect information and how the bid would have been different if the information had been correct. These proofs are not difficult for the contractor to demonstrate. However, the contractor must also reasonably interpret the contract indications. The contractor's reliance on the information may be reduced by other contract language, site visit data, other data known to the contractor, and previous experience of the contractor in the area. If these reduce the contractor's reliance on the indications, the contractor will experience more difficulty in proving the interpretation.

Owners seek to reduce their exposure to unforeseen conditions by disclaiming responsibility for the accuracy of the soil report and related information. Generally, this type of disclaimer will not be effective. The disclaimers are often too general and nonspecific to be effective in overriding the DSC clause —particularly when the DSC clause serves to reduce the contractor's bid.

Type II Conditions

A Type II DSC occurs when the physical conditions at the site are of an unusual nature, differing materially from those ordinarily encountered and generally recognized as inherent in work. The conditions need not be bizarre but simply unknown and unusual for the work contemplated. A Type II condition would be beyond the conditions anticipated or contemplated by either the owner or the contractor. As in the Type I DSC, the contractor must show that he or she was reasonably misled by the information provided.

The timing of the DSC may also be evaluated in Type II conditions. The contractor must establish that the DSC was discovered after contract award.

Development of Disputes

Disputes can develop for innumerable reasons. The client's engineer could be dissatisfied with the contractor's performance in some particular operation. The contractor may disagree with the directives of the client's engineer, or with his interpretation of the requirement under the contract. Methods of measurement and billing, disputes about whether something falls under a particular work-item or not, disagreement on whether the particular excavation can be termed 'unclassified or not, differing views on operational procedures, and hoards of other reasons can develop into disputes.

The constructors engineer and his team should prepare the background accurately and fill detail with all the supporting data, documents and photographs, where relevant. They should try to negotiate with the client's representative, keeping the management fully informed of the developments. This discussion, in an attempt to settle the dispute, may either crystallise into an 'agreement" or precipitate into a 'dispute", highlighting the views taken by both parties, and the quantities and costs involved. It would also decide whether to proceed with the work under protest without prejudicing the contractor's right for compensation in terms of time and cost under the terms of the contract and sound engineering practice. It is most essential to keep the minutes of all such discussions (along with the supporting documents), signed and dated by both the parties. It is a serious mistake if the contractor avoids formalising such minutes, feeling it might cause resentment with the client", representative. Such records are needed in the ultimate interest of both parties. Mature understanding of the respective responsibilities of both should be recognised. These procedures are required by duty, and are not governed by courtesy alone. These minutes will show whether the discussion lead to an agreement or culminated into a dispute. If the contractor feels his rights are being denied to him, the dispute degenerates and creates a claim, which the contractor must promptly and formally submit.

Creation of Claims

If his written request for settling the claim is rejected or not acknowledged within a specified time by the client's representative at the site, the contractor should then promptly address the written appeal to the client's higher authority within the period specified for such action in the contract. This is very important. An appraisal meeting should preferably be held at the time of submitting this appeal so as to bring home to the client the technicalities involved. This may result in a knowledgeable response through clearer understanding. It may be necessary to make successive appeals to each level of authority in the client's organisation to exhaust the bureaucratic barrier until all administrative remedies are tried respectfully. If the matter is still not resolved equitably or receipt of the appeal is not acknowledged by the client in the specified time, notice of the same should be given to the client, asking that the dispute be further considered at the time of settlement of all the outstanding.

The supporting documents for aft the claims should be meticulously prepared in detail and compiled carefully as soon as possible during the contract period. Depending on the case, the claim may involve the support services of experts in various disciplines, e.g. geotechnical, hydrological, structural, materials, quality control, electrical-mechanical-metallurgical, etc. Where the work is going to be 'covered' as the construction proceeds, due notice of the same should be given to the client to enable him to inspect and check what the experts might be investigating in support of the claim before the work is covered.

Seeking Settlement of Contractual Problems

Settlement through Direct Negotiations with the Client

If the actual work involved has been executed well and on time, the general atmosphere between the contractor and the client is bound to be cordial and satisfying and this should enable a more friendly accommodation and compromise in settlement of claims through direct negotiation. This is one reason for keeping the settlement of major claims open until the end of contract (assuming the work goes on well). The contractor should prepare all his claims accurately and in a chronological order along with all the supporting data and documents, including photographs and events diary. Having orchestrated the cases and their details among themselves, the contractors representatives should prepare an agenda for the negotiation and circulate it among the client's representatives together with very brief details of events, slightly prior to the actual meeting. Any personal clashes and conflicts should not be allowed to surface, even if provoked. Loud argumentation will only further prejudice the settlement. Give the client the feeling of superiority and help him to help you.

In any case, it is likely that a number of outstanding issues will be settled equitably at such contract settlement meetings. In the minutes of the meeting the contractor should mildly but firmly record those of the issues where he cannot agree to compromise or rescind. Promptly thereafter the contractor must address a letter to the client, detailing what transpired at the above meeting, what issues were settled (stating their effect in terms of money) and what issues remain outstanding. Through this letter the contractor should also communicate his release from the contract except for possible continuing guarantees, and state clearly that he reserves the right to pursue the settlement of the stated outstanding issues through arbitration and court of law, in that order, if and as provided for under the contract and the law of the land.

Settlement through Arbitration

Where the contract permits settlement through arbitration, it will also describe the procedure for selecting the arbitrator or arbitrators. Usually each party nominates one, and the two so selected may mutually elect a third one. Generally the decision of the arbitrators is laid down in the contract as binding on both the parties, and where this is not so, recourse can be taken by the aggrieved party (generally the contractor) to the court of law.

Governments may have their own arbitration laws and sometimes recourse might be taken to the Rules of Conciliation and Arbitration· of the International Chamber of Commerce, based in Luasanne, Switzerland. Arbitration proceedings approximate to court proceedings. All the proceedings discussions, occurrences, testimonies and decisions-should be recorded and written accurately and promptly. Witnesses should be presented and heard during the proceedings while the arbitrators hear both the contractor and the client. Arbitration is a much Lister procedure than court action. If the contract does not bar taking recourse to court action and does not restrict the decision of the arbitration to be final and binding, the aggrieved party can resort to court action if it feels that it has been wronged and has not received due justice.

Settlement through Court Action

Settlement through a court of law is obviously the last resort and this usually takes years and years and can be frustrating. Courts usually would like the party concerned to exhaust all other administrative channels for seeking redress (e.g. direct negotiations and arbitration) before hearing the case. Only if the contractor is extremely confident of his case and honestly believes that he has been wronged at the negotiations and at the arbitration levels, must he embark on court action. Court action will require engaging top legal minds (attorneys) who have some experience in construction industry feuds, and the ensuing long legal battle-which will also require prolonged attention and physical presence of the contractor's top personnel whose services will thus be diverted away from their usual business-can be extremely costly and demoralising.

Where the contractor is not a government-run enterprise (which ultimately draws upon unlimited support of the poor tax-payer to whom it does not feel accountable) but a private organisation answerable on its own strength to its shareholders, the decision to resort to court action should only be made when the gravity of the case is really enormous. Moreover, the trial of a construction case by a jury in the court may not prove to be very distilled and incisive because the usual public jurors with their divers~ backgrounds may not be in a position to appreciate the technical complications, since they may not be oriented and experienced in the technical problems on the anvil. The case might well involve understanding the finer aspects of some of the specialised topics such as subsoil and hydrological characteristics, materials, testing and quality-control, influence of the construction methods employed and of the plant and equipment deployed, ramifications of concrete mix design and hot or cold weather concreting, the effect of the spread of perishable and nonperishable materials and of falsework and plant at site prone to flooding or onslaught of tides and the elements, etc. The judge and the members of the jury are normally not familiar with such matters. It will therefore require exceptional effort on the part of the contractor and his non-technical attorney to explain the depth and detail of these points in the hope of receiving due justice and this is not easy.

Claim Preparation

Claim preparation involves the sequential arrangement of project information and data to the extent that the issues and costs of the dispute are defined. There are many methods to approach development and cost of a claim, but all require a methodical organization of the project documents and analysis.

Assuming that it has been determined that there is entitlement to a recovery, as determined by consideration of interpretation guidelines, the feasibility of recovery should be determined. Once these determinations are complete, claims are generally prepared by using either a total-cost approach or an actual cost approach.

An actual-cost approach, also called a discrete approach, will allocate costs to specific instances of modifications, delays, revisions, and additions where the contractor can demonstrate a cost increase.

Actual costs are considered to be the most reliable method for evaluating a claim. Permissible costs are direct labour, payroll burden costs, materials, equipment, bond and insurance premiums, and subcontractor costs. Indirect costs that are recoverable include labour inefficiency, interest and financing costs, and profit. Impact costs include time impact costs, field overhead costs, home office overheads, and wage and material escalation costs. Pricing the claim requires identification and pricing of recoverable costs. The recoverable costs depend primarily on the type of claim and the specific causes of unanticipated expenses. Increased labour costs and losses of productivity can occur under a wide variety of circumstances. Increased costs for bonding and insurance may be included when the project has been delayed in completion or the scope has changed. Material price escalation may occur in some circumstances. In addition, increased storage costs or delivery costs can be associated with many of the common disputes.

Equipment pricing can be complicated if a common schedule of values cannot be determined.

Total cost is often used when the cost overrun is large, but no specific items or areas can be identified as independently responsible for the increase. Stacked changes and delays often leave a contractor in a position of being unable to fully relate specific costs to a particular cause. The total-cost approach is not a preferred approach for demonstrating costs. A contractor must demonstrate that the bid and actual costs incurred were reasonable, costs increased because of actions by the defendant, and the nature of the losses make it impossible or highly impractical to determine costs accurately. Good project information management will improve the likelihood that the contractor can submit an actual-cost claim rather than a total-cost claim. However, due to the complexity of some projects, the total-cost approach may be the most appropriate method.

Dispute Resolution

Alternate dispute resolution (ADR) techniques have slowly gained in popularity. High cost, lost time, marred relationships, and work disruptions characterize the traditional litigation process. However, many disputes follow the litigation route as the main recourse if a significant portion of the claim involves legal issues. The alternatives — dispute review boards, arbitration & mediation and minitrials — are usually established in the contract development phase of the project. The traditional litigation process is the primary solution mechanism for many construction claims. This is particularly important if the dispute involves precedent-setting issues and is not strictly a factual dispute. The large expense of trial solutions is often associated with the cost of recreating the events on the project that created the original dispute. Proof is sought from a myriad of documents and records kept by contractors, engineers, subcontractors, and suppliers, in some cases. Once filing requirements have been met, a pretrial hearing is set to clarify the issues of the case and to establish facts agreeable to the parties.

The discovery phase of litigation is the time-consuming data-gathering phase. Requests for and exchange of documents, depositions, and interrogatories are completed during this time period. Evidence is typically presented in a chronological fashion with varying levels of detail, depending on the item's importance to the case. The witnesses are examined and cross-examined by the lawyers conducting the trial portion of the claim. Once all testimony has been presented, each side is permitted to make a summary statement. The trier of the case, a judge or jury, deliberates on the evidence and testimony and prepares the decision. Appeals may result if either party feels there is an error in the decision. Construction projects present difficult cases because they involve technological issues and terminology issues for the lay jury or judge. The actual trial time may last less than a week after several years of preparation. Due to the high cost of this procedure, the alternative dispute resolution methods have continued to gain in popularity.

Dispute review boards have gained an excellent reputation for resolving complex disputes without litigation. Review boards are a real-time, project-devoted dispute resolution system. The board, usually consisting of three members, is expected to stay up-to-date with project progress. This alone relieves the time and expense of the traditional document requests and timeline reconstruction process of traditional discovery and analysis. The owner and contractor each appoint one member of the dispute review board.

The two appointees select the third member, who typically acts as the chairman. The cost of the board is shared equally. Typically, board members are highly recognized experts in the type of work covered by the contract or design. The experience of the board members is valuable, because they quickly grasp the scope of a dispute and can provide their opinion on liability. Damage estimates are usually left to the parties to work out together. However, the board may make recommendations on settlement figures as well. Board recommendations are not binding but are admissible as evidence in further litigation.

Arbitration hearings are held before a single arbitrator or, more commonly, before an arbitration panel. A panel of three arbitrators is commonly used for more complex cases. Arbitration hearings are usually held in a private setting over a period of one or two days. Lengthy arbitrations meet at convenient intervals when the arbitrators' schedules permit the parties to meet; this often delays the overall schedule of an arbitration. Information is usually presented to the arbitration panel by lawyers, although this is not always the case. Evidence is usually submitted under the same administrative rules the courts use. Unless established in the contract or by a separate agreement, most arbitration decisions are binding. An arbitrator, however, has no power to enforce the award. The advantages of arbitration are that the hearings are private, small claims can be cost-effectively heard, knowledge of the arbitrator assists in resolution, the proceedings are flexible, and results are quickly obtained.

Mediation is essentially a third-party-assisted negotiation. The neutral third party meets separately with the disputing parties to hear their arguments and meets jointly with the parties to point out areas of agreement where no dispute exists. A mediator may point out weaknesses and unfounded issues that the parties have not clarified or that may be dropped from the discussion. The mediator does not participate in settlements but acts to keep the negotiations progressing to settlement.

Mediators, like all good negotiators, recognize resistance points of the parties. A primary role of a mediator is to determine whether there is an area of commonality where agreement may be reached. The mediator does not design the agreement. Confidentiality of the mediator's discussions with the parties is an important part of the process. If the parties do agree on a settlement, they sign an agreement contract. The mediator does not maintain records of the process or provide a report to the parties on the process.

A major concern that can be expressed about the ADR system is that it promotes a private legal system specifically for business, where few if any records of decisions are maintained, yet decisions may affect people beyond those involved in the dispute. ADR may also be viewed as a cure-all. Each form is appropriate for certain forms of disputes. However, when the basic issues are legal interpretations, perhaps the traditional litigation process will best match the needs of both sides.

Review Questions

1. What triggers claims and disputes in construction industry?
2. How bid proposal process can be a potential for disputes?
3. Who is the low bidder?
4. Explain what is different about "differing conditions"?
5. How can the risks associated with differing site conditions be reduced?
6. How late bid counts for dispute?
7. Explain various aspects of contract interpretation of dispute in construction industry.
8. What are some possible contractual problems?
9. Write short notes on:
    - Currency Fluctuation Effect
    - Ambiguity in Specifications and Drawings
    - Equitable Adjustment
    - Defective specification
    - Settlement through Arbitration
    - Settlement through court action
    - Dispute resolution
    - Mediation

6

Valuation Analysis

Meaning, Steps and Advantages

Valuation

Valuation is the art of determining present value of a property such as a building, a factory or other engineering structure. By valuation the present value of property is fixed. The present value of property can be determined on the basis of rent, or income it may fetch. It is also determined or decided by its selling price. The value of a property depends on its structure, life maintenance, location, etc.

Value means the saleable price (pre sect value) of structure and cost means the cost of construction or purchase. The value may be less or more than the cost.

There is a saying 'only things certain are death and taxes' and the same is not exactly true. No one can forecast his exact moment of death as also exact amount of taxes to be paid on his property which will vary with the variation in its fair market value. Valuation is a logically deduced estimate of value, on economic decision of the Valuer or Engineer or an Architect. He has to work out value of the existing property for various purposes because of this experience, training and expertise. Valuation is needed for wealth tax, estate duty or death duty and municipal taxation, etc.

Purpose of Valuation

Following are the main purposes of valuation:

1. **For rent valuation:** Valuation of a property is done to fix the rent. Rent means the amount it can fetch if it is let out. Rent fixed on the basis of certain percentage of the valuation. It is generally 6% to 10%.
2. **For buying or selling:** When a property is required to sell or buy, the valuation of the same is done both by seller and buyer in order to arrive at reasonable price. Every seller or buyer is willing to sell or purchase the property upto certain limiting price and that is only arrived at by valuation.

3. **For security of loans or mortgage:** When it is required to have loan against the security of any property, the person who is advancing the amount valued the property. Thus valuation is necessary.
4. **Acquisition:** When any property is compulsorily acquired by government, compensation is given to the owner for which valuation is necessary.
5. **For assessment of tax:** To determine the property tax. House tax, etc., valuation of the property is necessary.

Definition of Value Analysis

Value Analysis is a tool of management which attempts the question of saving cost from the point of view of, 'Value' or in other words the main aim is to study the relationship between the design function and cost of a part, keeping in view to reduce part and cost through change in design, modification in specification of the material used by changing the source of supply and so on. The Value Engineering is also known as value analysis.

Value engineering identifies the areas of excessive or unnecessary expenses and attempts to improve the value of the product. It provides the base for better performance at a lower cost while reducing neither necessary quality, reliability nor maintainability.

Value Analysis is a technique of cost reduction based on systematic and organised examination of every item of cost which goes into the manufacture of the industrial product in terms of the value or customer satisfaction it adds to the product.

Another definition of Value Analysis is "the systematic examination of all factors which contribute to the cost of a product, part or a material with the object of uncovering these possibilities of cost reduction which will not in any way reduce the quality, performance or any other attribute of the product which is sought by the customer."

Value Analysis means the organised and exhaustively critical study of a product in terms of the design, functions and costs with the object of cost reduction.

Other Definitions

1. **Market value or fair market value:** Market value is that value of property when it is put to open sale by auction at particular time. It differs time to time with the advancement of industries, fashions, transportation facilities in that area. The market value of property will depend upon a number of factors like:
   (a) User of the plot, its shape, size, area and nature of locality.
   (b) Whether the locality is a developed one or not.
   (c) Effects of the local authority's or other statutory bodies' rules and regulations controlling the development of the properties.
   (d) Means of transportation and easy accessibility.

(e) Loan facilities by government and semi-government bodies.
(f) Taxation policies.
(g) Demand and supply.
(h) Abnormal conditions.

Some of the factors given above are physical, some are social, some are economic and some are legal and properties are affected by one or more of the factor.

2. **Municipal taxes:** Municipality imposes certain taxes on the property to collect money to undertake the maintenance of public utility services such as roads, drains, water supply, etc. These taxes are known as municipal taxes. The taxes assessed on some percentage of the net income from property and varies from 10% to 25%. The taxes are usually less for small houses and high for big houses.
3. **Scrap value:** When any property loses its utility and it is not in a position to serve after repair, it is required to dismantle and reconstruct. The value of dismantled materials such as steel, timber, bricks, etc., is known as scrap value. The scrap value is generally 10% of the total cost of construction. In case of machinery value of metal is scrap value, it is also known as Demolition value or Junk value.
4. **Salvage value:** It is the value at the end of utility period without being dismantled. In the case of a machine, when its useful life of span ends and it becomes uneconomic, it may be sold and purchaser may use it for other purpose. The sale value of machine is salvage value. It does not include cost of removal.
5. **Book value:** It is the amount shown in account books after allowing necessary depreciations. The book value depends upon amount of depreciations. The book value of a property is the original cost minus the amount of depreciation upto the period passed. It continuously reduces year to year and at the end of utility period the book value is only scrap value.
6. **Outgoings:** The expanses, those are spent on the property by taking periodical repairs, are outgoings. It also includes taxes those are levied by Government or local bodies, i.e., property tax, etc., sinking fund, management charges, insurance charges and some other miscellaneous charges which are borne by the owner.
7. **Rateable value:** It is the net annual letting value of a property, which is obtained from the gross income after deducting the amount of yearly repairs. Municipal and other taxes are charged at a certain percentage on the rateable value of the property.
8. **Assessed value:** It is the value of the property recorded in the register of local authority. The assessed value is used for the purpose of determining the various taxes to be called from the owner of the property.

9. **Replacement value:** It is that value of a property calculated on the prevailing market rate to replace the same.
10. **Distress value:** When a property is sold at a lower price than that of an open market rate due to certain reasons (panic, war, etc.). It is said to have distress value.
11. **Obsolescence:** Sometimes a building though physically quite sound yet it becomes out of date in style, in structure, in design and thus it loses its functional utility. This is known as obsolescence. The obsolescence may be due to the reasons changes in fashions, changes in planning ideas, new inventions and improvements in design technique. An old designed machine becomes obsolete, though it may be in good running condition and its value will be less. Obsolescence is divisible into two parts functional and economic.
    (a) **Functional obsolescence.** The functional obsolescence may be on account of:
        (i) Advancement in the planned ideas.
        (ii) Extra high ceiling.
        (iii) Columns not property planned in godown.
        (iv) Ill-lighted and ventilated rooms or room having no window at all.
        (v) Baths and kitchens not finished and equipped with accessories of present day standard.
    (b) **Economic obsolescence:**
        (i) Residential building in a factory zone is said to suffering from economic obsolescence.
        (ii) Shifting of business centre and loss of trade.
12. **Gross income:** It is total income that can be fetched from property as rent or other source without deducting out going, operational and collection charges.
13. **Net income:** It is the amount left with the owner from the gross income after deducting outgoings, operational and collection expenses.
    Net income = Gross income - Outgoings.
14. **Annuity:** Annuity can be defined as the annual periodic payments for repayment of the capital amount invested by a party. These annual payments are either paid at the beginning of the year or at the end of the year, usually for a specified number of years.

If the amount of annuity is paid for definite number of periods or years, this is known as annuity certain.

If the payment of annuity begins at some future date after a number of years, this is known as differed annuity.

If the payment of annuity continue for indefinite period, it is called perpetual annuity.

Annuity may be paid by twelve monthly instalments or quarterly or half yearly instalments.

15. **Capital cost:** The total cost of construction of a project including land is called capital cost.
16. **Capitalized value:** This value of property is that amount of money whose annual interest at the highest prevailing rate is equal to net income from the property.
    Capitalized value= Net annual income X Year's purchase.
17. **Year purchase (Y.P.):** It is defined as the capital sum required to be paid once for all to receive an income of Rs 1.00 at certain rate of interest.
18. **Sinking fund:** Sinking fund is a fund which is gradually accumulated for the sole purpose of replacement or reconstruction of the building or structure at the end of its useful life. The sinking fund is regularly deposited in a bank or with some insurance agency so that at the end of the utility period of the building or structure, a sufficient amount is available for its replacement or reconstruction. The owner should not neglect this otherwise he will find difficulty to rebuild the building. The calculation of sinking funds depends upon the life of a building as well as upon the rate of interest. It is generally calculated on 90% of the cost of construction, as the owner will get 10% as scrap value of the building for the cost of old materials when the life of the building is over. The cost of land is not considered into account while calculating sinking fund as the land remains intact. If a property or building is owned or constructed by taking loan, a sinking fund must be created to repay the debt at the end of the term of loan. The amount which will be set aside, will be paid to lender directly or in instalments.

Types of Values

1. Cost Value

It is the cost of manufacturing a product or component.

2. Use Value

It considers work done, functions performed or service rendered and efficiency/effectiveness of the product.

3. Esteem Value

It involves the qualities and appearance of the product which attracts persons and creates a desire in them to posses the product.

4. Exchange Value

It considers the properties or qualities which will remain attractive enough to other people to permit market resale in the future.

Maximum value is obtained when essential function is achieved for minimum cost.

This statement can be expressed mathematically as follows:

$V_{max} = F/C_{min}$

where, V_{max} = Maximum value, F = Functions, C_{min} = Minimum cost

Functions of a Product in Value Analysis

Functions of a product can be classified into following two categories:

1. **Basic (or Primary) Functions**

These are the functions without which the product would virtually lose all of its value, and in some cases even its identity. For example, if lead is removed from an ordinary wooden lead pencil, it would not only eliminate its basic function of pencil but will leave us with not more than a stick.

2. **Secondary Functions**

These functions support the basic functions, although they may or may not be essential functions. For example, pencil without lead may still remain the esteem value by just being seen in the pocket.

Steps in Value Analysis Approach

Sequence of steps for systematic approach of value analysis is:

1. **Orientation**

   Familiarization with needs, specifications and customer desire.

2. **Information**

   All those facts which have bearing on the problem should be gathered. A typical list of information gathered is given hereunder:

   (a) Engineering information
   (b) Procurement information
   (c) Reliability
   (d) Materials information
   (e) Cost information
   (f) Tooling
   (g) Manufacturing information
   (h) Quality control—rejection, tolerance.
   (i) Customers experience
   (j) Packaging and preservation
   (k) Testing
   (l) Scheduling

3. **Creativity**

Use of imagination and brain storming. Adopt the process of blast, create and then refine.

4. **Evaluation (Analysis)**

   Estimate value of ideas and explore best.

5. **Planning**

After selecting few alternatives or combination of alternatives, each of them investigated thoroughly. On the basis of final outcome, detailed planning is carried out and a report is prepared for approval.

6. Execution (Implementation)

After approval of the proposals, its recommendations are implemented. Value engineers are expected to see that the approved recommendations are implemented and hindrances, if any can be sorted out.

Unnecessary Costs and Value Analysis

Unnecessary costs are those costs which neither contributes to function nor the appearance of the product. Value engineering/analysis attacks unnecessary costs to overcome or minimize them.

Some of the important reasons for such costs are:

1. Poor design of the product.
2. Over design of the product by providing it with features that the user just does not use.
3. Unrealistic or too tight specifications.
4. Lack of standardization.
5. Advances in technology not considered, and old practices continued,
6. New cheaper and better sources not considered.
7. Lack of production control.
8. Non-consideration of changed circumstances.
9. Inability to utilize specialist knowledge/available talent.
10. Wrong belief, habits and attitudes.

Value and Cost

Normally selling price depends upon the benefit which the customers expect to drive by using the product. This benefit is translated by the customer in terms of money which he can pay or in other words, this is the value which the customer places on the article for its purchase.

This amount is entirely based on the wishes of the customer and has nothing to do with the money spent on producing the article. In some products, we usually observe that, price paid by the customer is very high even if the cost of production is small. The reason behind this is that he is very anxious to purchase it and has rated its value very high.

Thus it is necessary for the manufacturer to incur costs in such a way so as to get good response from the customers and the product will be appreciated by them. So as to ascertain the value of the product, the businessmen make the systematic efforts. The process of making the systematic efforts to evaluate the value of the product is known as 'value analysis'.

In the manufacture of industrial products cost is incurred to create value, costs and values have input-output relationships. The scope to optimise the input of costs in order to produce a given set of values is always there. Value is what the product is worth to the customer or user.

Value differs both from price and cost. It can be defined as the cost proportionate to the utility, i.e.

Value = Utility/Cost

Thus the value of a product can he increased either by increasing its utility with the same cost or by decreasing its cost for the same utility. Users have different concepts of value depending upon time, place and availability of substitute items. Value is adjudged by cost value, use value, esteem value and exchange (re-sale) value.

The price at which the product sells or can sell on the market is a measure of its value to the customer or user.

Cost comprises of the:

(i) Money value represented by the material used to make the product;

(ii) Wages of the workers; and

(iii) Other indirect costs attributable to its manufacture.

The excess of value or market price over the costs incurred in producing the product represents the margin of profit of the manufacturer. In order to make his product more profitable, the manufacturer should make a constant search for reducing the manufacturing costs.

Reduced manufacturing costs means one or more of the following advantages:

(i) Increase the margin of profit.

(ii) Reduce the loss.

(iii) Enable a reduction in selling price and thus achieve a greater demand for the product.

Customer or user pays the assigned price for the product because he considers the satisfac-tion derived from its use worth that price. He is generally unconcerned about what it costs the manufacturer to produce the product.

Team Approach for Value Analysis

Value Analysis requires the cooperation of all the functional deptts in an enterprise. As all deliberations must be based on the ultimate satisfaction which the customer wants from the product, the sales and market research Deptt. should be intimately connected with value analysis exercises.

The Design Deptt. and Engg. Deptt. which translates the Customer requirements into the physical embodiment of the product must also be intimately concerned as they alone would be in a position to know as to why certain parts with certain properties must be assembled to make the product.

As a major portion of the product cost is accountable by materials, the purchase Deptt. which translates the design into reality and the costing Deptt. which possesses the factual data about the different cost elements as also their interaction with the others which must necessarily be associated with Value Analysis.

This organised exercise is unlikely to succeed without the enlightened cooperation from all the functional areas. The team must be consisting of senior representatives of the above 5 functional Deptts and they should possess special expertise in their own area which they represent so as to enable to influence the changes consequent to the recommendations of the Value Analysis team.

In large enterprises, Value Analysis should be a full time job for the team members. The Value Analysis team must have a position directly reporting to the top management. It is only, when an enlightened support from the top is assured than only the Value Analysis team will be able to implement its suggestions to the advantage of the enterprise.

Value Analysis team must keep an open mind to any suggestion, and all the suggestions should be noted down in writing and should not be rejected without consideration even if they appear prima facie unworkable.

Subcontractors and suppliers of material can come forward with reliable suggestions as regards alternative materials and their advice must be explored. Representatives of other Deptts like those from the inspection, packing, shipping and store keeping can be associated when points pertaining to their specialised areas being discussed.

The actual explorations in cost reduction are based on the familiar method of questioning.

Following are these questions:

1. Can the part be eliminated?
2. Can you combine it with other parts?
3. Can you separate it into simpler parts?
4. Can you use a standard part? Generally standard parts are economical.
5. Can you use standard materials?
6. Can you use cheap materials? Cheaper materials need not always mean inferior materials.
7. Can you use less materials? There is scope for considerable saving in materials, particularly in sheet and plate work by preplanning the jobs.
8. Can you buy more cheaply?
9. Can you reduce rejects?
10. Can you ease limits? Unnecessary tight tolerances are costly to obtain and many times, they serve no useful purpose.
11. Can you save on finish? Critical examination often reveals possibilities of saving on surface finish, which includes, paintings, electroplating etc.
12. Can you use better methods? This question is particularly important for parts having high labour cost content.
13. Can you reduce the risk of error? Could the reliability of the process be improved.
14. Can you do anything else? Any other possibility of cost reduction in the peculiar circumstances of the best-usage may occur to the team.

Objectives of Value Analysis

(i) To simplify the product
(ii) To use cheaper and better material
(iii) To use efficient and economical process
(iv) To reduce the cost of the product
(v) To improve the product design
(vi) To increase the utility of the product
(vii) To increase the profits.

Stages in Value Analysis

Value analysis is carried out in the following stages:

1. What the Customer Wants
   In first stage customer's requirements are recognised.
2. Finding Out the Best Method of Performing the Work to be Done
   Full use should be made of modern knowledge relating to production. Standardisation, simplification, work study, operation research, network analysis and other techniques should be employed to determine the most economic methods.
3. To have the 'appropriate cost' for the 'appropriate performance', i.e. to have the correct relationship between cost and performance.
4. Search for better ways of performing work and consider the functional use of each part of a product. The vital parts of value analysis are to eliminate the unnecessary function.

Requirements of a Product in Value Analysis

Generally there are two requirements, namely functional or aesthetic, i.e. a product should be useful and should be pleasing in appearance.

Thus Value Engineers look them in the following ways:

1. Use

It is the function for which the part is manufactured. For example, what is a pen supposed to do? The answer is that it should write and hold ink.

2. Esteem

Actually this is something wider than the aesthetic side of design. The fact that a product is pleasing to look at, may induce people to buy but there are other factors. In short, the esteem aspect is concerned with getting the customer to buy.

For example, if the pen has gold plated body or other features, the customer is tempted to purchase more readily.

When value analysis is done, then it is clear that the use function cannot be avoided because it is a main function. But it is not necessary in the case of esteem function. Any benefit added on to a product, over and above those required to meet the use function, must add to the cost causing a higher price.

When analysing, it is essential to consider how much be added in order to sell the part. The esteem function cannot be eliminated altogether, but it should not be over emphasised. Now suppose the cost of gold plated pen is Rs.1000.

This cost is further divided as:

Use = Rs. 15

Esteem = Rs. 985

Now the question is whether Rs. 985 is justified for esteem purpose. Do the additional features add to the value? This matter is closely related with the profit making.

The reasonable price of a product can be obtained by comparing with a competitive product. When this comparison is done the possibility of substitution is examined, e.g. whether a pen with a plastic body? Can the method of filling be simplified? Does the cap have to be made of expensive material and so on?

Advantages of Value Analysis

There are many advantages beside the normal advantages of cost reduction-lower cost, lower prices and higher profits.

Further, the following advantages may be possible when value analysis is employed in the business:

1. The most suitable products are manufactured because a careful study is made to determine the desirable feature of each product in terms of customer's requirement.
2. Each product should be manufactured at the lowest possible cost because special attention is paid for simplification, standardisation and improved methods of production.
3. Quality is maintained at desired level because there is no question to reduce cost at the expense of quality.
4. Value Engineering is based on the principle that management effectiveness can be measured in terms of cost saving. Any saving in cost is treated as increase in efficiency.
5. The constant search for improvement will lead to greater all-round efficiency.
6. Suggestion box method may be employed and any idea can be looked into by the Value Engineers.
7. Easy to repair or replace any part of the product.
8. Lighter in weight.
9. Easy in packaging to protect the product till it reaches to the user.

Review Questions

1. Give a meaning of valuation.
2. What is the purpose of valuation?
3. Give different definitions of value analysis.
4. What are various stages being carried out in valuation analysis?
5. What are the Advantages of Value Analysis?

7

Construction Labour, Legislation and Arbitration

Labour is defined as human efforts or services which are directed towards production. As labour is human services, it has characteristics that differ from services of land and capital. Services of human beings are similar to those of land and capital. The services of all factors are perishable. Those services which are not used in the present can never be used. This applies to human services as well as those of inanimate objects. It follows that unemployment is very wasteful whether it is unemployment of workers or of land and capital goods.

Immobility of Labour

Labour cannot easily be shifted from place to place or from occupation to occupation because of the following reasons:

1. **Difficulty of acquiring knowledge:** For certain occupations, it is difficult to acquire the necessary knowledge and skill. To become a civil engineer, for instance, a person must attend high school, college and engineering institution. He must have sufficient ability of course to master the required subjects. Finally, he must have a period of apprenticeship in minor positions before he can assume major responsibilities. In a similar way, other professions also require equally long periods of preparation.
2. **Leaving accustomed surroundings:** The individuals become accustomed to their jobs and communities. As time passes, they become reluctant to change either work or residence and in many cases both of these factors operate. The differences in wage rates offered for the same type of work often are found in different communities because labour does not move freely from place to place.
3. **Reluctance to enter certain occupations:** Labour is often reluctant to enter certain unpleasant occupations. Many persons prefer white-collar

jobs or the professions even when larger remuneration could be earned in some of the trades. This reluctance to enter certain occupations restricts the supply of labour and thereby becomes a factor in determination of wage rates.

Of the three reasons, the last two refer to mental attitudes of the workers. The first cause of immobility often is beyond the control of the individual worker.

Theories of Wages

Various theories explaining why wages are to be paid have been put forward. We will briefly discuss the following three important theories of wages:

1. Subsistence theory of wages
2. Wages-fund theory of wages
3. Bargaining theory of wages.

1. **Subsistence theory of wages:** This theory is based on the famous Malthu's theory of population which states that population would increase faster than the food supply. This theory was developed in the early nineteenth century when England was becoming industrialized.

On the basis of Malthusian theory of population, Ricardo formulated a theory of wages as follows:

In the long run, wages will be set at a level that will just give the workers a bare subsistence and enable them to perpetuate their race without increase or diminution.

Ricardo's principle of wage determination soon came to be referred to as the iron law of wages. Its implications aroused dispair and perhaps contributed to class hatred. The leaders of workers felt that if this law was sound, labour had no chance of improving its lot and it would be condemned to virtual economic slavery. This theory, however, called attention to the following two truths:

(i) No one can for long work for wages that are below subsistence.
(ii) When labour is abundant, wages will be lower than when this condition does not prevail.

The subsistence theory of wages has been criticised on many grounds. In the modern World, population shows less and less tendency to press upon the means of subsistence and hence, this theory has little to offer in explaining how wages actually are determined. Furthermore, the implication that there is an iron law or natural price of labour encouraged the exploitation theory which in turn caused much misunderstanding between labour and management. It is quite clear that this theory does not give an adequate explanation of the determination of wages. Rangawala,. (2012).

2. **Wages-fund theory of wages:** This theory is based on the assumption that the fund allotted to labour by employer is fixed. The wage of an

average worker is obtained by dividing this fixed fund among the workers. The total fund divided by the number of workers will determine the annual wage rate.

It is true that the employers in planning production set up budgets and determine the amount that they can afford to spend for labour. But on close examination, it will be seen that the problem is not so simple in actual practice. In fact, the wage rate actually determines the size of the fund, or the amount of the budget allotted to labour. The fund, as assumed in this theory, does not determine the rate of wages. The truth is that wage rates are established in the market and with these rates in mind, the employer determines the fund for labour.

3. **Bargaining theory of wages:** This theory attempts to give a practical explanation of what takes place in the market. Labour is considered a commodity just like any of the other factors of production. The employers are the buyers of labour and the workers are the sellers of it. The buyers and sellers enter the market and bargain to determine how much is to be given for labour. The buyer of labour has a maximum limit above which he will not go. The subsistence is the minimum limit below which the workers cannot afford to work. The actual rate of labour will be somewhere between these two limits and will depend upon the relative bargaining power of buyers and sellers. Rangawala,. (2012).

Generally the buyer or the employer stands in a strategic position because he controls other factors of production. If labour organizes and presents a solid front, it has an advantage and perhaps can force employers to pay maximum wages.

Thus, buyer or seller which is more effectively organized will have ultimate supremacy in the bargaining price. If buyers and sellers are more or less of equal strength, the wages will be settled at about midway between the maximum and minimum levels.

Nominal Wages and Real Wages

The wages paid in cash to workers are known as nominal wages and by these wages, they are able to buy the necessaries of life such as food, cloth, etc. However, in certain industries, some types of free services are given to the workers. The workers on an agricultural field can get the corn at a low price. The dispensary may be run for the benefit of workers of an industry. The children of the workers may be supplied with free books and fees for education. The principal of a college may be provided with a furnished residential bungalow. In all these cases, indirect free services are given to the workers and if they are not supplied, the workers will have to spend from the nominal wages to get the benefit of such services. These are known as indirect wages and in a mathematical form, real wages can be expressed as nominal wages plus indirect wages in terms of free services.

Psychic Income

The psychic income can be defined as the pleasure or satisfaction derived from an occupation.

The individuals are willing to work for low-wages in certain occupations that provide a high degree of satisfaction to them because they obtain non-pecuniary compensation. It is found that where psychic income is a significant factor, the money wage is often low than might be expected for such occupation.

The psychic income is especially high in some of the professions. For instance, a chemist or a botanist will work for longer hours than he is supposed to work because he finds satisfaction in his research. In a similar way, a sincere teacher will spend more time to teach his students because he gets mental peace and pleasure from his work. Rangawala,. (2012).

System of Wage Payments

In construction projects, the following two systems of wage payments are adopted:

1. Time wage system
2. Piece wage system.

1. **Time wage system:** This is the simplest method of wage payment and it adopts time as basis of labour remuneration without specifically linking the output of the worker. The employer buys the time of the worker in units which may be hour, day or month. The employer will however indirectly watch the performance of worker while deciding his fixed rate. The inefficient worker will be discontinued and the efficient worker will be offered higher fixed rate.

Following are the advantages of this system of wages:

(i) It gives the workers a sense of security in job and he therefore does not lose interest in work.

(ii) It grants simple procedure for the calculation of wages.

(iii) It is suitable for jobs which cannot be measured.

(iv) It is willingly accepted by trade unions because it does not create any distinctions in the ranks of workers on the basis of efficiency.

(v) The worker is guaranteed a fixed rate on time basis and as he is not concerned with quantity of work done, the work of better quality is produced.

Following are the disadvantages or drawbacks of this system of wages:

(i) It creates difficulties for employer to calculate the labour cost on unit of production.

(ii) It deals with par the efficient as well as inefficient workers and hence, there is no appeal for improving the quality of production.

(iii) It is not possible to put an effective cost control.

(iv) It requires constant and regular supervision which also adds to cost of production.

2. **Piece wage system:** In this system, the worker is paid for the amount of work done by him rather than time consumed in doing it. Thus the payment in this system is linked to the production and the efficient worker is rewarded as larger production results in higher remuneration.

Following are the advantages of this system of wages:

(i) It ensures effective cost control.

(ii) It ensures fairness to everyone by relating wages to the performance.

(iii) The cost per unit of production can easily be calculated.

(iv) The idle time is minimized.

(v) The relations between employer and labour are well maintained.

(vi) The strict supervision is considerably reduced.

Following are the drawbacks of this system:

1. If output is affected by some action of the management, there are chances of dispute between employer and employees.
2. In the process of over-speeding, the workers may damage the machinery or may injure their health.
3. It is opposed by the labour unions on the ethical grounds.
4. It lays too much stress on quantity of production rather than its quality aspect.
5. It requires elaborate production records and pay rolls.
6. The wages are paid on the basis of output and hence, there is a sense of insecurity among the workers.

The piece wage system is mostly adopted in the construction industry because of the following reasons:

1. The individual output can be readily measured and priced
2. There is sometime urgency in completing the job.
3. The work is of a standardised and repetitive nature.

However, the following two modified forms of piece wage system may be adopted to minimize its drawbacks:

1. Balance and debt system.
2. Standard hour system.

1. **Balance and debt system:** Under this arrangement, the worker is paid on piece wage basis. But, if on a particular day, his wages according to the piece rate system fall below the daily wage on time basis, he would be paid the latter with the condition that if on subsequent days, his wages on piece rate system exceed the daily wage on time basis, he will permit to recover the excess wages paid earlier to him. Thus, the worker is guaranteed the wages on time basis. But at the same time, the extra amount, if any, paid to the worker is recovered in due course of time.
2. **Standard hour system:** This arrangement is also sometimes referred to as one hundred per cent gain sharing system and the following procedure is adopted:

(a) A standard time in hours is allowed to complete a particular job.

(b) If the worker completes the job in standard time or less, he will be paid for his standard time at the specified rate per hour even though the completion period is less than the standard time.

(c) If the worker takes more than the schedule time to complete the job, he will be paid at the guaranteed time rate for the actual time spent by him on the job.

(d) In the absence of any provision for the guaranteed time rate, the worker will be paid only for the standard time.

Incentive Plans

Both the systems of wage payments, namely, time wage system and piece wage system, have their own applications. To combine their good aspects and to form systems which would result in the increase of productivity, various incentive plans of wage payment with time and speed blended together have been formed. Following are some of the well-known incentive plans of wage payment: Rangawala,. (2012)

1. Halsey plan
2. Rowan plan
3. Taylor's differential piece wage plan
4. Merrick differential or multiple piece rate plan
5. Emerson's efficiency plan
6. Gantt's task and bonus plan
7. Bedeaux point premium plan
8. Hayne's plan
9. Priestman bonus system
10. Barth variable sharing plan
11. Group incentive plans.

The basic principle of all the above plans is that the workers who exceed the standards of output set through scientific methods must be rewarded. The main features of each plan will now be briefly described.

1. **Halsey plan:** This plan was originated by F.A. Halsey in America. The standard time required for a particular job is determined beforehand and the workers who perform the job in less than the standard time, are rewarded with a bonus. But the workers who take more than standard time to complete the job are not punished and they are paid on time wage system.

The calculation of the wage is made as follows:

Wage of worker = W + B

Where W = Wage as per actual time

B = Bonus which is the money value of 33.3 per cent of time saved in case of standard time set on previous experience and 50 per cent of the time saved when the standards are scientifically set.

This plan is suitable for standardized jobs where the time can be determined accurately beforehand.

It grants the following advantages:

(i) If ensures every worker a feeling of security in the sense that he is entitled to receive a minimum hourly wage.

(ii) It is easy in calculation.

(iii) The efficient workers are rewarded.

(iv) The plan lays emphasis on time saved and hence, it proves to be popular among workers.

2. **Rowan plan:** This plan was introduced in 1901 by James Rowan of England.

It is similar to Halsey plan except that the calculation of bonus is calculated as follows:

Bonus = Time saved × Time Taken × Hourly rate/Standard Time

The major drawback of Halsey's plan was that the worker may spoil the quality of product by over speeding to get more bonus. This drawback is removed by this plan. Following two points should be noted:

(i) As the saved time increases, the time taken decreases and hence, the bonus is always a decreasing proportion of an increasing figure.

(ii) Under this plan, the bonus earned by a worker cannot exceed 25% of the standard time. Rangawala,. (2012).

3. **Taylor's differential piece wage plan:** This plan was devised by F.W. Taylor in 1880. In this plan, the standard of output is fixed and two-piece wage rates are laid. The workers whose output just exceeds or becomes equal to the standard one, are paid at the higher rate and for those whose output is lower than the standard one, a lower rate is prescribed. Thus, the worker is encouraged to produce as much as the standard output.

4. **Merrick differential or multiple piece rate plan:** This plan is the modification of the Taylor's plan. But it lays down three wage rates - one for *the* beginners, the second for the developing workers and the third for the efficient workers.

The standard output is decided and the wage rate is calculated as follows:

(i) Those who attain output upto 83% of the standard one are paid the lowest rate.

(ii) Those whose output is more than 83% but less than the standard one are paid at the second rate which is higher by 10% of the first rate.

(iii) Those whose output exceeds the standard one are paid at the third rate which is higher by 10% of the second rate.

5. **Emerson's efficiency plan:** As per this plan, the efficient workers are paid a bonus. The efficiency is calculated by comparison of performance of worker either with the standard output per unit of time or the standard

time for a job. For instance, if the standard output is 90 units per day, the efficiency of workers producing 45 units, 60 units and 81 units would be as under:

For 45 units, efficiency = 45/90 × 100 = 50 %

For 60 units, efficiency = 60/90 × 100 = 66 × 2/3%

For 81 units, efficiency = 81/90 × 100 = 90%

Emerson proposed the following scale for the calculation of bonus:

(i) No bonus should be payable if the efficiency is less than 66 × 2/3%.

(ii) At 66 × 2/3 % efficiency, a bonus of 1 % on wages earned according to time rate should be paid and this rate of bonus would increase as efficiency rises so that at 90% efficiency, the bonus would be 10%.

(iii) At 100% efficiency, the bonus would rise to 20%.

(iv) Beyond 100% efficiency a bonus of 30% should be paid.

This plan is easy to understand and it can be applied to individual workers as well as groups of workers. However, the incentive of workers beyond the point of standard efficiency becomes too mild.

6. **Gantt's task and bonus plan:** This plan was devised by H.L. Gantt and it is aimed to moderate Taylor's plan. It is similar to Emerson's plan and under this plan, wages on time basis are guaranteed to every worker. A good standard of performance under the best conditions is determined and the efficiency of worker is also measured accordingly. No bonus is payable to a worker whose efficiency is less than 100%. A bonus of 20 per cent is paid to worker showing efficiency of 100%. For instance, if 20 hour job is done in 21 hours, the efficiency of worker will be about 95.24% and even then, he will not be paid a bonus. But if the job is done in 20 hours, he will be paid wages for 20 hours plus a bonus of 20 per cent. Thus this plan is basically a piece rate system with a guarantee of wa15es on time basis.

The plan provides for security and incentive. It is simple to understand and it can be easily introduced. It also encourages planning and better supervision. It satisfies the workers as the bonus is handsome.

7. **Bedeaux point premium plan:** In this plan, the value of time saved is shared between workers and foreman in the ratio of 75% and 25% respectively. This is evident from the fact that a worker cannot show good results unless his foreman is fully co-operative with him. The standard time is predetermined for a job in minutes with unit mentioned as B. Thus, each B represents a point equal to one minute. Rangawala,. (2012).

The standard B units of various jobs done by a worker per week are calculated and they are compared with the actual time. If the actual time exceeds the standard time, the worker gets only wages for the actual time on time basis. If the actual time is less than the standard time, the difference is divided by 60 to get hours saved and then, 75% of the wages of the time saved will go to the concerned worker and 25% to the foreman.

8. **Hayne's plan:** This plan is similar to Bedeaux's plan. The unit adopted in this plan is mentioned as MANITS (short form of man-minutes). The standard time for each job is decided before- hand and from the actual time taken by a worker, the saving in time is calculated in MANITS. Dividing by 60, the saving in time is obtained in hours and by applying time rate, the amount of time saved is obtained. The benefit of time saved is distributed as follows:

(i) If the work is standardised, 5/6th goes to the worker and 1/6th goes to the foreman.

(ii) If the work is non-standardised, 5/10th goes to the worker, 1/10th goes to the foreman and 4/10th goes to the management.

The standardised work means work which is of a repetitive nature and non-standardised work means work of different nature every time.

9. **Priestman bonus system:** This plan takes into consideration the productivity of all workers as a whole. If the output during a particular year has gone above the standardised output or the output of the previous year, the benefit is given to the workers in the same ratio.

This plan may encourage to build up team spirit among all the workers. But as it does not consider individual efficiency of worker, there will be lack of initiative at individual level. Rangawala,. (2012).

10. **Barth variable sharing plan:** In this plan, the wage of a worker is calculated by applying the following equation:

Wage of worker = √(standard hours × actual hours) × rate per hour.

The wages of a worker under this plan rise steeply though at a decreasing rate.

It can be adopted for encouraging less skilled workers. But it cannot be recommended where high production is desired from the skilled workers. It is also difficult to explain the method of calculation to the workers.

11. **Group incentive plans:** It is found that certain pieces of work can only be done by groups of workers and under such circumstances, it is desirable to implement group incentive plans for efficient completion of such tasks.

The major drawback of these plans is that all the members of a group would receive bonus on the same basis even if some of them are highly efficient and others have not contributed effectively to the completion of the job. This may lead to dissatisfaction among certain members of the group. To overcome this difficulty, the following arrangement is made:

1. A fixed piece rate is decided for the entire job for all the workers.
2. The wages of various workers are fixed according to skill and time devoted by each worker.
3. The bonus earned by the group is divided amongst the members in the ratio of wages calculated on time basis.

Payment by Result

When the payment to the worker is performance based, the system of payment is known as payment by result. It may take up various forms as follows:

1. **Differential piece rate:** The calculations are made as per Taylor's plan or Merrick's plan or Gantt's plan.
2. **Group bonus system:** The calculations are made as per group incentive plans.
3. **Piece rate with a guaranteed time wage:** In this case, an hourly or daily rate of wage is guaranteed, if the earning on the basis of output is less than the wages payable on time basis.
4. **Premium bonus plans:** The calculations are made as per Halsey's plan or Rowan's plan or Bedeaux's plan.
5. **Straight piece-rate:** The rates of wages are fixed per unit of output.

Profit-Sharing

For some of the industries, the management provides for the share in profit by the workers in addition to their regular wages. Thus the worker under the profit-sharing scheme will get his share in net profit of the enterprise at the end of six months or one year. Rangawala,. (2012).

1. **Procedure:** The management has to adopt the following procedure for implementing the plan of profit-sharing:
   (i) **Basis of sharing:** The basis on which profit-sharing is to be introduced should be decided. It may be on individual basis, departmental basis, unit basis, etc.
   (ii) **Net profit:** The net profits are usually calculated by applying the ordinary principles of commerce.
   (iii) **Share of workers:** The share or percentage of profit which will go to the workers should be decided. In most of the schemes, the workers share profits only and losses are to be borne by the management.
   (iv) **Share of individual worker:** The method of working out the share in profit of individual worker should also be decided. It is usually on the basis of hours worked or wages earned during the period for which profit is calculated.
   (v) **Form of distribution:** It should also be decided how the profit is to be distributed among the workers. It may be paid in cash or transferred to pension or provident fund.
2. **Advantages:** Following are the advantages of profit-sharing:
   (i) It does not require strict supervision as the workers co-operate warmly.
   (ii) It grants stabilisation of labour force as the worker has to remain in the concern for a certain period to get the benefit of profit-sharing.

(iii) It improves the industrial relations.
(iv) It increases the productivity of the unit.
(v) It is a positive step in the direction of industrial peace.
(vi) It results in additional earnings for the workers.
(vii) It serves as a measure of social justice by giving some status to the workers in an industrial enterprise.

3. **Disadvantages:** Following are the disadvantages of profit-sharing:
   (i) It does not distinguish between the efficient and the inefficient workers.
   (ii) It does not maintain relationship between effort and reward. In strict economic sense, profit accrues only to the enterpreneur.
   (iii) It results in industrial unrest during the period of low profits or losses by the concern

Morale

Every worker is basically a human being. The degree of enthusiasm and willingness with which he works for the organization is known as morale. The conditions of high morale prevail when the employees work with full spirit of co-operation and have very few grievances. On the other hand, in the conditions of low morale, the employees lack the co-operative spirit and they have many intense frustrations. Rangawala,. (2012).

The organization, if it has to fulfil the tasks satisfactorily, must possess a high morale. It is evident that the duties of workers are performed effectively only when they pour their hearts into the work. In a similar way, the office should also have a high sense of morale for its better functioning.

Alexandar Leighton has enumerated the following five factors determining the morals:

1. confidence of individual members in the purpose of the group;
2. confidence of the different members of the group in the leadership and the ability of the leadership;
3. confidence of the individual members in their colleagues;
4. organizational efficiency; and
5. conditions in the group and working conditions.

It is also observed that for small, less complex and decentralised establishments, the morale is high and for big, more complex and centralised establishments, it is on lower side.

Measurement of Moral

The morale indicates the human behaviour and it can be judged by various indirect ways such as:

1. behaviour of individuals and work groups;
2. changes in turnover rate;
3. effects on productivity of employees;

4. regularity in attendance to work;
5. remarks by workers; etc.

Various scientific methods have been found out to measure morale of an organization.

Following two such methods are very popular:

1. Interviews
2. Opinion survey.

1. **Interviews:** The individual workers are personally interviewed by an agency - usually an outside agency such as a consultant. The interviews are carried out in one of the following two ways:
   (i) **Guided interview:** In this case, each employee is asked a series of questions which he is supposed to reply in yes or no or a single word or phrase.
   (ii) **Unguided interview:** In this case, the employee is asked general questions and he is encouraged to express himself as he likes.

The results of interviews are suitably assessed to determine the prevailing morale in the organization. Sometimes, special interviews are arranged with employees who are leaving the service of the company. Such type of interviews is known as exist interview and the data obtained from such interviews may also prove helpful in measuring morale of the company.

2. **Opinion survey:** In this method, the questionnaires containing various matters related to the organisation such as company policies, working conditions etc. are framed. The usual practice is that the employee is supposed to write single response such as good or bad, yes or no, like it or do not like it, etc. against each item in the questionnaire. The opinion survey thus conducted provides enough data to assess the prevailing morale in the organization.

Trade Unions

The idea that labour by organising can secure higher wages has led to the formation of unions known as trade unions or labour unions. Thus a trade union indicates a continuous association of workers formed for the purpose of maintaining and improving their conditions of employment.

1. **Functions:** The functions of trade unions can be broadly divided in the following *two* categories:
   (i) Fraternal or mutual-help functions
   (ii) Militant or fighting functions.

The fraternal or mutual-help functions include arrangement of lectures, cultur.il activity, indoor games, outdoor games, running of hospital, running of school, welfare activities, etc.

The militant or fighting functions include fight for higher wages, gherao, go slow grievances, strike, work to rule, etc.

2. **Advantages:** Following are the advantages of trade unions:

(i) **Bargaining power:** It is generally recognised that if unions are able to limit the supply of skilled workers, they can raise both the wage rate and the marginal productivity of the workers in any given trade.

(ii) **Economic share:** The trade unions are sometimes able to offset the monopolistic position of an employer and they force him to pay the full economic share that should accrue to labour.

(iii) **Grievances of workers:** The employer comes to know the grievances of workers through trade union and he may act accordingly to solve them.

(iv) **Guarantee of industrial peace:** The successful working of trade unions guarantees industrial peace.

(v) **Improvement in skill:** By a programme of education, the union may be able to improve the skill of its members and thus assist in increasing their average productivity and wages.

(vi) **Sense of confidence:** They create a sense of confidence among the workers who become bold enough to raise their voice before their employers.

(vii) **Voluntary settlement:** The disputes of workers may be settled voluntarily through trade union leaders who possess strength, power and authority.

3. **Disadvantages:** Following are the *disadvantages* of trade unions:

(i) **Artificial shortage of labour:** By demanding only the union workers to be employed, the trade unions may develop artificial shortage of labour.

(ii) **High wages:** The trade unions may demand very high wages and other benefits for their workers.

(iii) **New methods of production:** The trade unions may show hostile attitude to new and novel methods of production requiring less labour force.

(iv) **Non-co-operation with employer:** If the trade union is unable to co-operate smoothly with the employer, it will result in decrease in national income, unemployment, less production, etc.

4. **Trade union and wages:** The trade union can succeed in raising the wages of the workers by various methods such as:

(i) by controlling labour supply;
(ii) by increasing labour demands;
(iii) by increasing standard wage rates;
(iv) by raising marginal productivity of worker;
(v) by stopping exploitation; etc.

It is now recognised that labour unions have a distinct place in our economic system. But it is often observed that the purposes of such organizations are not clearly understood by either - organizers or employers. Even though the labour is organized and even though it possesses enough power to bargain effectively with the employers, there is a point above which wages cannot go without serious harm to both labour as well as employer. This maxim is not determined by the employer nor by labour. But it is decided by the public which purchases the commodities or services which the employer sells and which labour helps to produce.

When union leaders understand the economic effects of forcing up wage rates and act with moderation, trade unions are desirable and they perform a useful function. However, if the bargaining theory is accepted and forced without qualifications, the demands may be made which, in the long run, are detrimental to the members of the union because the reduction in employment may more than offset the increase in wage rates.

Labour Welfare

To look after the welfare of labour in any production organization, the labour welfare officers are appointed by the government and they perform the following activities:

1. financial assistance to needy workers;
2. hearing general grievances of workers;
3. observing service conditions;
4. protection of seniority;
5. provision of provident fund and insurance;
6. regular payment of salary on specified date;
7. supervising labour relations with employer; etc.

In addition to the above specific duties, the labour officers also look after the following general activities associated with the welfare of labour:

1. accommodation requirements;
2 canteen facilities;
3. co-operative store facility;
4. drinking water arrangements;
5. education facilities;
6. existence of proper lighting and ventilation;
7. medical facilities;
8. parking facilities for vehicles of workers;
9. recreation facilities;
10. risk covering schemes for workers;
11. safety of workers;
12. sanitary arrangements;
13. training of workers;
14. transport arrangements; etc.

Characteristics of Indian Labour

Some of the important characteristics of the Indian labour can be enlisted as follows:

1. Indian labour is the cheapest in the world. This is due to the fact that the labourers maintain their lives with the minimum requirements. There are large number of labourers and hence, there is perfect competition in the labour market.
2. Indian labour is mostly illiterate and uneducated. It is found that only 20 per cent of the Indian labour has got the technical knowledge.
3. Indian labour is not properly trained and it has no knowledge of what is happening in the outside world.
4. Indian labour is more or less idle and has no incentive to carry out the duties in the most satisfactory manner. It is due to the hot climate prevailing in most of the parts of India. Due to heat, labourer gets tired and he is not able to carry out the work.
5. Indian labour is mobile in nature. The villagers in large number leave their homes in search of work in towns and cities.

Some of the important factors which are responsible for the low efficiency are as follows:

1. **Absentism:** The workers are inclined to remain absent from the work or to change place without proper reason.
2. **Factory condition:** The factories are over-crowded and they possess ill-ventilation.

   They also lac the proper disposal of dirty water.
3. **Housing:** The housing conditions of the Indian labour are very unsatisfactory.

   There are no arrangements for good drinking water, sanitation and other essential facilities. The surroundings are also bad. It results into all kinds of diseases.
4. **Indebtness:** It is found that most of the workers are in debt. The labourers have to pay excessive rates of interest and once they are under the shadow of debt, it is difficult for them to come out.
5. **Migrating character:** In developed countries like U.S.A. and Britain, the industrial labour stays in the cities. Indian labour, on the other hand, returns to the village frequently and when a worker returns again, he is in search of a new job. Thus, he is lacking with the attachment to the job.
6. **Standard of living:** The standard of living of the Indian labour is the lowest among the civilized countries.
7. **Wages:** The wages paid to the Indian labour are the lowest in the world. These meagre wages cannot satisfy the needs of labour. The labourer cannot even purchase the primary needs of life and he cannot think of having a luxurious life.
8. **Working hours:** The working hours of the Indian labour are long and it results into fatigue. The constant working tires the muscles of the human being with the result that the labourer feels a positive pain in short time.

With the above discussions, a number of remedial measures may be suggested to improve the efficiency of the Indian labour. Some of such measures are as follows:

1. The degree of absentism among the workers should be discouraged.
2. The factory buildings should be well planned with respect to light, ventilation, disposal of dirty water, etc.
3. The labourers should be supplied with good houses having enough privacy of family, sanitation, ventilation etc.
4. The workers should be instructed and advised to remain away from undesirable habits, unnecessary expenses, etc. and they should be encouraged to have saving for future.
5. The workers should be trained for their jobs. The arrangement should also be made for the education of the workers.
6. The attempts should be made to bring up the standard of living of the workers.
7. The wages of labourers should be suitably adjusted so that they can pass a good life.
8. The labourers should be allowed to work for a particular number of hours and during this time also, they should be given sufficient rest in the form of recess between working hours.
9. The labourers must be given hopes that if they are found useful and efficient, there is hope for them to advance in life.
10. The efficiency of employer also affects considerably the efficiency of workers.

The raw materials should be purchased at right time and delivered at right place in the hands of right labourer.

Present Scenario

More than 50 million construction workers in India build our homes and cities. An average construction worker earns just over INR 350 a day (US$ 4.70) with almost no perks or benefits. Many workers are their families' primary earners. For the high risk of the jobs they are involved in, they are seldom provided a safety net.

Approximately 87.4 percent of workers in the construction industry are categorised as casual labour, and they make up a majority of the informal workforce in our country. In 1996, the Indian parliament enacted two laws which had the potential of providing protection and dignity to his workforce:

- The Building and Other Construction Workers (Regulation of Employment and Conditions of Service) Act, 1996, (BOCW Act or BOCWA).
- The Building and Other Construction Workers' Welfare Cess Act (Cess Act).

In building and other construction works more than eight million workers are engaged throughout the country. These workers are one of the most vulnerable segments of the unorganised labour in India. Their work is of temporary nature, the relationship between employer and the employee is temporary, working hours are uncertain. Basic amenities and welfare facilities provided to these workers are inadequate. Risk to life and limb is also inherent. In the absence of adequate statutory provisions to get the requisite information regarding the number and nature of accidents was quite difficult and due to this to fix responsibility or to take corrective measures was not an easy job. Although the provisions of certain Central Acts were applicable to the building and other construction workers yet a need was felt for a comprehensive Central Legislation for regulating the safety, welfare and other conditions of service of these workers. In pursuant to the decision of the 41st Labour Ministers Conference held on 18th May, 1995, the Committee of State Labour Ministers had expressed its consensus for the Central Legislation on this subject. In order to regulate the employment and conditions of service of building and other construction workers and to provide for their safety, health and welfare measures the Building and Other Construction Workers (Regulation of Employment and Conditions of Service) Ordinance, 1995 (Ord. 14 of 1995) was promulgated by the President on 3rd November, 1995 as the Parliament was/not in session. To replace this Ordinance a Bill was introduced in the Lok Sabha on 1st December, 1995. Since the Bill could not be taken up for consideration it lapsed. On 5th January, 1996 the President promulgated the Building and Other Construction Workers (Regulation of Employment and Conditions of Service) Ordinance, 1996 (Ord. 3 of 1996). To replace this Ordinance a Bill was introduced in the Parliament which could not be passed and the President promulgated the Building and Other Construction Workers (Regulation of Employment and Conditions of Service) Second Ordinance, 1996 (Ord. 15 of 1996) on 27th March, 1996. As this Ordinance could not be replaced by an Act of Parliament, the President promulgated the Building and Other Construction Workers (Regulation of Employment and Conditions of Service) Third Ordinance, 1990 (Ord. 25 of 1996) on 20th June, 1996. In order to replace this Ordinance the Building and Other Construction Workers (Regulation of Employment and Conditions of Service) Bill was introduced in the Parliament.

Statement of Objects and Reasons

It is estimated that about 8.5 million workers in the country are engaged in building and other construction works. Building and other construction workers are one of the most numerous and vulnerable segments of the unorganised labour in India. The building and other construction works are characterised by their inherent risk to the life and limb of the workers. The work is also characterised by its casual nature, temporary relationship between employer and employee, uncertain working hours, lack of basic amenities and inadequacy of welfare facilities. In the absence of adequate statutory provisions, the requisite information regarding the number and nature of accidents is also not forthcoming. In the absence of such information, it is difficult to fix responsibility or to take any corrective action.

Although the provisions of certain Central Acts are applicable to the building and other construction workers yet a need has been felt for a comprehensive Central Legislation for regulating their safety, health, welfare and other conditions of service The State Governments and Union Territory Administrations have been consulted in the matter and a majority of them have favoured such a legislation. Also, in a meeting of the Committee of State Labour Ministers constituted pursuant to the decision of the 41st Labour Ministers' Conference held under the Chairmanship of the then Union Labour Minister on the 18th May, 1995, a general consensus had emerged on the need for the proposed Central Legislation.

In view of the circumstances explained above, it has been considered necessary to constitute Welfare Boards in every State so as to provide and monitor social security schemes and welfare measures for the benefit of building and other construction workers. For the said purpose, it has been considered appropriate to bring in a comprehensive legislation by suitably amplifying the provisions of the Building and Other Construction Workers (Regulation of Employment and Conditions of Service) Bill, 1988 which was introduced in the Rajya Sabha on the 5th December, 1988. It has also been considered necessary to levy a cess on the cost of construction incurred by the employers on the building and other construction works for ensuring sufficient funds for the Welfare Boards to undertake the social security Schemes and welfare measures.

As Parliament was not in session and in view of the urgency felt by the Government for meeting the longstanding demand for the aforesaid legislation, the President was pleased to promulgate the Building and Other Construction Workers (Regulation of Employment and Conditions of Service) Ordinance, 1995 (Ordinance 14 of 1995) along with another Ordinance for the levy of a cess on the 3rd November, 1995.

A Bill, namely the Building and Other Construction Workers (Regulation of Employment and Conditions of Service) Bill, 1995 was introduced in Lok Sabha on the 1st December, 1995 to replace the aforesaid Ordinance by an Act of Parliament. Since the said Bill could not be taken up for consideration in the Winter Session, 1995 and the Budget Session, 1996 of Lok Sabha, fresh Ordinances, namely, the Building and Other Construction Workers (Regulation of Employment and Conditions of Service) Ordinance, 1996 and the Building and Other Construction Workers (Regulation of Employment and Conditions of Service) Second Ordinance, 1996 (Ordinance 15 of 1996) respectively were promulgated on the 15th January, 1996 and the 27th March. 1996 with a view to provide continued effect to the legislative protection. With tile dissolution of the 10th Lok Sabha, the Building and Other Construction Workers (Regulation of Employment and Conditions, of Service) Bill, 1995 has lapsed. By virtue of Articles 123 (2)(a) of the Constitution the Building and Other Construction Workers (Regulation of Employment and Conditions of Service) Second Ordinance, promulgated on 27th March, 1996 would have also ceased to operate with effect from 4th July, 1996. In view of the urgency attaching to the matter and as Parliament was not in Session, the President was pleased to promulgate the Building and Other Construction Workers (Regulation of Employment and Conditions of Service) Third Ordinance, 1996 (Ordinance 25 of 1996), on 10th June, 1996 in order to give continued effect to the legislative protection envisaged in the said ordinance.

The Building and Other Construction Workers (Regulation of Employment and Conditions of Service) Third Ordinance, 1996, *inter alia,* provides for the following matters, namely: (i) provision to cover every establishment which employs or had employed on any day of the preceding twelve months, fifty or more workers in any building or other construction work.

Understanding these laws and what they mean

These laws mandate that construction companies pay a minimum of one percent cess on the construction cost. This money is then directed to the welfare of construction workers registered under the BOCW Act. The act ensures the safety and welfare of these workers and their families. Benefits include pension, education and maternity assistance, loans for purchase of tools, and accident and medical expenses.

Despite this, data released by the Ministry of Labour and Employment in early February 2019 showed that only about 35 million of the 50 million construction workers were registered under the BOCW Act. Even here, up-to-date information is unavailable on the number of active registrations. For example, Delhi has 5,39,421 workers registered totally, but only 1,28,394 are active as per the state website. It is not clear how many of these registered workers are local and how many are migrants, because none of the systems we have come across captures this.

Since the inception of the Cess Act, states and union territories have collected an estimated INR 52,000 crores from construction companies. However, until 2019, less than 40 percent of the corpus was spent on welfare of the construction workers. Here, there are sizeable differences among states:

- As many as 21 of the 37 states and union territories have spent less than 30 percent of their collected funds.
- On the other hand, Kerala was the only state to have spent more than the collected funds (121 percent).
- Importantly, some states, which collected the highest amount of cess, have spent the least on the welfare of construction workers. For

instance, Maharashtra spent only five percent of the INR 7,400 crore that it had collected. Similarly, Delhi collected INR 2,190 crore, but spent only nine percent of it.

Increasing BOCWA registrations: What needs to be done

1. Adoption of a single window system

The current system allows for registration and support in one state without any portability option. This is incompatible with the reality of migrant labourers who are often working in multiple states in a given year. For example, Delhi has seen 87 percent fall in registered workers since 2015—largely because people dropped off the system when they migrated elsewhere for work.

Consider Mr. X and his family. They moved from Jhansi to Jodhpur to make a living as construction labourers. They are registered under BOCW Act in Jhansi; however, because of the move, their registration has most likely shifted to an 'inactive' status, leaving them unable to avail any benefits. They share, "We have not got any benefits of that registration yet. We are not aware as to what the benefits are, and how to avail them."

Given this reality, states need to have mechanisms that ensure seamless portability of the informal workforce. A Model Welfare Scheme has been outlined by the Ministry of Labour and Employment following the Supreme Court directive, where each worker will be allotted a unique identification code that can be used across the country. As Jan Sahas recommends in their report, states should expedite the implementation of the new model that will help streamline the registrations and benefits in a single window.

2. Easy registration

Currently, workers are required to fill out a detailed form to register under BOCW Act. For instance, in Delhi the four page form was turned into a 12 page form. Filling out the form with all necessary details is usually a big task for workers as they have low literacy levels and little access to digital systems.

Additionally, the onus of registration is currently on the workers. It would be fruitful to shift this onto contractors or developers. In the short-term, employers will be able to easily register workers. They have the resources to do so, and they already collect details such as PAN, Aadhaar, and bank account details. In the long-term, employers could then formalise these employment relationships through official contracts and registration to authorities.

3. Relaxation in proof of employment certification

Under this act, workers need to submit 90 days certification of proof of employment in the last 12 months. In reality, the industry has a high rate of worker turnover. For example, at our Godrej Properties Limited construction sites, workers typically stay with us for about 45 days. This number is far lower at smaller construction sites. As workers keep moving, they find it difficult to have a 90 day documented proof of employability, especially from contractors that employ them for short projects. This rule should therefore consider relaxing the certification of work to 90 non-continuous work days annually.

4. Partnering to increase registrations

A large number of construction workers work at the *nakas* as daily wage earners. It is important to reach out to this segment and get them registered. To this end, a number of steps can be taken.

Firstly, state BOCW Act boards can partner with nonprofit organisations that work with migrant workers to effectively register daily wage workers at worker facilitation centres. Additionally, information counters can be set up at state transport bus stands, railway stations, and *nakas* for workers to easily access information. Even a central worker helpline number will be useful. For example, the Delhi government is holding a registration drive during the lockdown to get more workers in the system. Maharashtra is also replicating this idea, and it can be quickly taken up by other states. However, support from developers and the construction industry is crucial to boost this drive.

In addition, corporates can widen the social protection net by proactively engaging with suppliers, vendors, and partners to formalise registration and licensing, set standards for employee protection, and socialise linkages to public schemes. Stakeholders can be incentivised or penalised and progress can be monitored.

C.P.W.D. Contractor's Labour Regulations

(Ref. General Conditions of Contract- 2019- Construction Works)

1. Short Title

These regulations may be called the C.P.W.D./PWD (DA) Contractors Labour Regulations.

2. Definitions

1. Workman means any person employed by C.P.W.D./PWD (DA) or its contractor directly or indirectly through a subcontractor with or without the knowledge of the Central Public Works Department/PWD (DA) to do any skilled, semiskilled or unskilled manual, supervisory, technical or clerical work for hire or reward, whether the terms of employment are expressed or implied but does not include any person:
   (a) Who is employed mainly in a managerial or administrative capacity: or
   (b) Who, being employed in a supervisory capacity draws wages exceeding five hundred rupees per mensem or exercises either by the nature of the duties attached to the office or by reason of powers vested in him, functions mainly of managerial nature: or

(c) Who is an out worker, that is to say, person to whom any article or materials are given out by or on behalf of the principal employers to be made up cleaned, washed, altered, ornamental finished, repaired adopted or otherwise processed for sale for the purpose of the trade or business of the principal employers and the process is to be carried out either in the home of the out worker or in some other premises, not being premises under the control and management of the principal employer.

No person below the age of 14 years shall be employed to act as a workman.

2. Fair Wages means wages whether for time or piece work fixed and notified under the provisions of the Minimum Wages Act from time to time.
3. Contractors shall include every person who undertakes to produce a given result other than a mere supply of goods or articles of manufacture through contract labour or who supplies contract labour for any work and includes a subcontractor.
4. Wages shall have the same meaning as defined in the Payment of Wages Act.

3. 1. Normally working hours of an adult employee should not exceed 9 hours a day. The working day shall be so arranged that inclusive of interval for rest, if any, it shall not spread over more than 12 hours on any day.

2. When an adult worker is made to work for more than 9 hours on any day or for more than 48 hours in any week, he shall be paid over time for the extra hours put in by him at double the ordinary rate of wages.
3. (a) Every worker shall be given a weekly holiday normally on a Sunday, in accordance with the provisions of the Minimum Wages (Central) Rules 1960 as amended from time to time irrespective of whether such worker is governed by the Minimum Wages Act or not.

   (b) Where the minimum wages prescribed by the Government under the Minimum Wages Act are not inclusive of the wages for the weekly day of rest, the worker shall be entitled to rest day wages at the rate applicable to the next preceding day, provided he has worked under the same contractor for a continuous period of not less than 6 days.

   (c) Where a contractor is permitted by the Engineer-in-Charge to allow a worker to work on a normal weekly holiday, he shall grant a substituted holiday to him for the whole day on one of the five days immediately before or after the normal weekly holiday and pay wages to such worker for the work performed on the normal weekly holiday at overtime rate.

4. Display of Notice Regarding Wages etc.

The contractor shall before he commences his work on contract, display and correctly maintain and continue to display and correctly maintain in a clear and legible condition in conspicuous places on the work, notices in English and in the local Indian languages spoken by the majority of the workers giving the minimum rates of wages fixed under Minimum Wages Act, the actual wages being paid, the hours of work for which such wage are earned, wages periods, dates of payments of wages and other relevant information as per Appendix 'III'.

5. Payment of Wages

1. The contractor shall fix wage periods in respect of which wages shall be payable.
2. No wage period shall exceed one month.
3. The wages of every person employed as contract labour in an establishment or by a contractor where less than one thousand such persons are employed shall be paid before the expiry of seventh day and in other cases before the expiry of tenth day after the last day of the wage period in respect of which the wages are payable.
4. Where the employment of any worker is terminated by or on behalf of the contractor the wages earned by him shall be paid before the expiry of the second working day from the date on which his employment is terminated.
5. All payment of wages shall be made on a working day at the work premises and during the working time and on a date notified in advance and in case the work is completed before the expiry of the wage period, final payment shall be made within 48 hours of the last working day.
6. Wages due to every worker shall be paid to him direct by contractor through Bank or ECS or online transfer to his bank account.
7. All wages shall be paid through Bank or ECS or online transfer.
8. Wages shall be paid without any deductions of any kind except those specified by the Central Government by general or special order in this behalf or permissible under the Payment of Wages Act 1956.
9. A notice showing the wages period and the place and time of disbursement of wages shall be displayed at the place of work and a copy sent by the contractor to the Engineer-in-Charge under acknowledgment.
10. It shall be the duty of the contractor to ensure the disbursement of wages through bank account of labour.

11. The contractor shall obtain from the Junior Engineer or any other authorised representative of the Engineer-in-Charge as the case may be, a certificate under his signature at the end of the entries in the "Register of Wages" or the "Wage-cum-Muster Roll" as the case may be in the following form:-
12. "Certified that the amount shown in column No has been paid to the workman concerned through bank account of labour on at........."

Fines and Deductions which may be made from Wages

1. The wages of a worker shall be paid to him without any deduction of any kind except the following:
2. Fines
3. Deductions for absence from duty i.e. from the place or the places where by the terms of his employment he is required to work. The amount of deduction shall be in proportion to the period for which he was absent.
4. Deduction for damage to or loss of goods expressly entrusted to the employed person for custody or for loss of money or any other deduction which he is required to account, where such damage or loss is directly attributable to his neglect or default.
5. Deduction for recovery of advances or for adjustment of overpayment of wages, advances granted shall be entered in a register.
6. Any other deduction which the Central Government may from time to time allow.
7. No fines should be imposed on any worker save in respect of such acts and omissions on his part as have been approved of by the Chief Labour Commissioner.
8. Note: An approved list of Acts and Omissions for which fines can be imposed is enclosed at Appendix-X.
9. No fine shall be imposed on a worker and no deduction for damage or loss shall be made from his wages until the worker has been given an opportunity of showing cause against such fines or deductions.
10. The total amount of fine which may be imposed in any one wage period on a worker shall not exceed an amount equal to three paise in a rupee of the total wages, payable to him in respect of that wage period.
11. No fine imposed on any worker shall be recovered from him by instalment, or after the expiry of sixty days from the date on which it was imposed.
12. Every fine shall be deemed to have been imposed on the day of the act or omission in respect of which it was imposed.

Labour Records

1. The contractor shall maintain a Register of persons employed on work on contract in Form XIII of the CL (R&A) Central Rules 1971 (Appendix IV)

2. The contractor shall maintain a Muster Roll register in respect of all workmen employed by him on the work under Contract in Form XVI of the CL (R&A) Rules 1971 (Appendix V).
3. The contractor shall maintain a Wage Register in respect of all workmen employed by him on the work under contract in Form XVII of the CL (R&A) Rules 1971 (Appendix VI).
4. Register of accident: The contractor shall maintain a register of accidents in such form as may be convenient at the work place but the same shall include the following particulars:
5. Full particulars of the labourers who met with accident.
   (a) Rate of Wages.
   (b) Sex
   (c) Age
   (d) Nature of accident and cause of accident.
   (e) Time and date of accident.
   (f) Date and time when admitted in Hospital,
   (g) Date of discharge from the Hospital.
   (h) Period of treatment and result of treatment.
   (i) Percentage of loss of earning capacity and disability as assessed by Medical Officer.
   (j) Claim required to be paid under Workmen's Compensation Act.
   (k) Date of payment of compensation.
   (l) Amount paid with details of the person to whom the same was paid.
   (m) Authority by whom the compensation was assessed.
   (n) Remarks
6. The contractor shall maintain a Register of Fines in the Form XII of the CL (R&A) Rules 1971 (Appendix-XI)
7. The contractor shall display in a good condition and in a conspicuous place of work the approved list of acts and omissions for which fines can be imposed (Appendix-X)
8. The contractor shall maintain a Register of deductions for damage or loss in Form XX of the CL (R&A) Rules 1971 (Appendix-XII)
9. The contractor shall maintain a Register of Advances in Form XXIII of the CL (R&A) Rules 1971 (Appendix-XIII)
10. The contractor shall maintain a Register of Overtime in Form XXIII of the CL (R&A) Rules 1971 (Appendix-XIV)

6. Attendence Card-cum-wage Slip

1. The contractor shall issue an Attendance card-cum-wage slip to each workman employed by him in the specimen form at (Appendix-VII)
2. The card shall be valid for each wage period.
3. The contractor shall mark the attendance of each workman on the card twice each day, once at the commencement of the day and again after the rest interval, before he actually starts work.

4. The card shall remain in possession of the worker during the wage period under reference.
5. The contractor shall complete the wage slip portion on the reverse of the card at least a day prior to the disbursement of wages in respect of the wage period under reference.
6. The contractor shall obtain the signature or thumb impression of the worker on the wage slip at the time of disbursement of wages and retain the card with himself.

7. Employment Card

The contractor shall issue an Employment Card in Form XIV of the CL (R&A) Central Rules 1971 to each worker within three days of the employment of the worker (Appendix-VIII).

8. Service Certificate

On termination of employment for any reason whatsoever the contractor shall issue to the workman whose services have been terminated, a Service certificate in Form XV of the CL (R&A) Central Rules 1971 (Appendix-IX)

9. Preservation of Labour Records

All records required to be maintained under Regulations Nos. 6 & 7 shall be preserved in original for a period of three years from the date of last entries made in them and shall be made available for inspection by the Engineer-in-Charge or Labour Officer or any other officers authorised by the Ministry of Urban Development in this behalf.

10. Power of Labour Officer to make Investigations or Enquiry

The Labour Officer or any person authorised by Central Government on their behalf shall have power to make enquires with a view to ascertaining and enforcing due and proper observance of Fair Wage Clauses and the Provisions of these Regulations.

He shall investigate into any complaint regarding the default made by the contractor or subcontractor in regard to such provision.

11. Report of Labour Officer

The Labour Officer or other persons authorised as aforesaid shall submit a report of result of his investigation or enquiry to the Executive Engineer concerned indicating the extent, if any, to which the default has been committed with a note that necessary deductions from the contractor's bill be made and the wages and other dues be paid to the labourers concerned. In case an appeal is made by the contractor under Clause 13 of these regulations, actual payment to labourers will be made by the Executive Engineer after the Superintending Engineer has given his decision on such appeal.

1. The Executive Engineer shall arrange payments to the labour concerned within 45 days from the receipt of the report form the Labour Officer or the Superintending Engineer as the case may be.

12. Appeal against the Decision of Labour Officer

Any person aggrieved by the decision and recommendations of the Labour Officer or other person so authorised may appeal against such decision to the Superintending Engineer concerned within 30 days from the date of decision, forwarding simultaneously a copy of his appeal to the Executive Engineer concerned but subject to such appeal, the decision of the officer shall be final and binding upon the contractor.

13. Prohibition Regarding Representation through Lawyer

1. A workman shall be entitled to be represented in any investigation or enquiry under these regulations by:
   (a) An officer of a registered trade union of which he is a member.
   (b) An officer of a federation of trade unions to which the trade union referred to in clause (a) is affiliated.
   (c) Where the employer is not a member of any registered trade union, by an officer of a registered trade union, connected with the industry in which the worker is employed or by any other workman employed in the industry in which the worker is employed.
2. An employer shall be entitled to be represented in any investigation or enquiry under these regulations by:
   (a) An officer of an association of employers of which he is a member.
   (b) An officer of a federation of associations of employers to which association referred to in clause (a) is affiliated.
   (c) Where the employers is not a member of any association of employers, by an officer of association of employer connected with the industry in which the employer is engaged or by any other employer, engaged in the industry in which the employer is engaged.
3. No party shall be entitled to be represented by a legal practitioner in any investigation or enquiry under these regulations.

14. Inspection of Books and Slips

The contractor shall allow inspection of all the prescribed labour records to any of his workers or to his agent at a convenient time and place after due notice is received or to the Labour Officer or any other person, authorised by the Central Government on his behalf.

15. Submission of Returns

The contractor shall submit periodical returns as may be specified from time to time.

16. Amendments

The Central Government may from time to time add to or amend the regulations and on any question as to the application/Interpretation or effect of those regulations the decision of the Superintending Engineer concerned shall be final.

Indian Legislation in respect of Labour

India has a federal structure of governance. The constitution has divided the subjects on which laws are passed between Central government and State governments. There are some subjects where both Central and State governments can enact a law; these are said to be on concurrent list.

There are three types of legislations

1. Acts
2. Rules
3. Codes - Indian Penal code (IPC), Civil procedure code (CPC) and
4. Criminal procedure code (CrPC). Etc.

Generally Acts are passed by Legislature and outline what is desired. There are corresponding rules for the act that outline the procedure to achieve intent of Act. The rules specify the parameters to be achieved, structure of machinery to regulate the working and fees/fines under the act.

In addition to the above codes, there are codes that are formulated by professional bodies and mandatory in working.

Salient features of the acts/rules and codes applicable for construction work are listed for ready reference. Even though the title of act/rule denotes a year of enactment, the latest amendments need to be taken into account. It should be noted that normally Central rules set the basic parameters, whereas individual states can fix the parameters more stringent than Central rules depending on the conditions in the state. It is important to have a library of applicable acts, rules and codes at site for ready reference.

Atomic Energy Act 1962 and rules

An Act to provide for the development, control and use of atomic energy for the welfare of the people of India and for other peaceful purposes and for matters connected therewith. Radiography source emits radiations and needs to be handled by authorised person only. Storing and transportation of radiographic source should be as specified by this act.

The Air (Prevention and Control of Pollution) Act and Rules

An Act to provide for the prevention, control and abatement of air pollution, for the establishment, with a view to carrying out the aforesaid purposes, of Boards, for conferring on and assigning to such Boards powers and functions relating thereto and for matters connected therewith.

Arbitration and Conciliation Act 1996

An Act to consolidate and amend the law relating to domestic arbitration, international commercial arbitration and enforcement of foreign arbitral awards as also to define the law relating to conciliation and for matters connected therewith or incidental thereto.

Battery Management and Handling Rules, 2001

These rules shall apply to every manufacturer, importer, re-conditioner, assembler, dealer, recycler, auctioneer, consumer and bulk consumer involved in manufacture, processing, sale, purchase and use of batteries or components thereof.

Responsibilities of consumer or bulk consumer

1. It shall be the responsibility of the consumer to ensure that used batteries are not disposed of in any manner other than depositing with the dealer, manufacturer, importer, assembler, registered recycler, and reconditioner or at the designated collection centres.
2. It shall be the responsibility of the bulk consumer to:
   (a) Ensure that used batteries are not disposed of in any manner other than by depositing with the dealer/manufacturer/registered recycler/importer/reconditioner or at the designated collection centres,- and
   (b) File half-yearly return in Form VIII to the State Board.
3. Bulk consumers to their user units may auction used batteries to registered recyclers only.

The Bonded Labour system (Abolition) Act, 1976

An Act to provide for the abolition of bonded labour system with a view to preventing the economic and physical exploitation of the weaker sections of the people and for matters connected therewith or incidental thereto.

Building and other Construction worker's Act 1996

An Act to regulate the employment and condition of service of buildings and other construction workers and to provide for their safety, health and welfare measures and for other matters connected therewith or incidental thereto.

Central Sales Tax Act 1956

Taxes on sales or purchases of goods in the course of inter-state trade or commerce were brought expressly within the purview of the legislative jurisdiction of Parliament.

Restrictions could be imposed on the powers of State legislatures with respect to the levy of taxes on the sale or purchase of goods within the State where the goods are of special importance in inter-state trade or commerce.

The Child Labour (Prohibition and Regulation) Act, 1986

An Act to prohibit the engagement of children in certain employments and to regulate the conditions of work of children in certain other employments.

Companies Act 1956 and 2013

The Act in a comprehensive form purports to deal with relevant themes such as investor protection, inclusive agenda, fraud mitigation, internal control, director responsibility and efficient restructuring. The Act is also quite outward looking and in several areas attempts to harmonize with international requirements.

Contract Labor (Regulation and Abolition) Act 1970 and Rules

An Act to regulate the employment of contract labour in certain establishments and to provide for its abolition in certain circumstances and for matters connected therewith.

Employees Provident Fund Act and Rules

An Act to provide for the institution of provident funds, pension fund and deposit-linked insurance fund for employees in factories and other establishments.

Employees State Insurance Act 1948

To provide for certain benefits to employees in case of sickness, maternity and employment injury and to make provision for certain other matters in relation thereto.

Environmental (Protection) Rules- "Environmental Standards"

The Central Pollution Control Board (CPCB) has developed National Standards for Effluents and Emission under the statutory powers of the Water (Prevention and Control of Pollution) Act, 1974 and the Air (Prevention and Control of Pollution) Act, 1981. These standards have been approved and notified by the Government of India, Ministry of Environment & Forests, under Section 25 of the Environmental (Protection) Act, 1986.

Equal Remuneration Act, 1976

Payment of remuneration at equal rates to men and women workers and other matters.

The Explosive Act and Rules

To regulate the manufacture, possession, use, sale, transport and importation of explosives.

The Explosive Substance Act

The expression "explosive substance" shall be deemed to include any materials for making any explosive substance; also any apparatus, machine, implement or material used, or intended to be used, or adapted for causing, or aiding in causing, any explosion in or with any explosive substance; also any part of any such apparatus, machine or implement.

Factories Act 1948

An Act to consolidate and amend the law regulating labour in factories.

Gas Cylinder Rules

Filling, possession, import and transport of cylinders: (1) No person shall fill any cylinder with any compressed gas or import, possess or transport any cylinder so filled or intended to be filled with such gas unless:

(a) Such cylinder and its valve have been constructed to a type and standard specified in Schedule I as amended from time to time by an order issued by the Chief Controller;

(b) The test and inspection certificates issued by the inspecting authority in respect of cylinder and its valve are made available to the Chief Controller and prior approval of the said authority is obtained.

The Hazardous Wastes (Management and Handling) Rules

Unscientific disposal of hazardous and other waste through burning or incineration leads to emission of toxic fumes comprising of Dioxins & Furans, Mercury, heavy metals, causing air pollution and associated health-related problems. Disposal in water bodies, or in municipal dumps leads to toxic releases due to leaching in land and water entailing into degradation of soil and water quality. The workers employed in such unscientific practices suffer from neurological disorders, skin diseases, genetic defects, cancer etc. Hence, there is a need for systematic management of hazardous and other waste in an environmentally sound manner by way of prevention, minimisation, re-use, recycling, recovery, utilisation including co-processing and safe disposal of waste.

Indian Contract Act 1872

It determines the circumstances in which promises made by the parties to a contract shall be legally binding on them. All of us enter into a number of contracts everyday knowingly or unknowingly. Each contract creates some rights and duties on the contracting parties. Hence this legislation.

Income Tax Act

An Act to consolidate and amend the law relating to income-tax and super tax.

Central Excise Act 1944

An Act to consolidate and amend the law relating central duties of excise on goods manufactured or produced in India.

Customs Act 1962

An Act to consolidate and amend the law relating to Custom duty.

Indian Boiler Regulation Act and Rules

Indian Boiler Regulations are the standards m respect of materials, design and construction, inspection and resting of boilers and boiler components for compliance by the manufacturers and users of boilers in the country. These regulations are being updated regularly by amending them in line with fast changes in boiler technology by the Central Boilers Board.

Indian Electricity Act 2013 and Rules

The act covers major issues involving generation, distribution, transmission and trading in power.

Indian Stamp Act 1899

A fiscal statute laying down the law relating to tax levied in the form of stamps on instruments recording transactions & Stamp duties on instruments specified in Entry 91 of the Union List (viz. Bills of Exchange, cheques, promissory notes, bills of lading, letters of credit, policies of insurance, transfer of shares, debentures, proxies and receipts) is levied by the Union. Stamp duties on instruments other than those mentioned in Entry 91 of the Union List above are levied by the States as per Entry 63 of the State List.

The Industrial Disputes Act 1948

An Act to make provision for the investigation and settlement of industrial disputes, and for certain other purposes.

Industrial Employment (Standing Orders) Act

An Act requiring employers in industrial establishments formally to define conditions of employment under them. It is expedient to require employers in industrial establishments to define with sufficient precision the conditions of employment under them and to make the said conditions known to workmen employed by them.

Inter-state Migrant Workmen (Regulation of Employment and Conditions of Service) Act 1979

This Act protects the dignity of life and the interests of the workers whose services are requisitioned outside their native states in India.

Limitation Act 1963

An Act to consolidate and amend the law for the limitation of suits and other proceedings and for purposes connected therewith.

Minimum Wages Act

An Act to provide for fixing minimum rates of wages in certain employments. These rates differ in states and depend on the location of site. The rates are published in gazette.

Mines and Minerals (Development and Regulation) Act, 1957

The major objective of the amendment is to generate employment in the mining sector and enhance the contribution of the mining sector in the total GDP of the country.

Tribal Communities: Several tribal communities and Particularly Vulnerable Tribal Groups (PVTGs) fall into the mining zones. Their residence is also threatened by an increase in mining. Their rehabilitation & compensation is another major issue.

India should focus on generating employment too. The sector has huge potential for job creation at local level which can counter the migration and thus prevent cities from getting overburdened.

Motor Vehicle Act & Rules

Regulates all aspects of road transport vehicles. The Act provides in detail the legislative provisions regarding licensing of drivers/ conductors, registration of motor vehicles, control of motor vehicles through permits, special provisions relating to state transport undertakings, traffic regulation, insurance, liability, offences and penalties, etc.

National Building Code 2016

Issued by BIS is a comprehensive building Code, a national instrument providing guidelines for regulating the building construction activities across the country. The comprehensive NBC 2016 contains 11 Parts some of which are further divided into Sections totaling 26 chapters. The salient features of the NBC include the changes especially in regard to further enhancing our response to meet the challenges posed by natural calamities and reflecting the state-of-the-art and contemporary applicable international practices.

The Noise Pollution (Regulation & Control) Rules, 2000

Sound emitting construction equipment shall not be used or operated during night time in residential areas and silence zones.

Payment of Bonus Act

An Act to provide for the payment of bonus to persons employed in certain establishments on the basis of profits or on the basis of production or productivity and for matters connected therewith.

The Payment of Gratuity Act 1972

The Act provides for the payment of gratuity to workers employed in every factory, shop & establishment or educational institution employing 1 O or more persons on any day of the preceding 12 months. All the employees irrespective of status or salary are entitled to the payment of gratuity on completion of 5 years of service. In case of death or disablement there is no minimum eligibility period. The amount of gratuity payable shall be at the rate of 17 days wages based on the rate of wages last drawn, for every completed year of service.

Formula is - Last Wages *15*No. of years of services/26

Payment of Wages Act 1964

The Payment of Wages Act was enacted with a view to ensuring that wages payable to employed persons covered by the Act were disbursed by the employers within the prescribed time limit and that no deductions other than those authorised by law were made by them.

The Petroleum Act and Rules

An Act to consolidate and amend the law relating to the import, transport, storage, production, refining and blending of petroleum.

The Private Security Agencies (Regulation) Act, 2005

An Act to provide for the regulation of private security agencies and for matters connected therewith or incidental thereto.

The Professional Tax and Rules

Professional Tax is the tax charged by the state governments in India. Any one earning an income from salary or any one practicing a profession such as chartered accountant, lawyer, doctor etc. are required to pay this professional tax. The maximum amount payable per year is Rs. 2,500/- (and in line with salary.)

The Public Liability Insurance Act and Rules 1991

An Act to provide for public liability- insurance for the purpose of providing immediate relief to the persons affected by accident occurring while handling any hazardous substance and for matters connected therewith or incidental thereto.

Radiation Protection Rules 2004

These rules shall apply to practices adopted and interventions applied with respect to radiation sources.

Service Tax Act 1994

Principles for determining the modalities of levying the Service Tax by the Central Govt. and collection of the proceeds thereof by the Central Govt. and the State.

Shops and Establishment Act

The object of this Act is to provide statutory obligation and rights to employees and employers in the unauthorized sector of employment, i.e., shops and establishments. This Act is applicable to all persons employed in an establishment with or without wages, except the members of the employer' family.

The Standards of Weights and Measures Act and Rules 1976

Establishment of the weights and measure based on the SI units. Provides to prescribe specification of measuring instruments used in commercial transaction, industrial production and measurement involved in public Health and Human safety. The specifications are given in the Standard of weights and Measures (General) Rules 1987.

The Water (Prevention and Control of Pollution) Act and Rules

An Act to provide for the prevention and control of water pollution and the maintaining or restoring of wholesomeness of water, for the establishment, with a view to carrying out the purposes aforesaid, of Boards for the prevention and control of water pollution, for conferring on and assigning to such Boards Powers and functions relating thereto and for matters connected therewith.

The Water (Prevention and Control of Pollution) Cess Act

An Act to provide for the levy and collection of a cess on water consumed by persons carrying on certain industries and by local authorities, with a view to augment the resources of the Central Board and the State Boards for the prevention and control of water pollution constituted under the Water (Prevention and Control of Pollution) Act, 1974.

Workmen Compensation Act - The Workmen's Compensation Act 1923

Was enacted to help workmen face the hardships resulting from accidents. These legal provisions apply equally to women workers also. An employer is liable to provide monetary compensation to a disabled workman, or to his dependents, in case of his death, if the disablement or death occurs "out of and in the course of employment."

Works Contract Tax Act 1994

Service tax on works contract was imposed on 1-6-2007. However, it was restricted works contract relating to immovable property. Its scope has been extended to works contract of movable property also w.e.f. 1-7-2012.

Elements of Arbitration

One cannot imagine a society in which there are no disputes and differences right from primitive days to space age. The process of arbitration as a modality came to the rescue of mankind to resolve their disputes and differences. It started in a rudimentary form in every clan and tribe having their own rules of settlement till it developed as a system, Even in present time different communities in our country do have some forum to settle the disputes between their members, the forum named as conciliation and arbitration boards. Such boards are modified forms of our old "Panchas" (five wise persons settling the disputes unanimously or by a majority).

Though arbitration is a law oriented field, yet it has kept its doors open for the architects and engineers for playing their constructive roles in this field because of expertise possessed by them on construction and valuation of immovable properties. So important is the arbitration in construction industry, that every contract will carry an arbitration clause. The I.I.A. form of contract enrolls two contracts in one, namely construction contract as well as arbitration agreement as contained in condition Nos.55 d 56 of the format.

In light of the above the architects and engineers are well advised to study.

1. The arbitration and Conciliation Act, 1996.
2. The Contract Act, 1872.

So that they can discharge their duties as arbitrators within the four comers of law.

Methods to settle disputes and differences:

Besides other methods available to settle the disputes and differences between parties, the popular modes have been: Arbitration

ADRM (Alternative Dispute Resolution Methods) which will cover the modes of Conciliation and Mediation. The mode of conciliation has been given a privileged position along with arbitration in the title of the Act itself like the Arbitration and Conciliation Act, 1996 with section 61 to 81 dealing with the subject of conciliation whereas "mediation" as a mode of settlement finds place in Sec.30 of the said Act of 1996.

Thus it is essential to know core contents of each of the method.

Arbitration

It is a method of settling the disputes and differences between two or more parties whereby they appoint one or more persons to adjudicate upon the said disputes and differences that have arisen or that may hereafter arise and agree to abide by the decision of the said one or more persons nominated for the purpose of adjudication.

OR

An arbitration is the reference of a dispute or difference between not less than two parties for determination after hearing the sides in judicial manner, by a person or persons other than a court of competent, jurisdiction (Halsbury).

Thus, the arbitration aims at the establishment of a domestic tribunal by the parties to adjudicate upon the disputes and differences and the Arbitration and Conciliation Act, 1996 helps the parties to enforce that decision besides laying down rules, regulations and procedures.

Mediation

Mediation is a process involving chosen intervention by a neutral person appointed by both the parties to settle their disputes and difference. This person is called Mediator, who enjoys the confidence of both the parties and possesses expertise in the subject.

Mediation is a term of the International Law which signifies the interposition by a Neutral and friendly state between two states at war or on the eve of war with each other, of its good offices to restore or to preserve peace. This term is sometimes used as a synonym for intervention, but mediation differs from it being purely a friendly act.

It is here that process of "Shuttle Diplomacy" come up and same gets transplanted in construction field also.

Both the parties lay bare facts before the mediator in the absence of another.

After hearing the parties, analyzing the facts, the mediator culls out the relative position of each party to the disputes in question. Like a roving ambassador he meets each party to the dispute holding a mirror before it to show its relative position, law applicable as well as its impacts. Besides the above he uses his power of persuasion as well as to some extent spirit of" Give and Take".

Besides on the opinion of the mediator the parties have two options available namely:

(a) Enter into further negotiation
(b) Settle the matter.

The mediator is not an arbitrator and signed settlement is not an award but a mere contract.

Complete process of mediation is without prejudice to the legal rights of the parties to the disputes unless otherwise agreed upon. Even the courts appoint mediators for some disputed matter with a condition that their opinions will be final and binding on the both the parties.

Selection of two mediators one by each side or a panel of mediators comprising of three mediators are the variants of the process of mediation. The aim of such variants is representation of the technical matters in rights perspective.

The drawbacks of mediation are:

1. Hearing each party in the absence of another.
2. Disclosure or information of one to another so that the parties can weigh their case and come to a settlement. This will rebound if the settlement fails and each party is in knowledge of the grounds of the other well in advance and can meet the case in court.

It is submitted that shuttle diplomacy ought not to find a place in ADRM and as the same forms a breeding ground of doubts as to the integrity of mediator due to the fact that he hears one party in the absence of another, he discloses information of one to another which contravene rules of natural justice.

In order to overcome the drawbacks of shuttle diplomacy, mediation in strict sense is being used by our courts as an alias of arbitration. The courts may appoint particular person as mediator to decide specific issues, parties to agree to the decision of the mediator and to treat it as final and binding without appeal. The parties given rights to lead all evidence and mediator given right to call upon the parties to produce needed evidence. This mediation is also neither cheap nor quick but it an alias of arbitration, and makes mediation an arbitration oriented one.

In conclusion, mediation is arbitration in fact but not in law. Again Mediation as process has to be absorbed by arbitration and in whatever form it may grow, the process of mediation which does away entirely with the control of courts is not possible and practicable for simple reason as to the removal of mediator being guilty of misconduct for which retention of the powers of court are absolutely essential.

Conciliation

Conciliation is a process alternative to arbitration for setting the disputes.

Though somewhat resembling to arbitration, yet it is not arbitration. Conciliator usually possessing relevant expertise can be a sole, appointed by both the parties or they can be joint, one appointed by each party. At times even a conciliation committee comprising of three persons finds a place in this process.

If there are two or more conciliators, they meet together after study with an open mind to ascertain the fields of agreement and/or disagreement. The result will be agreement on a number of issues whereby the field of disputes gets curtailed.

However, the meetings will prove failure if the conciliators:

(a) Are emotionally involved.
(b) Suffer from superiority complex.
(c) Are under the dictates of the parties.
(d) The meetings between the conciliators are without prejudice and it is not advisable for conciliators to rush for every step taken to the parties to ascertain their claim.

(e) reactions to the same.
(f) Section 61 to 81 of the Arbitration and Conciliation Act, 1996 lay down norm.
(g) for conciliation.

Difference between Conciliation and Mediation

The difference between conciliation and mediation can well be appreciated by taking an illustration of fair market value of land for the purpose of recovery of the unearned increment. According to one it has to be Rs. 2000/- per S.M. whereas according to other party it has to be Rs. 300/- per S.M. An intermediary step in both the modalities of conciliation and mediation, to decide the issue.

In case of conciliation the parties themselves desire right value and they have not locked up their stables. They desire to have recommendation or settlement proposal from the intermediary so that they become "reconciled".

In case of mediation the parties have closed their stables and are unwilling to negotiate face to face and it is the mediator who prepares his own findings and legal conclusions so that parties can come to a settlement. It comes halfway between conciliation and arbitration and can work provided parties agree to give up some of its rights and forget some of its grievances. The settlement will form a contract and not an award. If some difference arise in the said settlement of contract, remedy will be a regular suit on the contract as the contract entered into in the form of settlement is no enforceable as an award.

Kinds of Arbitration:

1. Arbitration based on arbitration clause in the contract.
2. Arbitration with the intervention of a court where there is no suit pending.
3. Arbitration in suits.

Arbitration: based on arbitration clause in the contract

It arises in case of arbitration clause of the contract whereby present or future disputes are referred to the adjudication of the arbitral tribunal. No suit is filed in this case and the arbitral award takes the form of a decree of the court.

The parties can have arbitration in this manner under Cl. 56 of Indian Arbitration Act I.I.A., form of contract or in any other form of contract which carries with it the arbitration clause. This arbitration clause will constitute as an arbitration agreement.

Arbitration with the intervention of a court where there is no suit pending

Where the disputes and differences have arisen between the parties to which the arbitration agreement applies, the parties have a dual choice:

1. Appoint arbitral tribunal and get an award, if not,

2. One of the parties may apply to the appropriate judicial authority with the original arbitration agreement or duly certified copy thereof, the judicial authority will refer the parties to arbitration.

Arbitration in Suits

In case of a suit pending in a court, all the interested parties at any time before the judgment is pronounced, agree that the disputes and differences be settled by arbitration and the judicial authority will refer the parties to arbitration.

In all the above cases the arbitration will be subject to the Arbitration and Conciliation Act, 1966 which is deemed to have come into force on 23rd October, 2015. The act deals with a number of categories of arbitration like:

1. Domestic arbitration.
2. International arbitration.
3. Foreign arbitration.
4. Institutional arbitration.
5. Statutory arbitration.

Out of the above categories of the arbitrations, the students are supposed to have an elementary knowledge of Domestic arbitration, which takes place in India and governed by Indian Law.

Main advantages of settling the disputes and differences by Arbitration:

1. Persons usually experts in the field are appointed and as such decisions will be fair and refined and not involving any element of guess work. In case of a dispute concerning building construction job, the person chosen as arbitrators are usually architects or engineers who possess expert knowledge of the subject and are fully qualified to act as arbitrators.
2. Legal formalities are cut short and decision can be reached in much shorter time as compared to the time consumed in a court case.
3. Savings in cost.
4. Time, date and place of meeting can be arranged so as to suit the convenience of all concerned.
5. Privacy as hearing is not public.
6. The arbitrator can view the subject at any convenient time.
7. Finality of the award.

The Arbitration and Conciliation Act 1996 deals with a number of terms like:

1. Arbitrator.
2. Arbitral Tribunal.
3. Arbitration Agreement.
4. Arbitral award.

Arbitrator

He is the person to whom the dispute and differences are referred for necessary adjudication. If there are more than one arbitrators, they are known as "Joint Arbitrators" and if there is only one arbitrator appointed for the purpose, he is known as "Sole Arbitrator". Unless a specific provision is made for two or more arbitrators, the reference will be always be to a sole arbitrator.

Appointment of Arbitrator and Arbitral Tribunal

1. Toe parties are free to determine the number of arbitrators provided such number is not an even number. The purpose of providing odd number of arbitrator is to finalise the outcome by a majority decision in case unanimous decision is not forthcoming.
2. In case of sole arbitrator the appointment will be by agreement of the parties In case of disagreement, the appointment will be made by the Chief Justice or his designate. This Sole Arbitrator alone by himself will constitute the Arbitral Tribunal.

Arbitral Tribunal

3. In an arbitration with three arbitrators, each party to appoint one arbitrator and the two appointed arbitrators shall appoint the third arbitrator. This third arbitrator is to act as the Presiding Arbitrator.

   All the three arbitrators will constitute the Arbitral Tribunal headed by the presiding arbitrator. If the two appointed arbitrators fail to agree on the third arbitrator, the appointment shall be made upon the request of a party by the Chief Justice or his designate. Thus Arbitral Tribunal means a Sole Arbitrator or a panel of arbitrators.

In case a controversy arises on the above points, the matter will be decided by the court.

Arbitration Agreement (Submission)

"Submission to arbitration" means a reference by the parties of a dispute between them to arbitration. This term is not defined in the Arbitration and Conciliation Act, 1996. However, Sec. 7 (1) uses the term like "parties to submit to Arbitration". Thus submission will mean referring the disputes for adjudication of the arbitral tribunal either by arbitration clause in the contract (like CL 56 of Indian Institute of Architects (I.I.A). form of contract) or by a separate agreement similar to reference.

Arbitration agreement is written agreement by the parties to submit to arbitration all or certain disputes which have arisen (present disputes) or which may arise between them (future disputes) whether the name of arbitrator is mentioned or not in the agreement. The main ingredients of the arbitration agreement are:

1. It must relate to some disputes or differences between the parties either present or future.
2. Parties must desire to settle the disputes by arbitration.
3. Parties must agree to abide by the decision of the arbitral tribunal.
4. It can be part of the main contract like Cl. 56 of I.I.A. form of contract or it can be in the form of a separate agreement.

Terms of an Arbitration Agreement:

1. Reference to be to an arbitral tribunal comprising of sole arbitrator.
2. If reference is to more than one arbitrator, the number of arbitrators has to be an odd one with presiding arbitrator and such a panel of arbitrators will constitute the arbitral tribunal.
3. Arbitral award to be final and binding.
4. Incidence of cost to be settled by the arbitral tribunal.
5. Arbitral tribunal may make an interim award.

Reference

Reference in case of arbitration is an agreement drawn up between the two contesting parties referring the matter for the adjudication of the arbitral tribunal after the actual disputes or differences have arisen. It contains the names of the arbitrator(s) and specific disputes over which the arbitration is required.

It is known as reference or reference to arbitration and at times the word "reference" is applied to actual hearing before the arbitral tribunal. Thus reference to arbitration would mean actual hearing of the case by the arbitral tribunal brought: about by the parties to dispute who would like a decision on the disputed matters of an agreement.

Difference between Arbitration clause and order of Reference:

Arbitration clause	Reference
1. Nucleus of arbitration agreement	Nucleus of arbitration agreement
2. Originates by agreement of the parties before disputes have taken birth.	Originates after the disputes have taken birth.
3. Relates to present or future disputes. It is likely that no disputes can take place in future and as such there will not be any arbitration.	Relates to present disputes alone and does not contemplate future disputes.
4. It does not contain name of arbitrator.	Mention the name of arbitrator.
5. Not possible to define specific future disputes.	Disputes are mentioned in Sufficient detail over which adjudication is required

Arbitral Award

It is the decision given by the arbitral tribunal after careful investigation of the case submitted to it for adjudication. It conveys the findings of the tribunal and is binding on both the parties.

The essential ingredients of the arbitral award are:

1. The award must be made without undue delay as otherwise the arbitrator's mandate can be terminated.
2. The award as a rule must be in writing and to be signed by the members of the arbitral tribunal and signatures attested.
3. The award must be self-explanatory without any ambiguity and must show nature of adjudication whether unanimous or by majority. The

decision by majority will form a binding force in case unanimous adjudication is not possible.

4. The award to be made (say in 2022) on Stamp paper of Rs. 100/- value.
5. After the arbitral award is made, a signed copy *to* be supplied to each of the party.
6. The arbitral award is to state its date and place of arbitration, claim adjudication in term of money contents, interest on amount awarded or part thereof at such rate as it deems reasonable for the period or part of period from the date on which cause of action arose to the date of making of the award.
7. The award must provide interest on the awarded amount from the date of award to the date of payment.
8. The arbitral award must provide for legal cost, cost of arbitration and other incidental expenses and the manner in which cost to be shared and paid.
9. The arbitral award has to be a reasoned one unless both the parties agree for a non-reasoned award.
10. The arbitral award has now been given the status of a decree in as much as if the award is not set aside by the court, it will be executable as a decree and now not necessary to make an application to the court for confirmation of the award.

Meaning of Reasoned Award

It is the decision of the arbitral showing how it has arrived at such a conclusion with discussion, consideration of the facts, evidence and materials before it depicting the trend of its thoughts. The tribunal is not required to write a detailed judgment setting out each logical step of its reasoning. It is sufficient if it indicates the trend of its thought process so that errors can be eliminated and arbitrariness avoided.

If it has complied with the above, the reasonableness of the reasons given by the Tribunal cannot be challenged.

Conciliation

It is a procedure wherein an independent person with expertise in subject and having no interest in parties or disputes endeavours to bring them together and assist them to reconcile their differences. He has to dig out salient facts on the basis of confidential oral discussion. After issues and differences have been concretised, he brings them together in an open discussion and shows the parties the impacts of settled law on the issue.

On disagreement, right course will be arbitration wherein the conciliator is not to take part.

Conciliators

Conciliators are the persons appointed by the parties to bring about an amicable settlement between them so that emotional harmony does not suffer and each party gets justice, their disputes and differences having been settled amicably.

There will be only one conciliator known as Sole Conciliator unless the parties agree to two or three, known as Joint Conciliators. If there are two conciliators, each party to appoint one. In case of three conciliators, each party to appoint one, and third conciliator to be appointed by the consent of both the parties, who will act as the presiding conciliator. In the appointment of the conciliators the parties can take the help of a suitable institution or person.

The conciliation proceedings will start when one party sends another a written invitation to conciliate with identification of the subject and other party accepts the same. If no reply is received within 30 days of the date of invitation or other extended date, the invitation stands rejected.

With the good offices of the conciliator if a settlement is reached between the disputing parties, the signed settlement agreement by both the parties authenticated by the conciliator will have the same status as an arbitral award on agreed terms on the substance of the dispute as if rendered by an arbitral tribunal.

The parties not to initiate during the conciliation proceedings any arbitral or judicial proceedings in respect of a dispute that is the subject matter of the conciliation proceedings except that a party may initiate arbitral or judicial proceedings wherein its opinion, such proceedings are necessary for preserving its rights.

The Act gives necessary protections to the views, admissions, suggestions, proposals, acceptance of a proposal of conciliation proceedings from being introduced as evidence or as materials for reliance in arbitral or judicial proceedings.

Duties of Conciliator

1. Request the parties to submit written statement setting out the general nature of the dispute and the points at issue including supporting evidence.
2. To assist the parties in an independent and impartial manner to reach amicable settlement.
3. He has to be objective, fair and just in his assignment.
4. He has to give consideration to:
   (a) rights and obligations of the parties.
   (b) trade usages.
   (c) circumstances of the case including previous business practice.
5. Conduct conciliation proceedings in a fair manner and can suggest proposals for settlement of disputes at any stage, may be oral and no reasons to be given for such suggestion.
6. Can take administrative assistance from suitable institution or person.

Factual information concerning dispute received from a party may be disclosed to the other party in the form of "substance". However if the party gives the information subject to a specific condition that it be kept confidential, the conciliator not to disclose that information to the other party.

He may meet or communicate with the parties together or with each one of them separately. However joint meeting is preferable.

Listen to the suggestions made by the parties for settlement as parties may also agree to drop some disputes.

Formulate the possible terms of settlement, carry out relevant amendments as are equitable and request the parties to draw up final settlement on the lines suggested or help the parties in drawing up final settlement agreement.

The final settlement agreement duly signed by both the parties to be authenticated by the Conciliator and terminate conciliation proceedings. Direct each party to bear and to deposit an equal amount towards the cost and fees of conciliator.

Not to act as arbitrator in the same conciliation proceeding on failure of such proceeding, as also cannot represent any of the parties or cannot be represented as a witness in the arbitral or judicial proceedings relating thereto unless both the parties agree.

Summarizing the above it will be observed that intention behind conciliation being no judicial enquiry and no evidence and hence great responsibility is cast on the conciliator to study the case, analyse it rationally from all angles viz. technical angle and legal angle. His expert opinion will have an impact on the minds of the parties and his opinion cannot be brushed aside easily unless one of the parties behave irrationally.

Arbitration and Building Contract

Condition Nos. 55 and 56 of the Indian Institute of Architects form of contract enroll two arbitrations for different matters like:

1. Sole arbitration of the architect on "Excepted matters" and the parties themselves agree to his decision in the absence of fraud or mistake.
2. Sole or joint Arbitration relating to all disputes and differences of any kind whatsoever arising out of or in connection with the contract or the carrying out of the work except for "Excepted matters".

Architect to be the Sole Arbitrator for "Excepted Matters"

The subject of arbitration has been dealt with in condition Nos. 55 and 56 of I.I.A. form of contract. Condition No. 55 has got the marginal note as "Excepted matters" and condition No. 56 as" Arbitrator". The eye catching marginal notes appear to be incomplete in as much as condition No. 55 besides providing for "Excepted matters" also provides for "Right of arbitration". Similarly, conditions No. 56 though it carries the marginal note of "Arbitrator" also provides for procedure for arbitration proceedings. Thus, for the subject of "Settlement of disputes" both the conditions are required to be read together. The suggested marginal note for condition No. 55 will be "Right of arbitration and Excepted matters" and that for condition No. 56 will be "Settlement of disputes - arbitration".

Condition No. 55 is a substantive condition for arbitration and condition No. 56 is a procedural one. Condition No. 56 provides for settlement all disputes and differences by arbitration whereas their natures and rights are dealt within condition No. 55. The nature of disputes and differences pertains to:

1. Decision.
2. Opinion.
3. Direction.
4. Certificate or valuation.
5. Refusal to issue certificates.

In all other matters except for "Excepted matters", coupled with right of arbitration and review.

Condition No. 56 besides providing for procedural aspect for arbitration specifically excludes certain matters from the field of arbitration and leave the same to be decided by the architect named in the contract who has to perform his duties like a sole arbitrator with justice to all. It is the condition No. 55 which collates all "Excepted matters" of eight compartments at one place.

Development of "Excepted Matter" in R.I.B.A. Form of Contract: A reference to R.I.B.A. form of contract 1909 and 1931 Edition shows that the following matters were left to the sole discretion of the architect and his word was final like:

1. Discrepancies in the drawings or bills of quantities and/or drawings and/or specifications.
2. Removal of improper materials and their substitution.
3. Removal and/or re-execution of any works.
4. Dismissal of any persons employed on works.
5. Opening up of any work already covered in for inspection.
6. Assignment or sub-letting.

However, R.l.B.A. form of contract, 1959 Edition has amended Arbitration clause No. 26 whereby even the matter or thing left by the contract to the discretion of the architect has been made arbitrible.

However, R.I.B.A. form of contract has maintained the "Excepted matter", beyond the pale of arbitration.

The question of "Excepted matters" creates its own problem from clause No. 56 which in reality deals with the parallel arbitrations, one by the designated authority of the architect and the other by the arbitrators appointed by the parties. This shows that the parties have agreed that not all the disputes are arbitrible and the decision of the architect is made final and binding on parties on "Excepted matters" given below:

Clause 5: Contract drawings and specifications and architect's interpretation of these documents.

Clause 9: Architect's instructions as regards variation.

Clause 19: Contractor's field organization and equipments including engineer-in-charge, necessary office accommodation, sanitary convenience and all tools and plants.

Clause 25: Assignment or sub-letting of the contract prohibited without the permission of the architect.

Clause 26: Intimation to the architect about the names of the sub-contractors and the contractor not to employ one for whom the architect or the owner objects.

Clause 36: Materials and workmanship to be of relevant code of B.I.S. specification and of approved type. Permission for using equivalent quality of materials, inspection and testing of materials and removal of defective ones.

Clause 40: Extension of time limit due to - (a) force majeure; (b) exceptionally inclement weather; (c) civil commotion strike, lock outs; (d) delay on the part of nominated sub-contractors or nominated suppliers which the contractor has taken all practical steps to avoid or reduce; (e) delay on the part of artists, tradesmen or others engaged by the owner in executing work not forming part of the contract.

Clause 48: Determination of the contract by the owner, due to default on the part of the contractor, bankruptcy of contractor or any malpractices by the contractor for obtaining the contract or in execution of the work.

Though condition No. 55 provides that the decision, opinion, direction, certificate (except for payment) with respect to all or any of the excepted matters as given hereinabove shall be final, conclusive and binding on the parties hereto and shall be without appeal but the same are subject to the following conditions:

(a) No fraud or collusion.
(b) No gross error of judgment.
(c) No pressure from the employer on the architect leading to deviation from his honest judgment.

A reference to "Excepted matters" shows that majority of the items pertains to the question of facts and for such question of facts, the architect is the best judge provided he gives his honest decision.

What is the Legal Position of "Excepted Matters"?

The arbitrators in practice may come across a situation where the claimants may plead "Excepted matters" as arbitrible. The claimants have to prove that the subject-matter is covered by the relevant conditions, they have obtained the decision of the architect on such matters, and that the said decision is a biased one, given in bad faith, contains error of judgment and devoid of honest judgment and given under pressure from the owner, if such conditions are satisfied, the arbitrators do get right to enter in the field of "Excepted matters" subject to terms and conditions of arbitration agreement.

To substantiate the above, case law on somewhat parallel line is given below:

To understand the case law the relevant exception clause from P.W.D. contract which was before the Madhya Pradesh High Court is given below:

1. The Engineer-in-charge shall have powers to make any alteration in, omission from, addition to or substitution for, the original specifications, drawings, designs and instructions and rates for such variations to be fixed by the Engineer-in-charge. In the event of a dispute the decision of the Superintending Engineer of the Circle shall be final.
2. In case of delay the contractor shall be liable to pay as compensation an amount equal to one percent or such small amount as the Superintending Engineer may decide (whose decision in writing shall be final)

Exception Clause

Except where otherwise provided in the contract all questions and disputes in any way arising out of, or relating to the contract shall be referred to arbitration.

The rule is well settled that where parties to building or construction designate a person who is authorized to determine questions relating to its execution, and stipulate that his determination shall be final and conclusive, such parties are bound by his determination of those matters which he is authorized to determine, except in case of fraud or such gross mistake on his part as would necessarily imply bad faith, or a failure to exercise an honest judgment, i.e. on the grounds of collusion or misconduct.

Before entering the field of "Excepted matters" the arbitrators have to examine:

1. Whether architect has given a decision or opinion. Decision will be a judicial one of binding nature whereas the opinion is not binding.
2. Whether parties were heard by the architect.
3. Whether the architect had acted judicially or administratively.
4. When the architect has given his certificate or decision that is whether before a dispute has actually arisen or after the dispute. If the architect has issued certificate or decision before the dispute has arisen, the same will be final and conclusive between the parties. However, if the architect comes on the scene after the dispute has arisen, certifies or gives his decision on the excepted matters, the same is not binding, his power to certify is destroyed and the jurisdiction of the arbitrators arises.

Procedural Conditions

Conditions precedent to Arbitration as per Condition No. 56 of I.I.A. (Excluding Excepted Matters): if one were to analyse condition No. 56 of I.I.A. one can cull out the following conditions precedent for invoking arbitration.

1. (a) There has to be dispute and differences of any kind whatsoever;
   (b) Arising out of; or

(c) In connection with the contract; or

(d) The carrying out of the works (whether during the progress of the works or after their completion and whether before or after the determination, abandonment or breach of the contract)

2. Such disputes and differences shall be referred to and settled by the Architect who shall state his decision in the form of final certificate or in any other form but in writing.

3. If either party is dissatisfied with the decision of the architect on any matter. question or dispute of any kind or as to the withholding by the architect of any certificate to which the contractor may claim to be entitled to, then either party within 28 days from the date of-receipt of the notice of decision of the architect, may serve a written notice through the architect to other party, that such decision of the architect is disputed and such disputed matter should be arbitrated upon.

4. Such written notice to the party shall specify the matters which are in dispute and to be arbitrated upon. No other matter can be referred to arbitration.

   Appointment of sole arbitrator if agreed by both the parties and two arbitrators one by each party and a presiding arbitrator to be appointed by the said two joint arbitrators.

5. The arbitrators and the Umpire have to be Fellows of the Indian Institute of Architects.

   The arbitrators thus enter upon the reference and have powers:

(i) To open up, review and revise any certificate, opinion, decision, requisition or notice save in regard to the excepted matters.

(ii) Determine all matters in dispute which shall be submitted to them and of which notice shall have been given through the architect.

(iii) Costs.

(iv) Award of the arbitrators final and binding.

(v) Arbitration to be as per the Arbitration Act. in Vogue

However, condition No. 56 provides one exception that is: withholding by the Architect of any certificate under clause 49 to which the contractor claims to be entitled, shall not be opened or entered upon until after the completion or alleged completion of the works or until after the practical cessation of the works arising from any cause unless with the written consent of the owner and the contractor. Provided always that the owner shall not withhold the payment of an Interim Certificate, nor the contractor, except with the consent in writing of the architect in any way delay the carrying out of the works by reasons of and such matters, question or dispute being referred to Arbitration, but shall proceed with the work with all due diligence and shall until the decision of the Arbitrator or Arbitrators to be given, abide by the. decision of the architect and no Award of the Arbitrator or the Arbitrators shall relieve the contractor of his obligations to adhere strictly to the Architect's instructions with regard to the actual carrying out of the works. The owner and the contractor hereby also agree that Arbitration under this Clause shall be a condition precedent to any right of action under the contract.

General

1. The Arbitrator is a judge of the choice of the parties not being bound by the technicalities and he is to follow the principle of natural justice and those procedures which are provided under the status. Thus, technicalities of Civil Procedure Code need not be adhered to.
2. Whether a matter should be referred to arbitration or not will also be a difference arising out of the agreement capable of being resolved by arbitration.
3. The question of construction of contract be a question of law, can be decided by the Arbitrator, as also question of breach of contract.
4. The arbitrator can decide the question of escalation of prices and allow the same.
5. The arbitrator has jurisdiction to decide regarding additional work which forms part of the principal work and the same partakes the form of extra works.
6. The arbitrator has to confine himself to the matter of Arbitration Agreement.
   He cannot go beyond the scope of such limitations.
7. The Arbitrator has to pass a speaking award unless agreement provides for non speaking award by consent.
8. Interest: There are two stages to grant of interest in arbitration proceedings.

Stage-1

9. Interest from the date on which cause of action has arisen to the date on which the award is made, at such rate and on such amount as the arbitrator adjudicates as fair and reasonable.

Stage-2

Interest from the date of award to the date of payment.

10. The architect/engineer has to charge reasonable fees when acting as arbitrator as arbitration is a social obligation and the architect/engineer must meet with the expectations of the society in which they live.

Review Questions

1. What do you understand by immobibility of labour?
2. Describe various theories of wages.
3. Explain two systems of wage payments.
4. What are the characteristics of labour?
5. What is the time wage system? Mention its advantages and disadvantages.

6. Write short notes on:
   (1) Wages-fund theory of wages
   (2) Bargaining theory of wages
   (3) Psychic income
   (4) Balance and debt system
   (5) Standard hour system
   (6) Bedeaux point premium plan
   (7) Priestman bonus system
   (8) Payment by result.
7. Explain the piece wage system and mention its advantages and drawbacks.
8. Mention the various incentive plans of wage payment.
9. Describe the following incentive plans of wage payment:
10. (1) Halsey plan
    (2) Rowan plan
    (3) Taylor's differential piece wage plan
    (4) Merrick differential plan
    (5) Barth variable sharing plan.
11. Discuss Emerson's efficiency plan.
12. Explain Gantt's task and bonus plan.
13. Describe the group incentive plans.
14. Mention the procedure the management has to adopt for implementing the plan of profit-sharing.
15. What are the advantages and disadvantages of profit-sharing?
16. What are the factors determining morale?
17. Explain the methods of measurement of morale.
18. What are trade unions? Explain their functions, advantages and disadvantages.
19. Differentiate between the following:
    (1) Nominal wages and real wages
    (2) Halse plan and Rowan plan
    (3) Bedeaux s plan and Hayn e's plan
    (4) Guided interview and unguided interview
    (5) Labour, land and capital
    (6) Time wage system and piece wage system
    (7) Standardised work and non-standardised work
    (8) Fraternal functions and militant functions.
20. Give the salient features of The Building and Other Construction Workers (Regulation of Employment and Conditions of Service) Act, 1996, (BOCW Act or BOCWA).
21. Briefly explain the C.P.W.D. Contractor's Labour Regulations-2019.
22. List out the Indian Legislation in respect of Labour.

23. What are various Methods to settle disputes and differences in construction industry?
24. Explain the meaning of Arbitration, Mediation and Conciliation.
25. What is Arbitral Award? Explain the meaning of reasoned Award.
26. Who can be appointed as Conciliators and what are the duties of Conciliator?
27. Explain the Legal Position of "Excepted Matters" under Indian Arbitration Act.

8

Risk and Insurance in Construction

Uncertainty

Uncertainties bring with them, elements of risk. The success or failure of a project largely depends upon the effectiveness and the efficiency with which the risks and uncertainties are managed. In fact, risk management has emerged, as the main function of project management.

Project risk management is the art and science of managing risks caused by unforeseen changes (uncertainties) which may require deviations from the planned approach and may therefore affect the achievement of the project objectives. It involves systematically identifying, analysing, planning and controlling risks. It provides greater insight into risks and their impact for making risk related decisions. Umesh Kohli & Krishan Kumar Chitkara,. (2012).

Project Risk Management Model

Processes	Tools and Techniques	Outcome
Risk Identification	Work breakdown analysis Management process analysis System flow analysis Brain storming Check-list templates Knowledge experts Judgment	Sources of risks Potential risk events Project risks checklist
Risk Assessment	Risk probability assessment Risk impact assessment Risk classification Risk ranking	Time-overrun Risk assessment Risk contingency assessment
Risk Response Planning	Risk mitigation planning (or strategy)	Risk response plan/baseline Responsibilities Allocation
Risk Response Control	Monitoring risks	Risk-related corrective action

Risk Defined

Risk is an abstract term. In a given situation, risk is viewed differently by different people. For example, the likelihood of rain on a cloudy day may be perceived as the risk of getting wet by a man walking in the street, a risk of slipping on the road by a motor cyclist and an opportunity to plant the seeds by a farmer. To some it may pose a hazard, danger or threat, while others may see it as an opportunity.

Risk generally signifies an uncertain event, situation, or condition which may occur. It may have either a positive or negative effect on the project objectives. Some risks may pose a threat to the achievement of project objectives while some other risks may enhance achievement of objectives. Favourable risk events are called opportunities, whereas unfavourable risks are termed as threats. Factors that induce unfavourable risky situations are called hazards. Risk increases with hazards but decreases with safeguards.

In the project context, BS 6079-3:2000 defines risk as the, uncertainty inherent in plans and the possibility of something happening that can affect the prospects of achieving business or project goals. PM BOK Guide, Third edition (2004) defines risk as, an uncertain event or condition that, if it occurs, has a positive or negative effect on a project's objectives.

A risk has three components:

- A situation leading to an event, the occurrence of which is likely to deviate from the estimated or forecast value
- The probability of occurrence of that event
- The likely impact of that event, i.e. loss or gain

Risk exposure is a function of the probability of occurrence and the consequences (amount of damage or gain) of not-achievement of a planned goal. Risk exposure is mathematically quantified by multiplying the probability of its occurrence with the financial impact of the occurrence of the risk event. Umesh Kohli & Krishan Kumar Chitkara,. (2012).

- Risk exposure = Probability of occurrence of risk × Financial impact on occurrence of the risk event

For example, risk exposure of encountering a hard soil, instead of the soft soil estimated during excavation in a building project, costing Rs. 20 per cubic metre extra, with 20% chances of its occurrence (due to lack of proper soil investigations), is worked out as:

Risk of excavation in hard soil= 20% × Rs. 20 per cm= Rs. 5 per cm. Thus, the project manager has to manage a risk exposure of Rs. 5 per cm.

If risk is affected by public perception, such as a fatal accident due to safety lapse resulting in a long duration strike by workers, then such risk can be defined as:

Risk exposure = Probability of occurrence x consequences of impact including public perception

Note: Risk impact is assessed at present monetary value.

Risks in the Construction Industry

In the execution of a normal construction job, the organisation's materials and equipment are susceptible to the risks of accidents, damage, deterioration and loss; and employees are exposed to accidental and occupational injuries and death. These risks may be arising out of and happening in the course of the following:

- Ocean marine and inland transit
- Storage
- Erection and/or other construction operations
- Erection testing and commissioning
- Operation of construction equipment
- Running of vehicles
- Employment of men
- Transit, holding, storing and usage of cash and valuable goods.

Besides, the organisation will have the legal liability for personal injury and death, property damage and loss to the owner, owner's employees and third parties including members of the public, caused by its own or its employees, or its subcontractors' negligent act, error or omission.

The most common perils to which a construction site is exposed to are:

- breakdown of equipment
- failure of safety devices
- hazardous work practice
- windstorm
- earthquake
- flood and subsidence
- fire and explosion
- burglary and theft fraud by employees
- strike, riot and civil commotion
- accidents to third parties by motor vehicles and mobile equipment
- personal injury, occupational disease and death.

Risk Management

Risk Identification, Analysis and Classification

The top management of the organisation must study the total working of each site separately with the help of a flowchart and identify the risks involved, affecting men and properties, both own and others. Then they must analyse and evaluate the probable risks with reference to causes and effects. Causes can be categorised as natural phenomena, breach of natural laws and human failures. Effects can be categorised as legal liability damages, property damage, bodily injury and death and loss of earning. The next step in the process is to classify the risks in the order of severity of the probable losses as follows: Umesh Kohli & Krishan Kumar Chitkara., (2012).

ISR = Insignificant risk
NR = Normal Risk, which would not affect the organisation's basic finances
SR-I = Serious Risk, which would necessitate raising additional finances by borrowing or by raising the paid up capital
SR-II = Serious Risks, which might lead to bankruptcy of this organisation

Risk Avoidance and Reduction Measures

By taking the appropriate risk reduction measures it might be possible to handle ISRs and NRs internally without transferring them to an insurer. It might also be possible to reduce the severity of SR-I and SR-II. By avoidance and reduction of risks, the organisation will have the advantage of avoiding delays, besides saving insurance premium. Accidents or other loss-producing events will result in not only direct cost but also indirect cost, delays in work and consequential losses. Umesh Kohli & Krishan Kumar Chitkara,. (2012).

The measures to be taken have to be decided by the organisation separately for every site depending upon the particular types of risks. A few examples are given below:

- Store petroleum products, inflammable and combustible good firewood and the like in the open, far apart and away from office -and store.
- If there is a rain water drain originating elsewhere and terminating at the site, ask the owner to divert it.
- If the organisation's own regular bank's nearest branch is situated, 11 a distant place, necessitating the carrying of large sums of cash IP the site exposed to the danger of robbery in transit, ask the bran Ii to open a pay office near the site or with that bank's permission open an account with another nearest bank.
- If the site's cashier is new, do not entrust him with large amount of cash.
- If a lifting tackle shows signs of weakness, do not put it to use before replacing the weak parts.
- Transfer certain types of risks to sellers of materials and equipment making the risk-covering clauses enforceable.
- If there is a dam nearby, request the authorities, in advance, to ensure proper water level control and timely warning so that no flood is caused by it.
- Provide sufficient watch and ward and floodlights.
- See to it that necessary fencing is done and that nobody is making an opening in it stealthily.
- Use a burglar-proof cash safe.
- Secure the office and store by strong doors and windows and regular locking habit.
- Provide sufficient fire-fighting arrangements and train the staff concerned, in their use.

- Where it is necessary and feasible, install fire detection systems.
- Provide escape outlets for use in the event of fire.
- Guard the major hazardous points.
- Ensure that mobile construction equipment and vehicles have good brakes, are generally in good condition and are operated by skilful operators or drivers.
- Make it compulsory that workmen wear the prescribed safety appliances and strictly observe the safety practices.
- Ensure quality control at every Stage of the construction work, reducing the chances of work failure.
- Follow safe work practices in erection at high elevations.
- Make staircases, walk-ways and working platforms strong and secure.
- Undertake cutting and welding with adequate fire safety precautions.
- Strictly adhere to statutory regulations in handling power connections and in the use of electric power.
- Inspect the premises thoroughly before closure for evening and weekend.
- Do not dispose off waste materials in a dangerous manner.
- Display safety-warnings and educative posters.
- Keep the site clean and tidy.
- Prohibit smoking and the presence of flames of all kinds at dangerous work spots and inside the store.
- Follow special storage instructions to protect goods from damage - causing exposure.
- Give awards for "no-accidents" and "suggestions" on accident -prevention and loss-reduction.
- Whenever any new equipment is added or technology introduced or work started, think of the additional risks it can cause and take the appropriate risk avoidance and risk reduction measures.
- Do effective internal audit and punish employees for embezzlement as a deterrent against fraud.
- Subcontract the most hazardous activities to experts in the field for safe operation without loss-producing incidents.

Self-Insurance

Giving due consideration to the probabilities of occurrence of los producing events and comparing the costs of risk reduction and insurance premium for risk transfer, the organisation may choose not to insure some classes of risks, but at the same time pay particular attention to their prevention. This decision shall be taken by or with the knowledge of the top management and be recorded in writing so that nobody is blamed later. There must be constant checks at various levels on the measures of prevention. Umesh Kohli & Krishan Kumar Chitkara,. (2012).

Risk Transfer

In addition to all preventive measures, certain types of risks shall be insured as a measure of caution. The nature of certain types of major risks is such that although the chances of their occurrence are remote in the unlikely event of the happening of any of them, its financial effect will bankrupt the organisation. There are certain other types of minor risks with comparatively low financial effect but with the probability & high frequency, like occupational injury and motor vehicle accidents. Such major and minor risks must be transferred to insurers although reasonable measures of prevention should still continue to be taken. Additionally, there are risks which are required to be insured statutorily, like third party liability under Motor Vehicles Act. Umesh Kohli & Krishan Kumar Chitkara,. (2012).

Risk Management Strategy

Project risk management strategy involves identifying, analysing, planning and controlling risks, which might affect the performance of the project. The risk identification process involves researching the project to determine the sources of risk, and connected potential risk factors that lead to risk events and thereby reducing the chances of overlooking any potential risk event.

Risk analysis or risk assessment aims at quantifying risk exposures to enable mitigation and development of project risk response plans for managing risks during execution of the project. Quantifying risks involves the estimation of the probability of occurrence and the cost of risk consequences of each risk element. This is followed by the ranking of risks.

The risk response process considers risk transfer, reduction and avoidance possibilities. A response risk plan ensures that the appropriate risk warning tools are in place, to handle risks efficiently. A predetermined risk response plan can minimise the consequences of possible adverse future events and can maximise the benefits of positive future events.

Risk control aims at controlling deviations to cut down risks and maximise the project value. It handles risks in a manner that achieves project objectives efficiently and effectively by monitoring and adjusting plans and evolving alternate plans to manage foreseeable risks, whenever necessary.

Risk management enables greater certainty of achieving the project goals by making the project plans realistic, reducing costly surprises, minimizing losses and delay and enabling exploitation of beneficial opportunities.

Risks Check-list

The risk factors and risk events are consolidated to form the risks check-list. The check-list will vary from project to project and agency to agency. A typical check-list showing risk events categorised under risk sources, which may be encountered in a turnkey construction contract, is given below. The list is indicative and is not limited to the following.

1. **Project Scope Risks:** High complexity, ill-defined project scope, frequently changing scope requirements, no project charter, no delegation of authority, ineffective control systems, no extra work control, no analysis of changes and problems in quality control.
2. **Design and Specifications Risks:** Inadequate design information, incorporation of new construction technology, unrealistic specifications, likelihood of design changes, difficulties in interaction of design with method of construction, non-standardisation of resources, designer's delays, poor design and shop drawings, non-conformity with national and local specifications.
3. **Quality Risks:** No quality assurance plan, no soil investigation, no method statements, poor quality materials, untrained manpower, absence of approved soil and material testing laboratories, unachievable quality specifications, problems in quality control and re-working of defects during construction.
4. **Time Overrun Risks:** Inaccurate activity time estimates, unrealistic time schedules, incomplete work breakdown structures, no formal sequencing plan, poor allocation of resources, incomplete assessment of project time, cost, resources and quality implementation plans, no database, ineffective control system, inflexible and unrealistic project plans, unsatisfactory conduct of status review meetings, inability to take timely corrective action.
5. **Cost Overrun:** Risks-Inaccurate cost estimates, inadequate cost planning and control, no extra work control and no analysis of changes, constantly changing market conditions, incomplete project closure.
6. **Leadership Risks:** No project vision, no team building, poor motivation of participants, high turnover of critical team members, indecisiveness, unreasonable stakeholders expectations, lack of senior management support, lack of team consensus over project plans, limited authority/ control of the project manager, poor communications, poor industrial relations, high rate of sickness and absenteeism, unsafe working conditions resulting in accidents and poor turnover, conflicts among staff and participating organisations, lack of co-ordination, insufficient liaison with public services, barriers in information communication.
7. **Organisational risks:** Inappropriate organisation network, poor assignment/allocation of tasks and responsibilities, lack of competent persons, no project manual/documented procedures/processes, project being too complex for the resources available, inadequate communications infrastructure, wrong selection of project management, no database, inflexible and unrealistic project plans, poor quality control, unsatisfactory conduct of status review meetings, inability to take timely corrective action, incomplete project closure.

8. **Physical Resources Mobilisation and Utilisation:** Risks-inadequate and low quality procurement of resources, non-availability of spares parts, special equipments and materials; transhipment delays, low productivity, bad weather and working conditions, non-availability of suitable sub-contractors, damage during construction due to negligence, transportation or storage vandalism, accidents, wastage, theft and fraud, lack of safety and hygiene measures and local requirements.
9. **Technology Risks**: Inadequate information on new technology, non-replacement of old technology, non-availability of competent and professional staff to use new technology, lack of managerial skills.
10. **Contractual Risks:** Non-standard and inconsistent conditions of contract, insufficient time to prepare bid tenders, delay in possession of site, errors or omissions in bills of quantities, payment problems, extra work variations, unrealistic tendered amount, no credit worthiness of contractor, high cost of legal decisions, insufficient insurance and surety, incorrect documentation of claims and disputes, unfamiliarity with local laws.
11. **Force Majeure and Ecological Risks:** Acts of God such as earthquakes, floods, landslides, ecological damages, epidemics, etc.
12. **Political, Legal and Social Risks:** Changes in government, policies, regulations, rules, laws, war, revolution, civil disorder, risks under criminal law and the law of tort, pollution, waste treatment, local regulations, constraints on the availability of labour, import-export restrictions and procedures, joint venture rules, inconsistency of regulations within the country, requirements for licenses and permits, crime and insecurity, bribes and corruption, religious and cultural conflicts.
13. **Financial and Economic Risks:** Investment risk, inflation, escalation of prices, availability and exchange rate fluctuations, local and national taxes, inadequate sources and availability of funds, cash flow problems, effect of time and cost overruns, default of stakeholders and suppliers, demand scenario and constantly changing market conditions.

Risks Assessment Methodology

Nature of Risk Assessment

Risk assessment involves overall ranking of risks using qualitative assessment approach and quantifying the risk exposures for mitigating risks and developing risk response plan. Umesh Kohli & Krishan Kumar Chitkara,. (2012).

Risks assessment approaches vary with the purpose and the project.

The choice of technique for risk analysis dependent on the nature of the problem being modelled, the amount and reliability of the information available and the nature of the output required. The particular needs of the client and the type of decisions to be made will determine the nature of the output that is required. For example, there are different approaches for time risk analysis, budgeting project contingencies, analysing risks in selecting a project out of alternate opportunities, analysing economic viability during project feasibility studies, sourcing and selecting contractors and input suppliers, and analysis of contingencies in completion cost estimates.

The Human Side of Risk Management

Risk analysis cannot be mechanised, and the most important characteristic needed to analyse and manage risk is an appropriate frame of mind or the risk-taking attitude of the project manager.

The risk-taking decision depends upon the utility, or the pleasure or displeasure one derives from the expected outcome. The utility theory explains how, not why, rational people sometimes prefer outcomes which do not necessarily have the highest monetary value. The equation that describes the utility curve is the utility function. Utility functions vary from person to person. Also, the utility function of an individual may not be identical to the utility function of organisation that employs him. Behavioural scientists have classified individuals with risk tolerance characteristic into three categories, i.e., risk averter, risk neutral and risk seeker.

In the case of risk averters, when cost increases, the utility rise is marginal as the slope of the curve decreases with the increase in expected monetary value. In other words, a project manager who is a risk averter is cautious and conservative, as his tolerance reduces when more money is at stake. More money increases his utility only very slightly. On the other hand, a risk seeker or risk taker may be willing even to pay a high penalty in order to gain a higher uncertain outcome. A risk-neutral project manager with linear utility curve is most likely to use the expected outcome as the decision criterion and act only when the expected outcome is positive.

It is important to note that the knowledge of risk analysis tools and techniques is necessary to manage risks, but it is not sufficient in controlling risk. The final piece of the jigsaw is the human being. Achieving this involves gaining a clear understanding of the human perception of project risk and uncertainties. It is also important that the risk-taking attitude of the project manager and the characteristic of the project should be compatible.

Role of the Project Manager in Managing Risks

A project manager's capability to accomplish the mission in a large-sized, modern construction project is a prime requirement of professional skills in risk management. His role in managing risks calls for the following:

- Estimating cost and time contingency allowances, and allocating these in commensuration with major risks and uncertainties that are identified.
- Regularly monitoring the risk response plan, and reviewing these with the concerned persons in order to reduce misunderstandings and ensure that the full spectrum of uncertainties is exposed.
- Adopting methods for allocating the remaining risks to the various parties in a way, that will optimise project performance.

- Recognising that risks and rewards go hand-in-hand and, that the allocation of a risk to a party should be accompanied by a suitable incentive.
- Keeping an open-minded approach to innovative solutions to problems.
- Making appropriate timely decisions.

The Benefits of Managing Project Risk

The overall awareness of risk exposure and the methods of handling risk, add to the effectiveness and efficiency of project management. This is due to several reasons. Umesh Kohli & Krishan Kumar Chitkara,. (2012).

- The risk response development process gives an insight into the project management process. Accordingly, the issues/problems of the project are clarified, understood and allowed for right from the start
- A pre-planned contingency plan provides clearer definitions of the specific risks associated with a project. It allows prompt, controlled and pre-evaluated responses to any risk that may materialize.
- The structure and definition of the project risk are continually and objectively monitored. This in turn reduces exposure to project risks.
- Risk response decisions are supported by a thorough analysis of available data.
- A fully documented risk management process, builds-up a profile of historical risk to allow better modelling for future projects.
- It encourages problem-solving and provides innovative solutions to the risk problems within a project.
- A risk reporting framework avoids sudden risk shocks.

It is sometimes argued that the risk identification process induces negative and cautious attitudes among the team or the project sponsors. Contrary to this, it can also be said that the risk identification process, if developed with the participation of the project team, enables effective management of projects by:

- Identifying the participants who can be entrusted with the management of risks.
- Creating environments for managing risks efficiently and effectively among the concerned people, as they become aware of the risky situations well in advance.

The very process of breaking a project down into its sources of risk and systematically analysing them ensures that managers develop a much more realistic feel for the project and its range of possible outcomes. Risk analysis supplements professional judgment. However, risk analysis is not a substitute for professional experience, judgment and the attitude of mind of the appropriate decision-makers. Project managers cannot stop the fast-changing, unstable, risk-prone environments but they must prepare themselves for managing the resulting impacts of risks to their projects. This in turn reduces exposure to project time, cost and performance risks.

Risk Management Programs

Risk management in the construction industry includes an organized effort to identify and quantitatively evaluate potential exposures, along with an advance plan designed to eliminate or mitigate the consequences of the risks. A comprehensive risk management -program throughout the life of the project continues the approach developed by initial plan.

The owner, the architect/engineer, the general contractor, the developer or construction manager, and the subcontractors all have their individual needs and objectives along with individual risk tolerance. Some of these risks can be covered by insurance and bonding. Others can be eliminated by contractually transferring the risk to another party. Some construction methods may be more risky than others but offer potentially greater rewards. Evaluation of anticipated frequency and severity of the risk compared to the contemplated reward can assist in choosing the optimum method.

Risk Identification

Risks can be divided into internal and external, predictable and unpredictable, technical and nontechnical. Some of the risks are best handled by the owner, are best handled by the designer, and some are normally assumed by the contractor. Assigning risks best assumed by one of the parties to other parties not in a position to control or mitigate the risk has been the root cause of much of the adversarial relationships and litigation now experienced in the industry. Umesh Kohli & Krishan Kumar Chitkara,. (2012).

Insurance

Glossary of Insurance Terms

Definitions for insurance and risk management terms follow:

Loss: The basis for a claim for damages under an insurance policy.

Primary insurance: Basic insurance coverage from first dollar to the stated insurance' policy limit.

Excess insurance: Insurance coverage applying only to loss or damage in excess of a stated amount or of a primary insurance policy.

Self-insured retention: The amount of each loss for which the insured party agrees he responsible before the primary insurance coverage begins to participate in the loss.

Loss-sensitive or retrospective policy: An insurance policy whose premiums are directly affected by the loss experience. Standard premiums can be increased or decreased based upon actual performance under the policy. This type of policy is most commonly used for worker's compensation insurance, where the frequency and severity of losses has a direct bearing upon the premium.

Subrogation: The right of subrogation preserves for insurers the right to sue, over damages from the party who caused the loss, unless that party is a named insured under the policy, or the right has been waived by contract.

General Principles

The basic principles of insurance, how they apply to the construction industry and how they have to be observed by the organisation in practical situations are discussed here.

Insurable and Measurable Interest

The first principle of insurance is that the organisation must be subject to a risk which is not an imaginary one. The beneficiary under an insurance policy may be either the organisation itself or its employees or the owner under a contract or members of the public under a legal liability.

But the liability must be the organisation's, either contractually or legally, if its proposal for an insurance against the same is to be accepted by an insurer Also, such insurable interest must be measurable in terms of money.

Indemnity

The organisation can recover from the insurer under a policy only such amount of money as is required to be paid to others in discharge of its legal or contractual liability or otherwise spent to make good the actual loss or damage to itself, caused by an accident or other loss-producing

Event. No profit nor any other undue benefit can be derived out of an insurance. In case of a property damage, the insurer will pay the cost of replacement or repairs to restore the property to its pre-accident condition. In other words, an insurer will indemnify the organisation only against actual loss or damage. But at the same time, the indemnity may not be full if there is an under-insurance. If the full value is not insured, the principle of average will apply. That means the amount of compensation will be proportionate to the insured value if full value is not insured.

Subrogation

If the organisation has any legal rights of recovery or action against carriers or any other third party responsible for the loss, such rights shall be transferred to the insurer when he pays the claim or reimburses the loss.

Contribution

Even if the organisation insures the same risk for full loss with more than one insurer, it cannot recover more money than the actual loss. It is forbidden under common law. But in that case, the insurer who pays the organisation shall have a right to recover from the other insurers the proportionate contribution towards the compensation. Each insurer shall be liable only for the rateable proportion of the claim, on the principle of "average".

Good Faith

An insurance contract is based on utmost good faith between the insurer and the insured. Therefore, at the time of making an insurance proposal, the organisation must disclose all the known material information about the risk proposed. Withholding of any such information or giving wrong information which might come to light later will be a ground for the insurer's refusal to pay compensation or even to render the policy void.

Starting Point of Insurer's Indemnity

An insurer will hold the organization insured and indemnified from the time he receives the first instalment of premium.

Cost and Expenses to Defend Claims

The cost and expenses to defend claims against the organisation shall be borne by the insurer.

Negligence

Negligence is an insurable risk, but it should not be 'wilful'. The owner under a contract will not be liable for the organisation's negligence. So also the organisation shall not be liable for the negligence of its subcontractors except that the organisation has to indemnify the owner against its own and its subcontractors' negligence. It is for this reason that the subcontractors should include in their work insurance policies the organisation as a co-insured. For the negligence of the subcontractor officially nominated by the owner, the organisation need not indemnify.

Onus of Proof

The responsibility to prove negligence will always rest with the party who has suffered. If the victim has died, the claimant should prove the negligence and his right to claim.

Insured to Employ only Competent Employees

The organisation should employ only competent employees for accident prone operations. Persons under 15 years of age should not be employed.

Excess or Deductible Franchise or Self-Insurance

Almost all classes of insurance policies, especially contract work insurance covers and motor vehicle policies, shall be subject to a minimum "excess" This means that an initial, specified amount of every loss or damage shall not be claimed from the insurer. Or from every claim of a larger amount the specified amount shall be deducted as the quantum of self-insurance by the organisation.

In addition to the above principles, there are the following condition and warranties which are more or less applicable uniformly to all classes of insurance.

1. *Losses or damage wilfully caused* by the organisation to itself shall not be entertained by an insurer.
2. *War risks* are always excluded from policies.
3. The organisation should exercise reasonable *care in selecting employer,* and supervising their work, from their competence and have all points of view; take reasonable precautions *to protect* the insured objects from perils and *prevent* and *minimise losses.*
4. The organisation shall give immediate notice to the insurer of any loss -producing event. Where a third party is involved or where the event of a theft or burglary or unlawful commitment or a death, immediate

police report shall be made and within seven days of the event (unless the policy stipulates a longer period) full details of the loss or damage with an estimate of the claim shall be given to the insurer.

5. The organisation shall protect the insurer's interest by sending timely claim notices to the party who has actually caused the loss or damage, like the carrier in the case of transit damage.
6. All claims shall be truthful and only for the actual losses or damage.
7. Disputes arising between the insurer and the organisation as regards the quantum of compensation or admissibility of a claim shall be referred to arbitration in accordance with the Arbitration Act of the country where the policy was issued.
8. The organisation shall not, without the consent in writing of the insurer, either repudiate a third party's claim or admit or negotiate or offer to pay or pay any claim; but shall only pass on all such claims, notices, summons and demands to the insurer to handle.
9. Under all policies, the maximum amount of compensation shall be the limit of indemnity declared in the policy, unless a larger amount is decided by the appropriate court or tribunal under common law.
10. Insurance policies cannot be transferred, but the insured property can be replaced and even co-insureds' names can be incorporated later. Bankers' or owner's names can be incorporated at any time as co-insured for their respective interests.
11. To almost every class of policy, extensions can be made to cover related additional risks at extra premium.
12. If an organisation does not file a suit against the insurer in a court of law within 12 calendar months from the date the insurer disclaims liability for any claim, such claim shall become unenforceable after 12 months - from the date of disclaimer.
13. Where a policy is issued as a matter of custom or practice and premium paid on the basis of an estimate of annual, turnover (as in cash-in-transit and workmen's compensation insurance), the organisation shall furnish actual turnover and adjust the premium within one month from the expiry of the insurance period.

Classes of Insurance Relevant to a Construction Organisation

Contract Work Insurance

There are two different classes of insurance for contract work. They are contractors' All Risks (CAR) policy and Erection All Risks (EAR) policy.

The basic difference between the two is that while CAR is mainly for civil engineering work, EAR is for mechanical and electro-mechanical erection work. EAR is known also by the term Storage-Cum-Erection (SCE) when it does not cover ocean, marine or inland transit risks. Joy P.K,.(2005).

Contract work insurance of either class can cover:

- the organisation's entire property at the site including conn, materials, construction equipment, tools, consumables and temporary work
- the owner's properties to be covered under contractual obligation
- subcontractors' and third parties' properties if they are required to be covered
- legal or vicarious liability to public in the course of:
    - ocean, marine and inland transit, storage, shifting, assembly, erection, cold testing, trial run, commissioning, guarantee test operations by the organisation under maintenance liability clause, clearance and removal of debris and any other operation connected with the execution of the contract covered by the policy.

For losses and/or damage arising out of the following perils:

- fire, including fire-fighting operation, lightning, explosion and common sources of fire or damage caused by a crashing aircraft
- strike, riot, civil commotion and malicious damage
- storm, hurricane, tempest, tornado, cyclone, typhoon, flood, inundation and subsidence
- earthquake, earth tremor, volcanic eruption, landslide or rockslide
- sinking of vessel or derailment of train
- burglary and theft
- negligence, lack of skill and faulty operation
- collision, impact, snapping, dropping, falling and similar accident
- electrical and mechanical breakdown, including short-circuiting
- water damage to works in rivers, sea, canals, etc.
- every other peril not directly or wilfully caused by the organisation

The insurance can cover also:

- express freight and overtime expenses for replacement or repair and escalation in replacement costs including fluctuations in tax' and duties.

The value insured will comprise the cost of materials, wage, supervision, freight, insurance premium itself, taxes and duties and owner's free supplies. Generally, the organisation will be required to take CAR or EAR policy only under turnkey contracts involving supplies and work. Where supplies are by the owner, he will take the work insurance. Then the organisation needs to insure only its own construction equipment, tools, consumables, temporary work, etc., for which different classes of insurance are available.

If the owner covers the marine, storage and erection risks, the organization shall cover its own construction equipment by a contractors' construction equipment policy or contractors' plant and machinery policy which akin to CAR or EAR policy, covering only the risks to plant and equipment used for carrying out the construction work. Joy P.K,.(2005).

In contractors' construction equipment policy, the exclusions also include the following in addition to those principal exclusions from CAR/which are listed in a subsequent paragraph:

- wearing parts such as packing, gaskets, fan-belts, ropes, chains and blades
- immersion in tidal water
- transit from one site to another
- working underground
- alterations, additions or improvement
- consequential loss

Construction equipment shall be insured for their replacement value rather than written down value.

A damage to an equipment is called a total loss when the replacement repair cost minus the value of salvage is more than the insured value.

Here the insurer will pay only the full insured value.

Contract work insurance policy shall be preferably in the joint names of the owner concerned and the organisation so that his liabilities to be indemnified by the organisation are properly covered and he is able to receive compensation directly. Subcontractors' legal liability shall be included in the organisation's public liability cover so as to be able to meet any possible claim against the organisation arising out of any sub-conntractor's negligence. This inclusion will not be necessary if every sub-contractor takes a public liability policy and gives its copy to the organisation. If the organisation covers them, a pro rata recovery of premium shall be made from them.

While perils covered by CAR policy will be peculiar to civil construction work, those by EAR will be peculiar to mechanical and electromechanical work.

The following are principal exclusions from contract work insurance:

- shortage or loss discovered only at the time of physical inventory
- normal wear and tear, rust, damage to painted or polished surface
- faulty design, defective material , and bad workmanship (but fault in erection is not excluded) (professional indemnity is a different insurance)
- loss or damage to files, drawings, bill, cheques, accounts and other documents (which can be covered under a different insurance policy but best taken care of by proper risk management)
- consequential losses
- war and warlike operations
- nuclear reaction, nuclear radiation, radioactive contamination
- wilful act or wilful negligence of the organisation
- shortfall in billing
- cessation of work whether partial or total
- contract work itself after it is handed over to the owner (but damage resulting from the organisation's operation or rectification work, etc., during the maintenance guarantee period is not excluded)

Premium rating for contract work insurance will take into account the extent or scope of cover needed. The organisation shall have the option to take combined work insurance policy covering all the risks or separate marine policy, storage-cum-erection policy, contractors' equipment policy, third party policy, etc., considering the actual requirements, convenient and overall economy. Even where the owner takes the work insurance, the organisation as the contractor executing the work will have, significant role to play in the operation of the policy. Therefore, it is desirable to get a copy of the policy and study its conditions. The owner should, in his own interest, give a copy. Joy P.K,.(2005).

Product Liability Insurance

If the organisation itself manufactures a good number of components and incorporates in a work contract, it may be advisable to take an insurance protection against any liability arising out of its manufacturing defect that may cause death, bodily injury, illness or even property damage to others. Damage to the product itself from intrinsic causes cannot be insured in the normal course. However, the need for a product liability cover will depend upon the nature of the components.

Professional Indemnity

Professional indemnity is meant to provide insurance protection to practising professionals including doctors, engineers, architects and accountants. It is possible to take a policy to cover any liability arising out of the negligence in the duties of any of these professionals employed by the organisation in the execution of a construction job. Usually the cover will be in respect of legal liability for property damage including damage to and defects in the object planned, designed or supervised. The indemnity cannot be covered in a CAR or EAR works insurance policy. The policy can be extended to also cover employees' dishonesty including fraudulent, criminal or malicious conduct which might cause financial loss to any client. This insurance is widely used by consultants and architect in the construction industry.

Separate Fire Insurance and Burglary and Theft Insurance

Fire, burglary and theft are normally included in CAR or EAR policy. Where there is no need for CAR or EAR insurance policy, the organisation shall take separate fire policy and burglary and theft policy covering its properties other than heavy construction equipment. Construction equipment policy will cover all risks including fire, burglary and theft as far as the equipment is concerned. The general fire policy and burglary and theft policy shall cover general stores, tools and tackles, instruments, temporary buildings, store racks, furniture, office equipment, etc. Fire policy will cover fire from all sources including firefighting operation, short-circuiting, lightning, explosion and a crashing aircraft. Burglary is the cover for losses and/or damage resulting from house-breaking or a forceful entry into locked store or office. Theft cover is an extension to the burglary over at an extra premium. The theft cover is for losses resulting from ordinary act of larceny or stealing which was immediately noticed and reported for which there is some clue of evidence and not for inventory shortage which might have been the result of a past unnoticed theft.

In fire, there are a host of exclusions especially of public liability.

Ocean Marine and Inland Transit Insurance

It has been seen that where supplies are involved, there is option either to take a combined policy covering marine (both ocean and inland), storage and erection risks or to take separate policies - one for marine and another for storage-cum-erection. Whether combined or separate, the basic conditions and procedures are the same.

The marine insurance is designed to protect the insured against accidental losses and/or damage only. It does not provide indemnity for material wear and tear and inevitable losses.

In ocean marine, there is a wide range of subclasses and choices varying according to the interest of the insured, valuation of the subject matter, term of the policy, manner of description and scope of coverage. In actual practice, should there be an occasion to take a marine policy, the advanntages of the different subclasses of policies should be studied brforehand. The organisation's usual insurer's marine expert should be called for detailed discussion. The staff dealing with marine insurance must study the uses of different types of marine insurance contracts, analyse the policy terms concerning description of the scope, period of coverage, perils covered, double insurance if any, subrogation, types of losses and damage covered and documents required to establish losses.

As a construction contractor, the organisation should be interested in insuring the materials against all perils and risks including customary peril during voyage and all types of losses so as to get the full replacement cost or repair charges and there should be provision for express freight charges of replacement.

For imported goods, the inland transit shall be an extension of ocean transit. But, as the carriers will be different and because the primary liability for transit damage and loss rests with the carrier concerned, the insurer will insist on a package-wise examination of the cargo for external damage before leaving the port, conducting of a survey where necessary and serving of claim intimation, to the ocean carrier so that the insurer claim is protected. Carriers will have their own independent insurance to cover their losses.

For inland transit perils, inland carriers are responsible. If the perils are insured, the insurer will follow up claims with the carriers but pay the organisation before the carriers' settlement is received by him. All types of inland transit perils shall be insured fully so that all replacements and repairs are paid for by insurers in full. Insurance is available for goods transit by passenger train, express, mail goods train, motor lorry,anima1 drawn carts, aircraft, inland water way and post. Transshipment is also covered without any additional premium. Although under inland transit also, there are several subclasses like those under ocean marine, a construction organisation, we are concerned with only "transportation risks policies on goods". Besides contract materials, the organisation's supplies of indirect materials, tools, construction equipment, furniture, office equipment, etc., will also need inland transit cover, mostly for single transit.

Both ocean marine and inland transit policies can be taken either independently for every shipment or as an open policy. Open policy is an open cover for a year or a contract period under which a declaration is necessary for every shipment or consignment for specific cover. Premium shall be calculated and collected separately for every declaration. Against the open cover there may be a very nominal deposit. Even in marine transit, joint insurance is possible.

The extent of risks covered, the nature of goods carried (for example for perishable and brittle goods the premium rate is high), the value of the goods, volume of business and for certain special equipment, the type and even dimensions of equipment are the factors influencing the premium rates. There are four different rates. They are as follows:

Uniform rates which are uniform for all risks of the same description.

Loaded rates with additional uniform loading for all risks of similar type to take care of additional hazards.

Discretionary rates which are left to the discretion of the insurer in a competitive market.

Merit rates specially offered to selected customers considering the volume of business and low history of claims as a result of better n management.

Contractors' Third Party Liability/Public Liability Insurance

During its operations at the construction site, the organisation shall have certain duty of care to members of the public. If losses or damage occur to public arising out of negligence on the part of the organisation or its employees while working, it shall be liable for such losses or damage. It has to indemnify the owner also against any expenses, liability, loss, claim or proceedings in respect of any injury or damage to third parties, arising out of its operations at the site.

Both the classes of liabilities mentioned above arise under *tort,* that is the breach of private rights of individuals.

Contractors' third party liability or public liability is usually included CAR/EAR policy. Where there is no need for CAR/EAR insurance, this class of liability can be covered by a separate policy. The amounts of compensation under this class are mostly decided by courts only. In some rare cases, tripartite agreements among the claimant, the organisation and the insurer may be reached. The limit of indemnity to be insured shall be decided by guess-estimates keeping in mind the probabilities of accidents andt the amounts of compensation granted by courts in similar cases in the recent past. Limits of indemnity are usually stated in public liability policies from every claim, all claims arising out of one cause and for all turns from all causes together during the policy year.

Liabilities arising out of the use of cranes and hoists are generally excluded from public liability insurance. But they shall be covered by the contractors' construction equipment policy.

The following are the other exclusions from liability policies because the organisation has no legal liability for them:

1. *Valenti Non fit Injuria:* If a person voluntarily consents to run a risk, he shall have no right of action against the organisation for the injury or loss he has suffered in consequence of his action.
2. *Inevitable accident:* For an accident which occurs in spite of the organisation exercising the usual care, caution and skill, there cannot be a claim against it.
3. *Act of God:* For events like flood, storm, earthquake, lightning, etc. which happen without human intervention, no claim can be brought against the organisation. That the organisation shall take extended cover for its own property is a different matter.
4. *Fire and explosion:* Claims arising out of fire and explosion are excluded from third party/public liability.
5. *Action in emergency:* If an employee of the organisation, in a moment of imminent danger does something to save his life and that action causes injury or loss to another, no claim can be brought against the organisation.
6. *Contributory negligence:* It the party who has suffered a loss or damage had played a wilful part in the happening of the accident-producing event, then the claim will stand reduced by the extent of his part of the negligence.
7. *Relief by contract:* If an activity involving a probable risk and the resultant liability has been subcontracted to another party and the other party was informed of the presence of the inherent risk, then no claim can be brought against the organisation in respect of that risk.
8. *Claim settled is once for all:* A claim once settled and discharged cannot be re-opened.
9. *Statutory limitation:* The organisation or its insurer will not entertain a claim after 3 years from the date of occurrence of the loss-producing event.

Employer's Liability Insurance

Employer's liability has two distinct parts. The first is the organisation's liability for damage, loss, injury or death caused to others by the negligence of its employees during the performance of authorised acts and the second is the liability towards the employees themselves to pay compensation in respect of injury, disease or death arising out of and in the course of employment. The first shall be covered partly by the public liability policy and partly by a motor vehicle policy and the second partly by a workmen's compensation poicy and partly by a group personal accident policy- the former for workmen and the latter for senior staff and officers.

Workmen's Compensation Insurance

The organisation as the employer is liable at law for accidental injury and death to its workmen in the course of employment. In the construction industry there is practically no occupational disease. However, that also comes under employer's liability. These liabilities may arise out of: negligence on the part of the organisation or negligence of the workmen themselves or their fellow employees; unsafe work practices and poorly maintained tools, tackles and equipment.

Rates of compensation for these liabilities are fixed under a statutory law - the Workmen's Compensation Act which, incidentally, is known by the same name in most countries.

Insurers give full cover for liabilities under the Workmen's Compensation Act. Premium rates would depend upon the nature and type of construction work undertaken. The rates are high for hazardous jobs. If subcontractors' workmen are to be covered, their total wages also shall be considered for the purpose of premium calculation. Policy shall be issued on the basis of an estimate of annual wages, subject to adjustment at actuals at the close of the year.

It is possible to extend the policy to cover medical and hospitalisation charges resulting from accidents.

For accidents directly attributable to the following, the organisation shall have no legal liability to pay compensation.

- a workman having been under the influence of intoxicating drinks or drugs
- the workman wilfully disobeying an order or rule expressly framed for the purpose of securing the safety of men
- the wilful removal or disregard by the workman of any safety guard or other device provided for securing his safety
- the wilful misconduct of a criminal or quasicriminal nature or wanton and reckless act.

No compensation will be due for a natural death while on duty.

Similarly, for an injury or death while on duty by an accident that is foreign to or unconnected with the organisation's work, there will be no compensation. For example, while engaged on a chimney erection if a workman stretches himself to pluck mangoes from a nearby mango tree and in that process falls and dies, no compensation will be payable for the death.

Valenti Non fit Injuria mentioned under Third Party/Public Liability Insurance is not applicable to workmen, while on duty.

It is clarified for the readers' guidance that a workman is not in the course of his employment from the moment he leaves his residence. The course of employment starts only from the time the workman reaches the work place or a point where he has to report.

If a workman dies, beneficiaries for the compensation are: his widow, minor legitimate son, minor legitimate daughter or a widowed mother, in that order.

Workmen's compensation insurance scheme covers the expenses for carriage of dead body and funeral, without extra premium.

In some states, employees' state insurance scheme operates. But that is not applicable to construction employees.

Under the Workmen's Compensation Act, in some countries, as in India, there is an obligation to file an annual statutory return of compensation payments. The organisation shall keep a note of the due date and file this return without fail.

What is said under "accident while on duty" in the organisation's personnel manual may also be read while dealing with insurance.

Group Personal Accident Policy

The Workmen's Compensation Act applies only to the categories of employees coming under the statutory definition of "workmen". Senior staff and officers above the grade of workmen also need the organisation's protection although there is no statutory requirement. They shall be covered by a group personal accident policy which is an annual accident policy. Its scope of coverage is almost the same as that of the workmen's compensation policy. It can be extended to provide for round the clock cover at a higher premiunrn. Premium is linked with the amount insured for each individual. To ward off the possibility of a deceased employee's beneficiary claiming a larger amount of compensation, it is advisable to let every employee know in advance the amount he is insured for. It has become more or less customary to insure an employee for a sum equal to his 60 months' salary.

Health Insurance

Workmen's medical treatment and hospitalisation resulting from accidents while on duty can be included in-workrnen's compensation insurance as an extension. Other medical treatments and hospitalisation of both the workmen and staff, if there is a provision in service rules to that effect, can be insured under a separate policy. This is purely a financial insurance. It may be tried for a year and then the overall economy re-examined before renewing. Many construction companies have tried and discontinued the health insurance and even the workmen's medical treatment extension, as they proved uneconomical. For a very few top personnel alone it might be beneficial. Joy P.K,. (2005).

Motor Vehicle Insurance

Every motor vehicle, whether light or heavy, running at the construction site, shall be insured. Although only third party risk is the statutory requirement, the organisation shall insure its vehicles for all risks under comprehensive policies. This is because of the high risk rate at the construction sites. The comprehensive policies cover: third party liabilities, damage to the vehicle itself, passengers in the vehicle; driver and the driver's legal liability. But they shall exclude the following:

- Use of the vehicle for any purpose other than those declared in the policy or in a later endorsement. For example, if a private carrier is used as a public commercial carrier, any accident occurring during that time will be outside the policy coverage
- Any liability covered by another insurance
- Use of the vehicle to tow another vehicle not insured with the same insurer
- Driving of the vehicle by unlicensed persons.

While the insurer will pay the organisation the compensation to repair or replace the insured vehicle without delay simply on the basis of estimates and surveyor's report or in the case of theft, on the basis of police report, any compensation to others will be paid only on the basis of either a motor accident claims tribunal decision or a court decision.

Every vehicle shall be insured for its replacement value. For risks to passengers, separate indemnity may be fixed in the policy at an extra premium. But larger amount may be paid towards public liability on the basis of tribunal/court decision. The organisation's own claims will be always subject to an 'excess'.

Motor vehicle insurance would cover medical treatment, ambulance, nursing, surgical and hospitalisation expenses for accident victims.

Although it is one of the principles of insurance that the insured shall not admit any liability on behalf of the insurer, in the event of motor vehicle accidents, the insured is allowed to provide at the insurer's cost, the immediate and necessary medical and surgical aid to the affected persons.

Motor vehicle insurance premium rates are not subject to bargain as they are controlled by "tariff". But fleet rating with special discount may be available if several vehicles are insured together under one block -policy with individual value for every vehicle.

If a vehicle is off-road for a long time, it may be worth discontinuing its policy unless it enjoys a substantial no-claim discount which might be discontinued on an interruption of the insurance.

While a policy cannot be transferred to another owner when a vehicle is sold, an owner can retain a policy substituting or replacing the vehicle.

Fidelity Guarantee Insurance

This class of insurance is a guarantee that if the organisation suffers any direct financial loss arising out of an employee's dishonesty, the insurer will pay such loss within the limitations prescribed by the policy conditions. Cashiers, store-keepers, local purchasers and other employees who handle the organisation's cash and valuable materials and documents shall be covered by this insurance.

Fidelity guarantee policy shall have a schedule showing the names, titles, duties and individual amounts of guarantee covered by it. Instead of names, this insurance is available by "positions" also so that changes in the persons holding the particular positions would not affect the cover.

The risks covered by fidelity guarantee are: acts of forgery, embezzlement, larceny and fraudulent conversion of money or materials for misappropriation.

The organisation's obligations under this class of insurance will include the following:

- not to allow accumulation of cash in the hands of the insured employee
- not to entrust cash and accounts together to one person without daily checking by another
- not to allow huge cash transactions
- to insist on the issuance of printed and consecutively numbered cash receipts with counterfoils, for all cash collections
- to involve at least two employees in every cash transaction
- to ensure internal and professional audit at the site at short intervals
- surprise cash checks by a director or chief executive at irregular but frequent intervals
- not to entrust a re-employed employee with cash and/or valuables, if he had previously left the organisation of his own accord.

This cover is against direct pecuniary loss only and not against consequential losses. The insurer's maximum liability in respect of every person or position will be the amount guaranteed against the individual or position.

Cash-in-Transit and Cash-in-Safe Insurance

The organisation shall take these two classes of insurance either jointly or separately. However, the risks are distinct and they are exactly what the names indicate. The limits of cover are: the maximum amount of cash carrying in any single transit and the largest amount of cash held in the safe on any single day, as declared, in the policy. Cash-in-transit policy is issued on the basis of an estimate of annual cash carrying and an advance against estimated premium, subject to adjustment when the year has passed. Premium for cash-in-safe must be fully paid in advance. Joy P.K,. (2005)

Cash-in-transit is a first loss cover providing for payment of full claim without bringing in the questions of under-insurance or average. Cash-in-safe insurance is strictly a burglary and house breaking cover and the loss is to be evidenced by cash book. Cash-in-transit will also cover 48 hours of cash keeping for disbursements after the cash reaches the site.

Principal exclusions are: shortage due to error or omission; loss arising from entrusting the cash to a person who is not an employee of the organisation; losses covered by other policies including a fidelity cover and loss caused by strike, riot and civil commotion for which a policy extension may be available.

Pedal Cycles Insurance

The organisation may cover all the bicycles used at the site for: loss or damage to the cycles themselves and legal liability for bodily injury or property damage caused by the use of the cycles, to members of the public.

Contract Guarantees

Householders' Comprehensive Insurance

If a director's or manager's service condition includes the organisation' responsibility to insure his household and personal effects, a multiperil policy which is a comprehensive domestic multirisks insurance shall be taken, after negotiation with the insurer.

War Risk Insurance

War risk is an exclusion in all classes of insurance policies. But the respective governments will directly provide insurance against war risk. In the event of necessity, the government concerned will entrust one of the departments with this responsibility or form a separate bureau to handle war risk insurance. It would accept proposals and premium and issue covers setting forth the applicable terms and conditions. It would contain guidelines to lodge claims and obtain compensation. When there is a war threat or promulgation of a state of emergency, one should keep a watch for a government notification concerning the war risk insurance. This should not be neglected.

Insurance Procedure

The procedure or the process of taking an insurance policy is as follows:

Proposal and Premium Rating

For different kinds and classes of insurance, different forms of proposal are used. They are supplied by insurers. A completed proposal form containing the prescribed information is an offer. Based on the proposal the insurer would quote his rate of premium. There are two classes of premium - 'tariff' and 'non-tariff'. Tariff rates are decided by a Tariff Committee consisting of members representing all the major insurance companies.

Tariff rates are usually fixed for most common and statutory insurance like fire, motor vehicle and public liability. In India, almost all classes of insurance are under the purview of the tariff committee. All insurance companies are required to charge the same premium rates for those classes of insurance. Of course there are discretions to offer discounts and special rebates. Also insurers offer special services while competing with others. Non-tariff rates can be subjected to negotiation between individual insurer and the organisation. The methods of premium rating are different for different classes of insurance. For cash-in-transit, it is a rate per mile (thousand) of the actual cash carried during the insurance year. For cash-in-safe, it is a percentage of the maximum cash holding. For workmen's compensation, the premium is a certain percentage of total actual wages paid during the insurance period. Current rates may be obtained from the organisation's insurers. Where premium is to be calculated at the end of the insurance year, advance premium of not less than 75% must be paid on the basis of estimate.

When the insurer accepts the proposal along with the first premium, the insurance contract would come into existence. The insurer may issue a cover note immediately on receipt of the premium to be followed by the formal policy.

Policy Forms

Policy forms have some common features and some special features. Common features are those which are common for all classes of policies. They are: definitions, principles, obligations and procedures. Special features are the scope and conditions that are special and peculiar to the type of risks covered in each class of policy.

A policy may have various endorsements or appendages or riders attached to it carrying extended covers and/or special features.

If any particular risk or liability is left uninsured, it should be made known to the top management in writing. Depending upon the severity, the top management may decide to insure such risks by an extension of the relevant policy at an additional premium.

Renewals

The organisation shall keep a register of insurance policies. Such register shall have one or more pages allotted to each policy and contain: policy No; insured object or risk; amount insured; premium; period of insurance; particulars of claims lodged and settlements made. The dealing officer shall go through this register regularly for the purposes of claims follow-up and policy renewal.

While renewing, the value insured shall be reviewed and, if necessary, revised. On renewal, it should be examined if there is an entitlement to a no-claim discount. For renewal of work insurance for an extended contract period, the extra premium should be nominal.

Usually insurers send renewal notices. But the organisation shall not wait for renewal notices for its renewal action.

Termination of a Policy

Either the insurer or the organisation can terminate an insurance contract before its expiry by sending by registered post to the other party a notice of termination setting forth the reason. If no notice period is stipulated in the policy, a seven days' notice will be adequate.

Claim Procedure

Initial Notice

Immediately on the happening of any loss-producing event, an intimation must be sent to the insurer with reference to the relevant policy number. If there is a separate declaration in respect of the affected object under an open policy, the reference of such declaration also should be given. Notice forms are usually supplied by insurers. If not, a simple letter will suffice.

Simultaneously, a claim notice should be sent to carriers or any other party responsible for the loss or damage and its copy given to the insurer. In many cases, this will not be relevant.

Detailed Statement and Estimate of the Claim

On receipt of the notice, the insurer will allot a claim number and send a claim form to the organisation. It should be filled up with full details of the accident and given back to the insurer with estimates or quotations or proforma bills for repairs or replacements, where those are relevant.

Survey or Investigation

The insurer might depute his own expert surveyor to inspect the spot of accident, examine the affected object, discuss with the organisation's representative and the repairers or suppliers and make a report. Where a police investigation is in progress, the insurer may insist on getting a copy of the final report of the police on the findings. In the case of a death, he will need a copy each of the death certificate and postmortem certificate. Where the claim is in respect of a serious bodily injury, a medical certificate would be asked for. Sometimes the insurer's medical advisor may be asked to be associated with the medical examination of the affected person.

The survey or investigation will be aimed at ascertaining and assessing the actual nature of the event and the extent of liability of the insurer.

Settlement

The insurer will be quick to restore the organisation's own affected property to the pre-accident condition. Also the organisation's own employees' claims will be settled fast if the documentation is complete. But the claims of third parties will be settled only on reaching a compromised settlement or after getting a court's or tribunal's decision, whichever is first.

Claim bills supporting orders, vouchers and certificates wherever available for repairs and replacements must be given to the insurer to enable him to make payment. Where there is a possibility for the insurer to recover the loss from a carrier or other parties, a letter of subrogation shall be taken by him from the organisation authorizing him to receive payment on its behalf.

When a claim is settled, the following also shall be given to the insurer:

1. For repairs to the organisation's own automobile or equipment, a satisfaction voucher signifying that the repairs are to its satisfaction.

2. For any major payment of workmen's compensation, an agreement of settlement signed by the claimant and the organisation's authorised officer and witnessed by the local government officer administering the Workmen's Compensation Act. For payment of normal compensation for loss of wages, formal receipt signed by the workman concerned and countersigned by the organisation's representative will suffice.
3. For discharge of a third party's liability, a legally worded discharge form signed by the claimant and witnessed by a respectable person indicating his full address. Discharge form will be given by the insurer.

Construction Insurance

This section will review a number of the more common and usual types of insurance which would be of general interest to owners, contractors, subcontractors, construction managers, developers and design professionals.

Worker's Compensation and Employer's Liability Insurance

Worker's compensation insurance is designed to provide the statutory benefits required by state law to an employee who is hurt or killed as a result of employment. Some states and the Canadian provinces have established mandatory funds which employer must utilize unless they can qualify as self-insured. Some states have established non-mandatory funds and in these jurisdictions the employer may choose be private insurance and the state program. In most states, however, the preponderance of worker's compensation insurance is provided by private insurance companies.

Initially, contractors without state experience enter the worker's compensation rate pool at the manual or entry rate. Based upon the contractor's own experience, the manual rate is adjusted up or down by an experience modifier which will result in lower rates for employers with excellent safety records and higher rates for those with poor safety records. In the retrospective plan, employers will initially pay based upon their experience modifier. Their accident loss record on the project will be tabulated and the contractor will either receive a credit or be charged an additional premium depending upon performance. Worker's compensation policies provide statutory limits to comply with the laws of the state.

Employer's liability insurance is usually written in combination with worker compensation insurance and provides broad coverage for work-related injury or death of an employee which is outside the scope of the worker's compensation policy. Joy P.K,. (2005).

Comprehensive General Liability Insurance

Comprehensive general liability insurance generally insures against liability imposed by law for negligent acts occurring in the conduct of the business which result in bodily injury or damage to the property of others. Typically, the basic policy can be endorsed to include coverage for owner's and contractor's protective insurance, product, and completed operations, blanket contractual, personal injury (libel, slander, etc.) and will frequently include coverage for liability arising from the insured's automobiles.

Contractors will frequently obtain excess liability insurance to protect themselves in the event of a catastrophic loss. Amounts of excess liability insurance are often dependent upon the perceived maximum risk as developed from a risk analysis program. Joy P.K,.(2005).

Contractual Liability Insurance

Contractual liability insurance protects the contractor when he assumes the legal liability of others, generally the owner, designer or other designated party. Most contracts will contain some form of indemnification clause or hold-harmless provision in which the contractor is required to assume the potential liabilities of others.

Professional Liability Insurance

Professional liability coverage is designed to provide protection to architects and engineers from liability based upon professional errors or omissions in performing design, construction management or other services. A number of public and private owners require that architect/engineers, other design firms, construction managers and sometimes other consultants carry a specified minimum amount of this type of insurance.

Builder's Risk Insurance

Builder's risk insurance covers the cost of damage of a physical nature to a building or other component of a construction project. Coverage applies to material and equipment not yet 'incorporated into the work when located on site or in transit to the site.

Current policies generally include the interests of the owner, the general contractor and all subcontractors and material suppliers. While policies may be written upon a "named peril" form which insures against only those perils specifically named, such as fire, windstorm, vandalism, etc., the more common form is "all risk" which provides protection from all perils except those specifically excluded. Most common exclusions include war, nuclear, faulty design, faulty materials, faulty workmanship, earthquake and flood. Some exclusions can be eliminated for an additional premium. Deductibles are commonly used, the amount depending upon the size and type of the work.

Equipment Floater Policy

The equipment floater provides coverage for damage to mobile and stationary construction equipment which is not generally subject to vehicle registration. The floater does not normally provide for liability and property damage insurance for cars, trucks and other equipment subject to the motor vehicle licensing laws, which are covered under other liability policies. It usually provides coverage for damage to the equipment whether located at the job site, in transit, or at the contractor's yard. Premiums are normally computed periodically based upon the value of the equipment for contractor-owned items and based upon the rental rate for rented equipment. Newly acquired equipment is insured upon acquisition and reported for premium purposes in forthcoming reporting period.

Payroll Taxes and Insurance

Other obligations of construction and other employers imposed by law include payroll taxes and insurance. Included is social security tax for both employers and employees. Other payroll based taxes include state unemployment tax, employment training tax, and other items depending upon individual country/state requirements.

Wrap-Up Insurance

Particularly in the private sector, a number of owners have established wrap up insurance programs to cover the owner, contractors, subcontractors and sometimes construction or program managers as well as the designers. Several major public programs including rapid transit, wastewater and other programs have utilized wrap-up insurance where state law does not prohibit their use. Basic policies can include comprehensive general liability insurance, excess general liability, builder's risk, worker's compensation, and occasionally errors and omissions insurance. Automobile liability insurance, equipment floaters and the deductible or self-insured retention normally continues to be the responsibility of the individual insured.

Considerable controversy exists within the industry regarding wrap-up programs. Certain large insurance brokers specialize in putting together wrap-up insurance pro grams for owners on major projects. Advantages claimed for the owner are lower insurance costs, control over the insurance program, standardization of risks and centralization of responsibility.

Contractors as a group are normally not in favor of the owner placing worker's compensation insurance and receiving the benefits of any savings on retrospective policies due to superior safety performance on the part of the contractor. Contractors with minimum experience in the state or those with poor safety records will pay more for this type of insurance than will experienced local contractors with excellent safety records. When this type of insurance is mandated by the owner to be excluded from contractor bids, the safe contractor with a favorable experience modification record will not normally benefit in his ability to bid lower than his less safe competition because of his favorable record.

The best-qualified contractors, as a group, generally prefer to place general liability insurance with their own broker, usually under continuing policies that automatically cover new projects as they are obtained. When the owner supplies this insurance, normal business practices can be disrupted and contractors must deal with a new set of administrators, insurance adjusters, safety inspectors and other representatives from several insurance companies. On the other hand, contractors often like to see the owner pick up builder's risk and excess liability insurance for the project providing subrogation rights are waived by the insurer. Joy. P.K,. (2015).

Construction Surety Bonds

Surety bonds provide a third-party guarantee for the performance of construction contract obligations. This section will discuss the type of bonds most often used in public and private construction projects, including bid bonds, performance bonds labor and material payment bonds.

Public Works

Form of Performance Security (Guarantee)

Bank Guarantee Bond-Format – I: (Ref. *General Conditions of Contract-2019-Construction Works)*

In consideration of the President of India (hereinafter called "The Government") having offered to accept the terms and conditions of the proposed agreement between..................................and .... (hereinafter called "the said Contractor(s)") for the work... (hereinafter called "the said agreement") having agreed to production of an irrevocable Bank Guarantee for Rs. .... (Rupees .... only) as a security/guarantee from the contractor(s) for compliance of his obligations in accordance with the terms and conditions in the said agreement.

Private Work

In private work, bonding requirements are at the owner's option. Some owners routinely require payment and performance bonds while other owners depend upon a program of prequalification to restrict bidding to financially sound contractors who have demonstrated that they have the required financial, resource, and experience qualifications for the proposed project. Bid bonds are also optional in the private sector. Some owners waive bid bonds and require that the contractor will furnish a payment and performance bond only if requested by the owner, in which case the owner will reimburse the contractor for the premium. Private owners often combine performance and payment bonds into a single instrument in the amount of 100 percent of the contract price. Joy. P.K,. (2015).

Foreign Work

On most foreign work, performance and payment bonds are not acceptable to governmental clients and are replaced by an irrevocable letter of credit guaranteed by an acceptable financial institution. This letter of credit offers protection to the owner who can draw on it in the event of default or nonperformance by the contractor.

Performance Bonds

The performance bond is not insurance. It does, however, guarantee performance of contract obligations to the owner as set forth in the contract. The amount of performance bond represents 100 percent of the contract price, which is guaranteed to be made available by the surety to complete the contract in the event default by the contractor. In the event of default, the surety is obligated to either complete the work or to make an arrangement with the owner to pay for the cost of completion of the work less the balance of the contract price previously unpaid.

The surety of course retains by agreement the right to recover costs from the contractor for his failure to complete. Therefore, contractors do not normally default on work unless they are financially unable to finish the work. In the event of default, the surety has several choices. It can provide additional funds to the contractor under an agreement to permit the original contractor to complete the project. It can take bids from other contractors or he can negotiate with a favored contractor to complete the work. Alternately, it can arrange for the owner to take bids and to reimburse the owner for any additional costs up to the value of the bond. A number of construction contracts give the owner the right to take over the work, including the contractor's materials, plant and equipment, in the event of default and to complete the work as he chooses. The contractor and his surety would remain responsible for the additional costs incurred by the owner up to the value of the performance bond. Joy. P.K,. (2015).

If the original contractor is capable of completing the work with financial assistance, bonding companies often find it less costly to provide prompt financial assistance rather than to assume the additional costs and delays resulting from shutting down restarting the work. When this method is feasible, all parties, including the owner may benefit.

Bid Proposal Bonds

Bid and proposal bonds provide that the contractor's surety will guarantee that the contractor will enter into a contract and provide required bonding if selected. Public agencies and some private owners will require this type of bond. The contractor normally does not pay any premium for bid bonds. Surety remuneration comes from furnishing the performance and payment bonds if the bid or proposal is successful. Amounts for bid bonds usually are between 5 percent to 20 percent of the amount bid, expressed as a percentage, which means that the bond can be issued prior to the bid date which helps to eliminate further complexities involving last-minute adjustments to the bid by the contractor.

Payment Bonds

Labour and material payment bonds provide for prompt payment to all those furnishing labour and/or materials to the work. This type of clause has been interpreted by courts to over unpaid bills to a subcontractor even though payment has been made to the subcontractor by the general contractor. To help mitigate this situation, prudent contractors can:

Pre-qualify and select subcontractors and materials suppliers with care; require subcontractors to furnish evidence that they have paid for labor and materials furnished; require subcontractors to furnish a performance and material payment bond; and require lien releases prior to making final payment.

Additional liabilities may be incurred through mechanics lien laws and stop-notice requirements, which vary from state to state.

Combination Payment and Performance Bonds

Private work often utilizes a combination payment and performance bond in amount of 100 percent of the contract price. As the project proceeds to completion, the exposure of the surety to complete the work in the event of default normally decreases with time, except in unusual circumstances. However, the amount of bond does not decrease with the amount of completed work, but remains con throughout the life of the contract. Premiums for the combined bonds are normally the same as for separate performance and payment bonds. Premiums for bonds for cost-reimbursable negotiated work where required are substantially smaller.

Mechanic's Liens and Stop Notice

Every state has mechanic's lien laws which convey a right to secure payment work performed in the improvement of real property. Lien laws are subject to strict compliance by the courts and familiarity with such laws in each individual state is required. Mechanic's liens are recorded in a manner similar to a mortgage and represent a claim or lien against the real property.

Stop notices are generally utilized on public property which is not subject to mechanic's lien. A stop notice tells the public owner that bills have not been paid. State statutes require that the owner must withhold payment to the prime contract until the obligation is paid. Some states exempt the owner from lien liability provided that the general contractor has furnished a payment bond. Architects, surveyors and engineers are given lien rights in some states, subject to certain conditions.

Where state law permits, construction contracts can require that the right to lien be waived by general contractors or others. Many construction contracts require that the contractor supply a signed release and waiver of lien prior to receiving final payment from the owner. Joy. P.K,. (2015).

Review Questions

1. Explain the meaning of uncertainty in construction industry.
2. What are the courses when risks may be arising in construction industry?
3. What Measures you would take for Risk Avoidance and Reduction?
4. What are the various Classes of Insurance Relevant to a Construction Organisation?
5. Write an essay on Insurance Procedure.
6. Write short notes on the following:
   - Insurable and measurable interest
   - Subrogation
   - Excess or deductible franchise or self insurance
   - Performance Bonds
   - Worker's Compensation and Employer's Liability Insurance
   - Combination Payment and Performance Bonds
   - Mechanic's liens and stop notices

9

Critical Factors for Success of a Project

The construction industry constitutes those individuals and organisations who are engaged in the designing, manufacturing, and maintenance of the construction projects. Construction Industry is one of the most vibrant and evolving sector of Indian Economy providing livelihood to number of people. In the past few years, this industry internationally has laid increasing dependence on accentuating the prominence of construction to the economy of the country. Engaging an extensive variety of practices and methodologies, the intention of the industry is to influence policy makers to inspire development of construction related activities or, at a least, to avoid implementing policy actions that would discourage the companies to get involved in construction projects.

Project Performance Measurement

In every walk of life, performance is required to be measured either in order to draw valuable inferences about an individual or to compare the competing individuals.

Measuring the performance of a project in terms of success or failure is, in fact, a difficult task. This cause success/failure has different meanings for different participants. Further, defining the success of project has always been full of ambiguity and there is no universal definition of success. Success is viewed from the different perspectives of individuals and the goals related to a variety of elements including technical, financial, educational, social and professional issues. Each industry, project team, or individual has a definition of success. Failures and successes are relative terms and they are highly subjective.

The definition of success or failure can even change from project to project. Success for one participant may be failure for another (Iyer and Iha 2004, 2005).

An architect may consider success in terms of aesthetic appearance; the engineer in terms of technical competence; the accountant in terms of rupees spent under budget; the human resources manager in terms of employee satisfaction; and the chief executive officer in terms of the stock market. For same project, notice the different yardsticks employed by different participants to measure the success.

The perception of success or failure is also time-dependent.

In spite of the difficulties involved, some researchers have tried to define the success of a project.

The project is considered an overall success if the project meets the technical performance specifications or the mission to be performed, and if there is a high level of satisfaction concerning the project outcome among: key people in the parent organization, key people in the project team, and key users or clientele of the project effort.

Traditionally, success is defined as the degree to which project goals and expectations are met and the project requirements are fulfilled. However, modern projects involving multiple designers, contractors, sub-contractors, construction managers, consultants and specialists from different disciplines, and increasing domain of project requirements have compounded the problem further, and understanding the success of a project has become all the more complicated.

From the above discussion, one can realize the difficulty in measuring the performance of a project in terms of success and failure. Nonetheless, there are certain criteria that are in vogue for measuring the performance of a project.

Criteria for Project Performance Evaluation

Criteria are the set of principles or standards by which judgement is made and are considered to be the rule of the game. Traditionally, project performance is evaluated using schedule, cost quality performances. Subsequently, different research have proposed different sets of success evaluation criteria in addition to the iron triangle.

Budget or Cost Performance

The project if completed at or under the contracted cost is characterized as a successful project. The co criterion could be measured in terms of cost over/under-run as a percentage of initial budgets.

Schedule or Time Performance

The project if completed on or before the contractual finish time is a successful project. The time criterion could be measured in terms of over/under-runs as a percentage of the initial plan.

Quality Performance whether Specifications have been met

If the completed project meets or exceeds the accepted standards of workmanship in all areas and co user's expectations, it is regarded as a successful project. In other words, 'the project must produce what it is required to produce. Quality typically includes such measures as the amount of rework required.

Safety

If the project honours health and safety rights of the people involved with the project by ensuring safe conditions, it is regarded as a successful project.

Dispute

If the project is completed with the least number of litigations resulting from disagreements among participants, it indicates that the project is a successful one. Dispute could be measured in terms of either the number of disputes or the monetary value involved in the dispute, or through qualitative measures such as whether there are minor or major disputes, and so on.

Stakeholder Satisfaction

If the completed project meets or exceeds the stakeholders' goals, it is termed as a successful project. The key stakeholders include architect, client, contractor, engineer, project manager and subcontractor/vendor. The project is also considered successful by employees if it has enabled personal growth for the employ of contractor and owner organizations.

Other Criteria

Some other criteria used to measure the performance of a project are:

- Whether the performance of constructed facility is as per the expectation and specifications, and whether the design goals have been met
- Whether the project has met the functionality requirement as expected
- Whether any technical innovation has been achieved during the implementation of project
- Whether the efficiency of project execution was as per requirement
- Whether managerial and organizational expectations have been met during the project implementation
- Whether the financial performance/profitability is as per the expectations of the stakeholder
- Whether the project has made any substantial and positive impact on the customer
- Whether the project implementation has influenced the involved organization's business favourably and whether it has helped the organization better prepare for the future
- Whether the project has benefited the customer, and whether it has led to favourable influence on the National infrastructure

Construction projects are vital for the growth of a nation and so is the need to make all-out efforts in ensuring the successful outcome of a project. In the next section, we discuss some of the attributes that are considered key to ensure the success of a project.

Project Performance Attributes

Project attributes are the variables that influence the outcome of a project. The attributes can be people (project participants and their traits), resources, technology, working environment and system, or tasks. Project success is repeatable and it is possible to find certain success attributes; has been the genesis of many research works in this area. Also, there are certain attributes termed as failure attributes, which when present lead to failure of the project. Finding the success attributes and maximizing them is as important as finding the failure attributes and minimizing them. Accordingly, researchers have their energy into identifying success attributes and failure attributes, with the common objective of enhancing the chances of project success.

Success Attributes/Factors

The success attributes are classified under some major heads purely for ease in reading. The attributes are also referred to as critical success factors (CSF). CSFs are those key areas of activity in which favourable results are absolutely necessary for particular manager to reach his or her own goals.

Failure Attributes

One study concludes that choice of wrong project manager, unplanned project termination and unsupportive management are the main reasons for project failure. Another study in 1986 identifies that projects fail because of improper basic managerial principles such as improper focus of the management system, rewarding the wrong actions and lack of communication of goals. Inability to recognize this may lead to wrong selection of personnel for the pre-project and project phases, and ultimately lead to the failure of a project. Researchers have identified the factors responsible for using schedule overrun, cost overrun and so on, in different parts of the World. These factors have been taken from the studies conducted by several investigators and are summarized thus:

- Poor contract management - this may be due to lack of adequate experience and training at the senior management level
- Inadequate technical manpower
- Very low level of productivity
- Inadequate finances for short-term and long-term purposes, and absence of specialization
- Unforeseen ground conditions and changes in site conditions
- Shortage of materials and plant items
- Design changes
- Price fluctuations
- Inaccurate estimates prepared by contractor
- Too many change orders (variations of works)
- Fraudulent practices and kickbacks
- Poor site management and supervision
- Slow speed of decision-making involving all project teams
- Poor scope definition

Case Study By Ramlee, N *et.al.*, (2016)
(Critical Success Factors For Construction Project)

In developing countries such as Malaysia, construction industry directly and indirectly related to a big portion of the economy. The construction industry plays a major role in the economy as its share of the Gross Domestic Product (GDP) and its interaction with other sectors of the economy. When the project complete on time, on budget and achieves the performance goals, it is considered as success. However, there are a few projects which were not completed on time.

Therefore, there is a need for the organizations that involve in construction projects to come out with the strategies on how to guide the project until it becomes success in the future.

The success of a construction projects is an important issue for most of the governments, users and communities. In modern construction projects there are significant challenges for both the clients and contractors to deliver the project successfully due to increasing complexity in design and the involvement of stakeholders. In the project management literature, project success has been widely discussed by many researchers. Most of the studies in project success have been focus on dimensions in how it is measured and other specific factors influencing the project success.

Project will be considered as success when the project is completed on time, within budget and the quality is satisfied by all. Success also can defined as much better results than the expected or normally obtained in term of cost, schedule, quality, and safety. The meaning of 'success' itself has undergone many changes due to involvement of so many stakeholders in nowadays complex project environment. The concept of project success is developed with criteria and standards to help project participants to complete projects with the most desirable results.

There are four separate dimensions of CSFs in project success. The first dimension is meeting design's goals which is refers to the contract that was signed with the client. The second dimension is the benefit to the end users which is refers to the benefit to the customers from the project end products. The third dimension is benefit to the developing organization and refers to the benefit gained by the developing organization as a result of executing the project. The last dimension is the benefit to the national technological infrastructure, as well as to the technological infrastructure of the firm that was engaged in the development process. The combination of all these dimensions gives the overall assessment of project success.

The project success is divided into four dimensions of time-dependent. The first dimension is the period during project execution and right after project completion such as completion on time and budget. The second dimension can be assessed after a short time when the project has been delivered to the customer and it is focused more about satisfaction, functional of product and technical specifications. While, the third dimension can be assessed after a significant level of sales has been achieved in one to two years and it will show the organization performances. The last dimension can only be assessed three to five years after project completion, it about preparing for the future.

The study of project success and the critical success factors (CSFs) are considered to be a means to improve the effectiveness of a project. The measurement of success factors has been proposed into five main groups namely project management action; project procedures; project related factors; human related factors; and external environment. The five main groups of success factors have discussed about cost, time, quality, management, technology, safety, organization and environment in the construction project.

The success criteria for a construction project is not only evaluate the cost, time and quality as a success factors but also including the successful of project management, organizational success and the customer satisfaction. Another perspective was stated were the importance of organizational's planning effort, project manger's commitment and safety precaution in order to complete the construction project complete by meeting cost, meeting on time, follow the schedule accurately and meeting the quality needed to ensure the project success. Reviewing of the relevant literature suggests that different criteria were hypothesized by different researchers. The literature reviews by different researchers have been summarized and shown in the table 9.1.

Table 9.1: Summary of Success Factors by Previous Researchers

Author	Critical Success Factors (CSFs)									
	Cost	Time	Quality	Satisfaction	Management	Safety	Technology	Organization	Environment	Resources
Sadeh *et.al.* (2000)	X	X	X	X	X		X			
Paul Steinfort (2007)	X	X	X	X	X		X	X		
Alzahrani and Emsley (2013)	X	X	X		X	X		X	X	X
Shenhar et.al (2001)	X	X	X				X	X		
Lim and Mohamed (1999)	X	X	X				X			
Dr. Parviz (2003)	X	X	X	X	X					
Chan et al. (2004)	X	X	X		X	X	X	X	X	
Sigurour (2009)	X	X	X	X	X			X		

Note: 'X' is refer to the conclusion of CSFs from previous researchers

Discussion

There are many factors that can make a project is success or cause it fails. Table 9.1 has shown the summary for project evaluation criteria by previous researchers. Table 9.2 has shown the percentage of Critical Success Factors that found from literature review. Almost all researchers stated that cost, time, quality and management as the Critical Success Factors of their literature review. From the summary of the literature review, it may be concluded that the project is success when it complete by meeting cost, meeting on time, follow the schedule accurately, meeting the quality needed and managed by the best team members. Technology also has been one of Critical Success Factors that contributes to a construction project success.

Table 9.2: Percentage of critical success factors

No.	Critical Success Factor	Total	%
1.	Cost	8	16.3
2.	Time	8	16.3
3.	Quality	8	16.3
4.	Satisfaction	4	8.2
5.	Management	6	12.2
6.	Safety	2	4.1
7.	Technology	5	10.2
8.	Organization	5	10.2
9.	Environment	2	4.1
10.	Resources	1	2.0
	Total	49	100

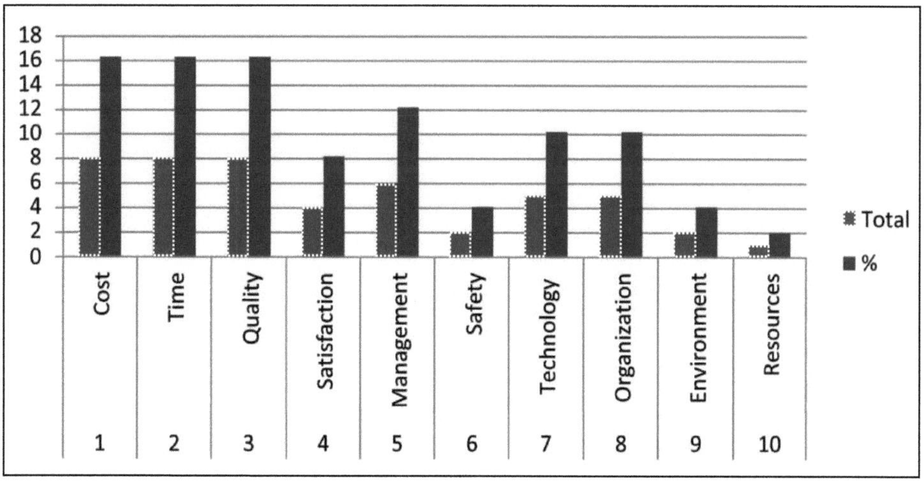

Fig. 9.1: The Percentage and Total of Critical Success Factors (CSFS)

Case Study By Afshin Pakseresht & Dr. Gholamreza Asgari,. (2012) (Determining the Critical Success Factors in Construction Projects: Analytical Hierarchy Process (AHP) Approach)

Critical success factors are those measures which result in the organization success. Of course since every organizations circumstances is unique, identifying the critical factors has been the subject of several studies.

So, by identifying and ranking the critical success factors, this research purpose is to give a solution for the executive planning of the project of Pars Garma construction and industrial company. It is obvious that identifying and ranking the critical success factors of this supreme company, which works in various fields like construction of dam, irrigation and drainage networks, road construction, bridge construction, tunnel excavation, construction of concrete and metal heavy building, and mass construction of the residential buildings in cooperation with Ministry of Defense. Roads and Urban Development, Energy, Oil, Social Security Organization, Organization of Construction Engineering and Tehran Municipality, could result in achieving reliability and more desirable record and lead to this organization's more success.

Thus, identifying the factors could not provide the organization's success solely.

The critical success factors are indicative of the subjects which could make the organization successful; if there is deficiency or lack in these fields, the organization would fail in achieving its goals.

The organizations should concentrate on the more efficient factors or so-called the critical success factors considering their restrictions.

In line with it, the main purpose of this research was to recognize the effective factors in the success of Saba Residential Tower Project and comparing it to the other projects of Pars Garma Company and also weighting and prioritizing them as the critical success factors of construction projects programming of this company.

In 1986, Ashley identified these seven factors as the success factors:

1. Construction activities programming
2. Design planning
3. Project manager commitment to the goals
4. Project team motivation
5. Project manager technical capabilities
6. Control systems
7. Definition of work and its field

He also identified theses six criteria as the success criteria:

1. Budget performance
2. Schedule performance
3. Employer satisfaction
4. Task-orientation

5. Contractor satisfaction
6. Project manager satisfaction

It is noteworthy that the differentiation between success criteria and success factors is also important. The success criteria are the measures based on which the project success or failure is judged; while the success factors are those entered into the management systems and result in project success directly or indirectly.

Then in 1987, Pinto and Slevin offered a project model and ten determining factors as below:

Project mission, scheduling, technical tasks, user consultant, user reception, feedbacks and reports, communications, fault detection, senior manager support, personnel (requirement, employment and training).

In 1996 Walid Belassi and Tukel Oya Icmeli also divided the critical success factors into four main groups in a new format:

1. Project dependent factors
2. Team members and project manager dependent factors
3. Organizational structure dependent factors
4. External environment dependent factors

Studying the results proves that the project manager, management skills, team members, teamwork, their technical filed, project properties, viable environmental factors could be considered as the organization's critical success factors; although the degree of these factors' significance varies from one industry to another one. While in the information system of management and construction projects, project managers, management skills are the most important key factors and the environmental factors influence and conduct the construction tasks.

In an empirical survey on the relationship between project planning and project success, Dov Dvir declared that there is a certain and positive relationship between the amount of effort in determining the project's goals, job requirements and product's technical features on one hand and project success on the other hand and no effort should be spared in the first stage of the project to define the receivable requirements and goals, and this demand is not met without involving the customer or the final beneficiary in the processes.

Adnane Belout and Clothilde Gauvereau considered the following four dimensions as the success factors in 2003:

1. Project Result
2. Customers effects
3. Commercial success and conduction and preparation for the future
4. Identifying the beneficiary groups (stockholders, managers, customers and personnel) which are significant as well as critical success factors until different people wants to see success in different ways.

In 2004 Nguyen et al. identified five critical success factors among the twenty factors of projects success:

1. Competent project manager
2. Providing adequate financial resources to the end of the project
3. Competent and multidisciplinary project team
4. Commitment to the project
5. Access to resources

Muhammad Sagib et al. researched about construction projects in Pakistan and succeeded to identify 77 factors in 7 different groups and finally achieved to the prioritization of 10 critical success factors in construction projects of Pakistan.

Research Conceptual Model

The conceptual model of this research is shown in figure 9.2.

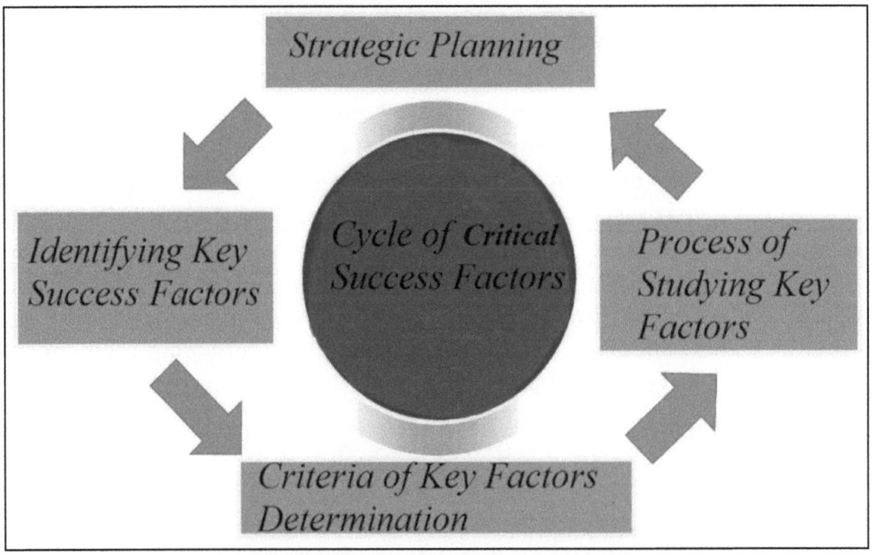

Fig. 9.2: Cycle of Critical Success Factors

Figure 9.2 shows that identifying the critical success factors is needed for strategic planning; identify the factors an appropriate strategy could be planned by determining the criteria and consequently by process of studying the critical factors.

Of course based on this approach and using the previous researches and experts opinion, to identify the critical factors the hierarchical structure consists of levels of objectives, criteria and choices as below:

First Level: Objective

Identifying the critical success factors and ranking them is the objective of this research.

Second Level: Criteria

To achieve this level's criteria, interview with project managers was done in addition to studying the principals of construction projects contracting and library resources. Finally seven criteria were selected:

1. Effective factors in project management
2. Employer related factors
3. Project logistics related factors
4. Consultant design team related factors
5. Contractor related factors
6. Project manager related factors
7. Environmental factors related to the project's business environment

Chart 1: Decision making hierarchical structure of selecting the critical success factors in construction projects

Third Level: Options

After examining the validity and reliability the Options (of the questions) related to every criterion, were also included in the questionnaire.

Methodology

This research is descriptive in terms of implementation and practical in terms of objective. The tools of gathering data are two questionnaires. The first questionnaire includes 32 questions in Likert type scale which identifies the critical success factors. The second questionnaire, which was distributed among the research experts after the first one, was in the form of AHP questionnaire and was designed for weighting the criteria and indexes related to the critical success factors.

The statistical population of this research consists of 100 people including the staff managers in the central office, project managers and technical experts in executing the company's projects, client's agent (employer and consultant) based in projects. The statistical population of Saba Residential Tower project consists of 35 people who are used for the critical success factors of Saba project and other projects of the company. Pars Garma Construction and Industrial Company currently have 7 active projects countrywide which human resources consists of 350 people directly and 700 people indirectly.

It is noteworthy that Cochran formula was used to calculate the appropriate and needed number. In Cochran formula:

N=population size, d=difference of characteristic real ratio in population the amount of which is considered 0.05 in this project and t=Z. The amount of variable in distribution is natural which is measured based on the table related to the considered probability level: considering the probability of 95%, it is considered equal to 1.96. Additionally, p=the distribution percentage of the characteristics in the population i.e. the ratio of the people who have the studied characteristic and it is considered equal to 0.9.

q=the percentage of the people who does not have that characteristic and it is considered equal to 0.1 in this research.

Analysis of the Finding

To analyze the obtained data of the first questionnaire, the software package of SPSS16 and to extract the weights related to the second questionnaire, the Expert Choice software was used. The data of the first questionnaire was examined by "z" test in the level of 95% and the factors the mean of which was less than 3 were omitted. Then they obtained criteria of the first questionnaire regulate in the form of an AHP questionnaire and the experts' opinions were collected and analyzed using the Expert Choice software. Table 9.3 shows the weights of the second level criteria of critical success factors. According this table, in the experts' viewpoint the project's logistics is the most important one. Project management is in the next place and paying attention to the business environment is in the last place.

It should be mentioned that the consistency rate of this model is 0.04

Table 9.3: The weights of the second level criteria of critical success factors

Name of criteria	Weight (percentage)	Priority
Project management	22.3	2
Logistics	24.1	1
Employer	13.2	4
Consultant design team	6.0	6
Contractor	12.0	5
Project manager	18.9	3
Business environment	3.5	7
Total	100	
Consistency Rate = 0.04		

The output results of the software (the weight of the third level sub criteria), which shows the prioritization of the effective factors in the success of Pars Garma Company's construction project based on weight percentage, are presented in table 9.4.

Table 9.4: The effective factors in the success of Pars Garma Company's construction projects

Overall Inconsistency=0.04

No.	Critical Factors	% Weight
(1)	(2)	(3)
1.	Technical and economic assessment of the project required resources	14.5
2.	Experience and executive records of the project manager	10.2
3.	Project strategic planning	8.8
4.	The executive experience of the contractor team about the project subject	6.5
5.	Ability of on-time decision making (employer's agent in project)	6.3
6.	Project control management	6.1
7.	Prioritization of purchasing the needed items considering the project schedule	5.0
8.	Mechanism of forming an experienced technical-legal team at the time of contract	4.6
9.	Authorization in Financial decision makings and cost control considering project type and size	3.7
10.	Mechanism of financial payment in relation to the project commitments and Plan	2.9
11.	Effective implementation of safety, health and environmental standards	2.8
12.	Experience of consultant design team	2.7

(Table Contd...)

(1)	(2)	(3)
13.	Ability of analyzing the project problems	2.6
14.	Mechanism of on-time decision making	2.4
15.	Ability of coordination and principled Agreements with consultant and employer's agent in meetings	2.4
16.	Providing a Safe Working Environment for Employees	2.4
17.	Executive Application of Indexes of Project Management Body of Knowledge (PMBOK)	2.2
18.	Mechanism of evaluating the and confirming the competence of the second hand contractors before signing the contract	2.2
19.	Clear and precise definition of project objectives by employer	2.0
20.	Using Experts Consultants to Design and Prepare the Project Executive Maps	2.0
21.	Mechanism of appropriate utilization to increase the efficiency of project machineries and equipments	1.7
22.	Mechanism of identification and management of project risks	1.6
23.	The amount of graduate studies in feasibility stage to decrease the future project problems	1.5
24.	Taking Inflation and economic crisis in financial programming into consideration	1.2
25.	The accuracy of plan analysis and evaluation by design team	1.0
26.	Reducing the errors and inconsistencies of the plans of project's second phase	0.7
	Total Weight	100

The critical success factors obtained from the 10 top effective factors in construction project success are presented in table 9.5.

Table 9.5: The critical success factor of Pars Garma Company's construction projects

No.	Critical Factors	% Weight
(1)	(2)	(3)
1.	Technical and economic assessment of the project required resources	14.5
2.	Experience and executive records of the project manager	10.2
3.	Project strategic planning	8.8
4.	The executive experience of the contractor team about the project subject	6.5
5.	Ability of on-time decision making (employer's agent in project)	6.3
6.	Project control management	6.1
7.	Prioritization of purchasing the needed items considering the project schedule	5.0

(Table Contd...)

(1)	(2)	(3)
8.	Mechanism of forming an experienced technical-legal team at the time of contract	4.6
9.	Authorization in Financial decision makings and cost control considering project type and size	3.7
10.	Mechanism of financial payment in relation to the project commitments and Plan	2.9

Based on the study, it may be inferred that:

Regarding the completed research, the critical success factors in Pars Garma Company and the obtained results of this research, the conclusions are as below:

- By more concentration on these factors, their effectiveness could be maximized.
- These factors assist the managers in utilizing the resources and facilities and consequently reforming their use pattern.
- They accelerate organization's success achievement by faster removing the existing barriers, especially in competitive conditions.
- They prevent probable perceptual errors and imaginations of the managers based on the importance of some factors to other factors and realize the factors prioritization and impede the realization of various errors.
- They help the organization to implement the strategic plans successfully.

The followings recommendations were made considering the obtained results of this study:

1. Spending enough time for the studies known as phase zero and ensuring the precise studies which leads to save the time, cost and quality of doing the projects.
2. Surveying the technical-economic briefing report and providing projects' credit by the employer or beneficiary.
3. Preparing the technical features, maps, and binding and technical documents
4. Prediction the necessary guarantees for doing the obligations of the contracting parties and considering the damages resulting from not accomplishing the commitments
5. Guaranteeing the payment of construction projects statements by finance and credit institutions and project insurance; of course, accomplishment of this task depends on the regulatory approvals and development of Iranian capital markets
6. The necessity to document the technical knowledge and experiences in the implementation of Iranian construction projects
7. Need assessment and education planning for empowering the effective and key employees, in line with strategy and critical success factors.

Holding course of knowledge management concepts and principles based on PMBOK standard is recommended.

8. Determining an especial team/committee, selected from the central office and senior authorities in every project, to survey and follow up the results of critical success factors in project (the recommended name: revision committee; it is offered to form in the periods of 6 months).
9. Forming a committee named CRM (Customer Relationship Management) in the central office, to follow the customers' feedback in order to increase the key customers' satisfaction.

Review Questions

1. What is the criteria for project performance evaluation?
2. What are Project Performance Attributes in terms of success and failure attributes?
3. Elaborate Critical Success Factors for Construction Project with the help of a case study.
4. How would you Determine the Critical Success Factors in Construction Projects by Analytical Hierarchy Process (AHP) Approach? Explain with the help of a case study.

10

Artificial Intelligence in Construction Industry

Today, Artificial Intelligence (AI) techniques are rapidly emerging as an important contributor to effective management. One of the greatest growth areas lies in the use of expert systems for supporting managerial decision processes. Expert system techniques, a major branch of AI have only recently begun to be applied to business problems.

Although AI has scarcely touched today's managers, it will begin to have dramatic and widespread impacts on their activities over the next few years. The application of AI play a major role in reshaping traditional notions of what organisations are, how they are managed, and how decisions are made. All of this, presents a major challenge and an important opportunity to today's managers and organisations. It will lead to tremendous increase in the productivity of managers and organisations. Those organisations in the forefront of applying Al methods to aid in management will have distinct competitive advantage over those that lag.

The construction industry is faced with many challenges that have hindered its growth and led to extremely low productivity levels when compared with other industries such as manufacturing. As a matter of fact, the construction industry is one of the least digitized industries in the world and most stakeholders acknowledge the age-long culture of resistance to change. The lack of digitisation and overly manual nature of the industry makes the management of projects more complex and unnecessarily tedious. The absence of adequate digital expertise and technology adoption within the construction industry has also been linked to cost inefficiencies, project delays, poor quality performance, and uninformed decision-making and poor performance in terms of productivity, health and safety. In recent years, it has become apparent that the construction industry must embrace digitisation and rapidly improve technological capacity especially with challenges of existing labour shortages, and the need to provide sustainable infrastructures. A foremost digital technology, Artificial intelligence (AI), has helped to achieve significant contributions to the improvement of business operations, service processes and industry productivity in recent years. The adoption of AI techniques has helped to enhance automated and provide better competitive advantages as compared to conventional approaches. The subfields of AI such as machine learning, natural language processing, robotics, and computer vision, optimisation, automated planning and scheduling, have been applied to tackle complex problems and support decision-making for real-world problems. For instance, in the manufacturing industry, the advent of the fourth industrial revolution, commonly known as Industry 4.0 is geared towards automation, data-driven technologies and the application of advanced AI techniques. It is evident that this revolution has led to significant process improvements, cost-efficiency, reduced production times.

Overview of artificial intelligence and its subfields (*Sofiat O. Abioye et.al., 2021*): The idea of developing machines exhibiting intelligence like humans traced back to several fields, which include philosophy, fiction, imagination, computer science, electronics and engineering inventions. Alan Turing's test for intelligence was a turning point in the field of AI as the test exceeds the traditional theological positions and mathematical conclusions about the possibility of intelligent machines. Sixty years later, intelligent machines are outperforming humans in so many domains such as learning by leveraging rapid advances in other technologies such as big data and computer processing power. The definition of *AI is the study of how to make machines do things, which at the moment, people do better"* captures the concept of AI perfectly. Fig. 10.1 shows an overview of the types, components and subfields of AI. There are three types of AI namely; Artificial Narrow Intelligence (ANI), Artificial General Intelligence (AGI), and Artificial Super Intelligence (ASI). ANI, sometimes referred to as weak AI, it is a form of AI where machines exhibit intelligence in a particular domain such as chess playing, sales prediction, movie suggestions, language translation and weather forecasts. AGI, sometimes referred to as strong AI, is concerned with making machines operate at the same level as humans. It refers to making machines that can solve a range of complex problems in different domains, control itself autonomously, with its own thoughts, worries, feelings, strengths, weaknesses and disposition. This remains a major goal of AI, but it has proved difficult to achieve and elusive. Artificial super intelligence (ASI) is concerned with building machines that exceed human capabilities across several domains.

The major components of AI, as shown in figure 10.1, are: (1) learning; (2) knowledge representation; (3) perception; (4) planning; (5) action, and; (6) communication. Some studies have also classified some of these components as tasks that can be carried out by AI as compared to human senses.

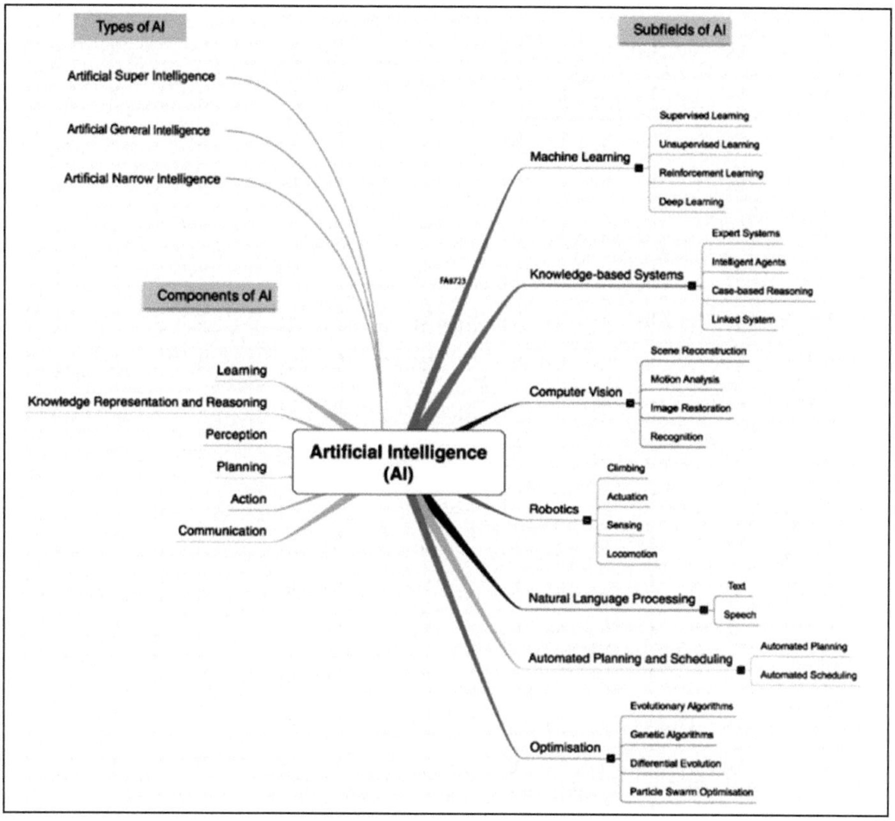

Fig. 10.1: Components, types and subfields of AI

To understand the current state of AI in the construction industry, it is pertinent to identify the major subfields of AI. Generally, the advances of AI application in industry has birthed various well known subfields of AI including: (a) machine learning (b) computer vision (c) natural language processing (e) knowledge-based systems (f) optimisation (g) robotics (h) automated planning and scheduling. An insight into each of these subfields is presented below.

(A) ***Machine learning*** Machine Learning is concerned with the design and use of computer programs to learn from experience or past data for the purpose of modelling, control or prediction using statistical techniques without being explicitly programmed. The Machine Learning methods include: (A) Supervised Machine Learning: this is concerned with the study of how machines make decisions based on what they have learnt from labelled datasets i.e. input and desired output pairings. It is categorised into classification and regression; (B) Unsupervised Machine Learning: this is concerned with making machines learn essential structure in unlabelled datasets. It is categorised into clustering and dimension reduction techniques and; (C) Reinforcement Learning (RL): this is defined as *"learning of a mapping from situations to actions so as to maximize a scalar reward or reinforcement signal"*. It is a computational approach involving learning from the outcome of interactions with the environment; and (D) Deep Learning: this is the current state-of-the-art in machine learning which has proven to give more accurate predictions than conventional machine learning techniques.

Computer vision Computer vision is a multidisciplinary field concerned with artificial simulation of the human visual system. To achieve the ultimate goal of making machines mimic human intelligence, computer vision seeks to enable high level understanding of digital and multidimensional images by capturing images via appropriate devices; processing them using state-of-the-art algorithms, and analysing images to facilitate decision making.

Automated planning and scheduling Planning is a subfield of AI concerned with enabling intelligent systems to achieve desired goals or objectives by carefully selecting and sequencing actions based on their expected outcomes. Scheduling involves the selection of plans and allocation of time and resources necessary to achieve the desired goals based on the total available resources. Planning and scheduling techniques are being adopted to provide solutions to complex applications that better fit the problem constraints and user needs. Planning, due to complexity, cost, and time-consumption is used in situations where its benefits outweigh the cost. Common heuristics and algorithms used for planning and scheduling include search techniques, optimisation techniques and genetic algorithms.

Robotics Robots are highly automated devices that carry out physical activities in the real world. Robotics is an interdisciplinary engineering activity that involves the design, manufacture, operation and maintenance of robots and other computer actions to mimic physical human actions. Robots are for highly specialised tasks, taking shapes most suitable for their use and not necessarily the humanoid shape, and they interact with the environment using sensors and actuators. One investigator explained that the majority of the learning problems in robotics are reinforcement machine learning problems.

Knowledge-based systems Knowledge-based Systems (KBS) is a branch of AI that is concerned with machine decision-making based on existing knowledge. Essentially, a KBS consists of a knowledge base, an inference engine and a user interface to interact with. The knowledge base is created from the storage of domain expert knowledge, past cases or experiences or other relevant sources. Its main advantage is the increased productivity and efficiency of easy access and interactions with large requisite domain knowledge. KBS makes inferences and arrive at conclusions that are heuristic, flexible and transparent, and provide the logic behind the advice given when necessary. KBS are classified into (A) Expert Systems: the knowledge base consists of task-specific knowledge from an expert in a particular domain to imitate decision-making in humans to solve specific problems; (B) Case-based Reasoning (CBR) Systems: the knowledge base of CBR systems is composed of past experiences or old cases which are used to explain, interpret, or critique new situations. For effectiveness in some domains, CBR systems must use expert knowledge to select the appropriate cases similar to the new project; (C) Intelligent Tutoring Systems: these systems use AI techniques to provide tutors with what they teach, who they teach and how they teach it; and, (D) DBMS with intelligent user interfaces and linked systems: are database systems with intelligent user interfaces driven by AI. Linked or Hypertext manipulation systems (HMS) allow easy traversal of complex information networks that enables writers and authors to easily link passages and references.

Natural language processing: Natural Language Processing (NLP) is a subfield of Artificial Intelligence that is concerned with creating computational models that mimic the linguistic capabilities of human beings. NLP has been applied in the fields of machine translation, natural language text processing and summarisation, user interfaces, multilingual and cross language information retrieval, speech recognition and expert systems. Tasks involved in NLP include part-of-speech tagging, chunking, named entity recognition and semantic role labelling.

Optimisation: Optimisation is concerned with making decisions or choices that provides the best outcomes given a set of constraints. According to one researcher; an optimisation problem is a construct of the problem of making the best choice from a set of choices. Optimisation is a lifelong phenomenon, originally known as a mathematical discipline concerned with finding an optimal solution to any given problem. With the birth of AI in the 1950s, a new family of metaheuristic algorithms known as evolutionary algorithms (EA) was created. Some of the recent algorithms belonging to the EA family include evolutionary strategies (ES), evolutionary programming (EP), genetic algorithms (GA), differential evolution (DE) and particle swarm optimisation (PSO).

Potential Benefits of Expert Systems

1. An expert system is able to provide timely advice when a human expert is unavailable. Unlike its human counterpart, an expert system can operate round the clock, all the seven days in a week, and every day of the year. An expert system does not get sick, take holidays, go on vacations or resign. It is not tied up in meetings, away on business.
2. An expert system can be readily replicated, it can be used simultaneously in many sites across the country or around the World. Once an expert system has been constructed, it is relatively inexpensive to distribute.
3. In some cases, human experts may already be in place providing advice. The introduction of an expert system that can offer comparable advice

should have a very positive impact on the human expert. It can reduce the demands on that human expert's time by insulating him from many kinds of consultation requests. They can focus on most challenging problems or concentrate on new creative activities. Human experts are normally a scarce resource. Their productivity can be increased by offloading consultation activity to expert system-thus human resources can be more effectively utilized.

4. Expert systems provide consistent, uniform advice. It is thorough and methodical. It does not have lapses that cause it to overlook important factors, slip steps or forget. It is not politically motivated, temperamental or biased.
5. Like a human expert, an expert system is able to explain the line of reasoning it uses for each problem it solves. A user can study the rationale on which the advice is based and is free to accept or reject it.
6. New expertise can be added by simply adding new rules or modifying existing rules. (S.Seetharaman,. 2017)

Limitations of Expert Systems

1. The over-importance of one individual expert in establishing the knowledge base, giving too strong a personal stamp.
2. The almost complete absence of any learning capability. An expert system does not improve itself by using its past experience, and except in a very few cases, cannot add to its knowledge base.
3. The restricted coverage of the field being studied.
4. The limited ability to provide explanation.
5. Inadequate interaction with the external World.

Needs of an Expert System in Construction Industry

Because of their characteristic attributes of combining factual knowledge with judgment being able to handle incomplete and uncertain data, and communicating with their users in the natural language like English, expert systems have a special appeal for the construction profession. The following typical features of today's construction environment show the need for an expert system - like technology for improving construction quality and productivity.

1. Unlike manufacturing, construction is non repetitive - each project being of different design layout, material used, construction methods, time, crews, weather and management. Ready algorithmic solutions are therefore not applicable to the day to day construction problems.
2. Construction environments are full of uncertainties in respect of labour and equipment productivity, market forces, material availability and changes, variations in regulatory agencies influence and variations in weather. Algorithmic computer programs cannot handle such uncertainties.

3. So far, not enough knowledge in the field of construction engineering and management has been formalized and encoded in text books. The industry runs on conventional knowledge and experience based judgement.
4. In many construction situations, there is not enough time to make a detailed analysis of all the influencing factors. Such decisions are often made on the spot so that the construction process is not interrupted. Expert systems can provide quick decisions in such situations.
5. Many construction decisions like safety management, labour relations, decision to bid, bid evaluation and risk management are qualitative and subjective in nature, needing heuristic approaches.
6. On a construction project, all the necessary information on a subject is almost never completely known and the decisions have to be taken on the basis of incomplete information. (S. Seetharaman,. (2017).
7. Expert-system building requires domain experts and there are highly knowledgeable experts in the construction industry who have successfully completed several projects and who are widely recognized in the profession.

State-of-the-art and future opportunities of AI applications in the construction industry *(Sofiat O. Abioye et al., 2021):*

Trends of AI application in the construction industry: In the 1960s, the use of artificial intelligence was still at its infancy with very few publications applying optimisation techniques. Over time, optimisation has been the foremost area of research interest in the application of AI subfields for the construction industry. This could be attributed to the age long struggle of the industry with low productivity levels. Another insight from the research trends over the years is that machine learning has overtook knowledge-based systems as a subfield of interest in the construction industry in the last decade. This can be attributed to the increased need to meet up with labour and skill shortages. Furthermore, robotics has also come to the fore of AI applications in the construction industry in the introduction of 3D printing, exoskeletons, UAV technologies for construction processes. However, natural language processing has been the least researched subfield in the construction industry. Over 60% of AI application research in construction was done in the last decade and has given rise or enhanced the emergence of advanced technologies including:

1. Quantum computing;
2. Internet of Things (IoT);
3. Cybersecurity; and
4. Blockchain.

Quantum computing has the capability to solve certain computational tasks more efficiently than classical (or traditional) computers. As a consequence of the special nature of quantum information, quantum algorithms are more effective than classical (or traditional) algorithms. AI can leverage on the capabilities of Quantum computing to accelerate problem-solving and optimise its solutions. The internet has evolved from just allowing humans to communicate with each other, to enabling communication between multiple objects and humans to create a smart environment. This transformation has been made possible by recent advances in technology such as sensors, actuators, wireless technologies (e.g. RFID), cloud computing, faster and cheaper devices with increased processing capabilities. In the construction industry, IoT has been integrated with AI in many ways including: energy-saving on demand for intelligent building energy monitoring; IoT-enabled building information modelling (BIM) platform to achieve real-time visibility and traceability in prefabricated construction; and to generate early warnings and alarms as dynamical safety barriers for hazard energy on underground construction sites. The numerous advantages precipitated by an increased access to the internet and interconnected systems all over the World is threatened by the challenge of ever-evolving cyber threats. Some examples of cyber threats that are common in the construction industry include malware, social engineering and phishing. It has been established by one investigator that although the construction industry does not figure much in findings about cyber security, the introduction of level 3 BIM with an increasing reliance on virtuality will increase exposure to cybercrime. The increasing integration of BIM with data from other external sources makes it particularly vulnerable to cyber-attacks. In fact, any digital technology used in the construction industry, AR/VR, IoT and even AI such as hard and soft robots are at risk of cyber-attacks if there is no proper network security and response plan. Since the invention of Blockchain in 2008, there has been a rapid growth in its adoption, especially in the areas of crypto currency, risk management, IoT, public, social and financial services.

Blockchain technology helps to ensure the legitimacy of a transaction, prevents double-spending, and allows for high-value transactions in a distrustful environment using cryptography and a consensus mechanism for verification. Some recent applications of blockchain in construction includes logistics of construction materials, integration with IoT and BIM to manage building lifecycle data and to enhance manufacturing supply chains in the composite materials industry. The applications of blockchain technology in the construction industry are non-trivial and it has the potential to solve most of the trust, communication and transparency issues in the industry. The applications of AI in construction could be integrated with blockchain technology to develop solutions that are secure, transparent and have a decentralised CDE.

Application of AI subfields in the construction industry: Advantages across all the subfields include increased cost and time savings, improved safety, better accuracy and overall increased productivity. Some of the limitations of AI subfields in construction includes incomplete data, high initial cost of deployment, data and knowledge acquisition issues.

Common application areas and future opportunities of AI in construction: Fourteen subdomains were identified with relevant state-of-the-art applications and potential opportunities for construction-specific issues. Some of the identified subdomains include resource and waste optimisation; value-driven services; supply chain management; health and safety, AI-driven construction contract analytics; voice user interfaces; and AI-driven audit system for construction financials. Under value-driven services, subdomains such as estimation and scheduling, construction site analytics, job creation, AI and BIM integrations with other industry 4.0 tools such as internet of things (IoT) are described.

Resource and waste optimisation: There is a growing amount of construction and demolition waste (C&DW) produced each year across the due to rapid continual development. These construction activities have adverse effects on the environmental, natural and human resources globally. One investigator observed a paradigmatic change in the approaches to waste management from waste intelligence, which suggests measures to reduce waste only after it happens, to proactive data-driven approaches i.e. waste analytics (WA), which minimizes waste through design. There has been an increasing use of BIM as a virtual, less-expensive and computational environment to enable construction design with a view to minimize waste generation. One researcher highlights the potential of advanced data analytics techniques to produce more detailed waste generation profiles. Therefore, the use of advanced data analytics could minimize waste extensively. Waste analytics depends on a diverse data from different sources such as building design, material properties and construction strategies, which allows for high-performance computation and real-time processing. The large amount of data requires advanced data analytics techniques to turn it into information relevant for waste minimzation. This calls for the adoption of artificial intelligence techniques for effective waste management. Particularly, the use of advanced AI techniques with BIM to optimise design for offsite construction, materials selection, reuse and recovery, waste-efficient procurement, deconstruction and flexibility.

Value-driven services: This section discusses a broad range of non-core services, which can be benefitted from the emerging trend of AI in the construction industry.

Estimation and scheduling: AI-based estimation (or prediction) models have wide applicability in various domains of the construction industry. Particularly, these estimation models are instrumental in the early prediction of construction cost and duration, which are key project success factors. Unreliable project cost and time estimates could have huge economic and financial implications. BIM is the current state-of-art in the construction industry, which is used to enhance reliability of cost and time estimates in the construction industry. The integration of time (4D) and cost (5D) to BIM and cost estimation. Nevertheless, in spite the huge benefits of BIM, evidences have also shown that the automation advantage brought about by BIM accounts for only the information in the model while disregarding external subjective factors, additional materials and resources respectively. Consequently, applying BIM for cost and time estimation still requires a lot of work for the cost estimator. This calls for the integration of advanced AI techniques such as deep learning with BIM for cost and time prediction to take advantage of better accuracy. Consequently, deep learning techniques can also aid in better predictions of other relevant factors such as bankruptcy, success, energy, carbon efficiency and even waste.

Construction site analytics: Construction sites are rapidly being transformed to smart working environments due to the increasing pervasiveness of IoT sensors and other digital technologies. Construction site analytics is concerned with the generation, collection, storage and analysis of construction site data to generate deep insights for visualisation. A large volume of images, videos and other forms of data is generated on construction sites such as reports, real-time equipment and site monitoring, which are mostly unstructured. The data collected can be aggregated in BIM and analysed using advanced AI techniques to optimise site performance in all key areas such as planning, design, safety, quality, schedule and cost. One investigator presented a theoretical foundation for visual data and predictive analytics for proactive project control on construction sites. One investigator discussed the potential of big visual data and BIM for modelling construction performance analytics. One researcher described the process of digitizing construction site layout planning process in BIM by applying rule-based model checking. There is a need for the development of a holistic site analytics tool using AI for real-time, cloud-powered analytics of data generated on the construction site. This will improve productivity, quality control and help to achieve set performance targets. A construction site AI chatbot to get real-time updates about site activities would prove to be very helpful for project managers and other relevant stakeholders.

Job creation: Construction jobs are the third highest jobs at risk of automation in the next decade due to the increasing prevalence of automation technologies such as AI and IoT. Most construction jobs that require medium to low education are at a high risk (38-45%) of automation by mid 2030s. However, the adoption of such technologies could also lead to the creation of entirely new roles to assimilate and reskill the displaced workers in the industry. For example, the advent of BIM has led to creation of new roles such as BIM project manager, director, coordinator and designer. The adoption of digital technologies like AI in the construction industry will give rise to new types of jobs such as construction AI researcher, trainers and engineers. The researchers will be responsible for bringing construction industry to the frontier of AI adoption by continually carrying out research and bringing innovation into the sector. Construction AI engineers will focus on developing and deploying state-of-the-art AI solutions suited to the construction industry. Finally, trainers and testers will be needed to ensure these solutions such as robots and autonomous agents carry out designated goals effectively and efficiently. For example, low to medium skilled workers that will lose their job to automation could serve as trainers and testers for those systems replacing them.

AI and BIM with industry 4.0 tools: BIM is the current state-of-art in the architecture, engineering and construction (AEC) industry being adopted Worldwide. For example, the UK government mandated BIM level 2 for all publicly procured projects. However, despite attempts to make BIM as the de-facto standard globally, its rate of adoption is still very low. Consequently, it is important to carry out relevant research on ways to improve the adoption of BIM. For example, some studies have combined BIM applications with AI subfields such as NLP to improve navigation of BIM interfaces. According to a study of construction project work carried out by group of investigators in 2006, who argued that nature of construction practise is largely dependent on direct verbal and symbolic communication for sharing expertise and information. Perhaps, this may be because speech is undoubtedly, one of the oldest and most natural form of communication. The integration of voice interaction with BIM will make it feel more natural and truer to the nature of construction. Other studies have also combined AI and BIM with other industry 4.0 tools such as Internet of Things (IoT), smart cities, augmented reality, blockchain and quantum computing.

Supply chain management (SCM): A recent study by a few investigators to explore factors affecting supply chain excellence and outcomes, revealed issues such as SCM knowledge education and supply chain culture. The major barriers found are lack of top-level buy-in and general understanding of supply chain. The literature also revealed that the high cost of advanced IT for SCM, lack of unique and regional specific performance measurement frameworks, lack of organisational trust and effective communication channels between partners had been a major hindering factor. AI techniques can play a significant role in solving these issues that are hindering excellence in the supply chain.

One investigator developed an IoT-based real-time risk monitoring system for controlling product quality and occupational safety risks in cold chains using wireless sensors network, cloud database services, and fuzzy logic. According to one researcher, real-time integration of in decentralised supply chains is at the top of the agenda for European institutions, industries and multinational consulting firms. One investigator developed a process specification language to improve supply chain communication for offsite construction. This study calls for the development of a decentralised supply chain knowledge management and monitoring system. The advantages of blockchain to ensure transparency and legitimacy of transactions, combined with the advanced analytics powered by AI has the potential to solve the organisational trust and communication issues in construction supply chain. For example, one investigator explored the applicability of blockchain to enhance manufacturing supply chain and results show a tamper-proof history of product manufacturing and storage. AI can be used to manage the whole supply chain process to detect potential issues and ensure timely and quality project delivery. Also, the creation of AI chatbots for supply chain management system to harness relevant information in a simple manner.

Health and safety analytics: Health and Safety analytics (HSA) makes use of advanced data analytics techniques to predict and prevent occupational accidents in the workplace. The construction industry records a significantly higher rate of occupational injuries and deaths than other industries. This is because construction work is highly risky due onsite dangers like working from heights, getting trapped, falling objects, equipments and tools, dust and toxic materials, and loss of hearing due to noise.

AI-driven construction contract comprehension: Construction contract management can be very complex, involving bulky and multiple contracts where mistakes often have huge implications. Some of these implications include litigation costs, claims, project delay and loss of reputation. A major challenge is that current contract management is heavily dependent on humans such as contract managers or legal team, which makes it prone to errors and mistakes. This calls for the automation of the contract management with AI to augment the efforts of contract managers or legal teams by ensuring processes are faster and more accurate. This will allow for easy querying of contracts to retrieve relevant information, contract summarisation, contracts drafting, identification of dispute prone clauses, contract negotiation and dispute-resolution, contract-obligation-enforcement and closure. One major challenge with developing an AI-driven tool for contract.

Challenges So far, this study has identified the opportunities as well as emerging trends in AI application in construction processes. To further strengthen this area of knowledge, it is important to identify and discuss the key challenges. Figure 10.2 shows the opportunities, emerging trends, challenges and open research issues of AI in the construction industry.

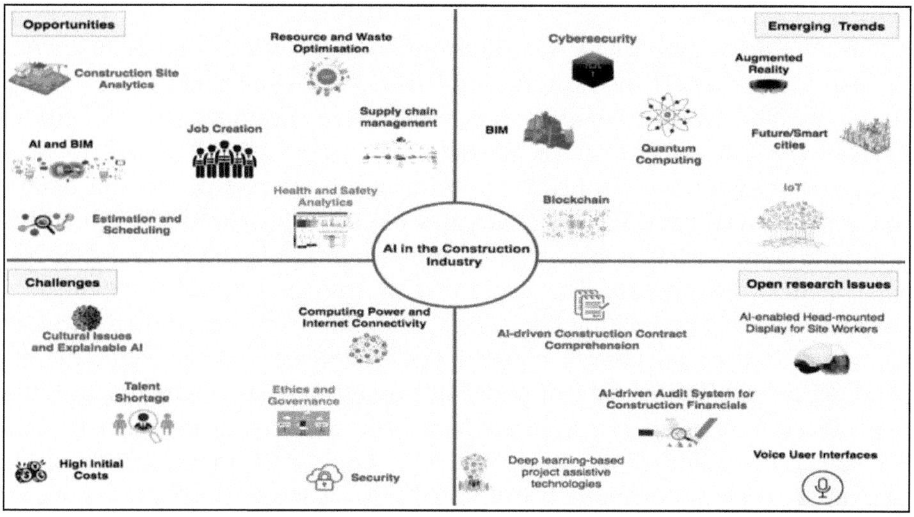

Fig. 10.2: Opportunities, emerging trends, challenges and open research issues

Six main challenge areas affecting the adoption of AI in construction are presented below.

Cultural Issues and Explainable AI

It is a known fact that the construction industry is one of the least-digitized industries, and slow to adopt new technologies. This could be due to the risky and costly nature of most construction processes, where little mistakes often have huge implications. In construction, traditionally known ways of doing things are favoured above the new but un-trusted technologies promising to deliver great rewards. This means that the construction industry is slow to adopt innovative technologies. Unlike sectors like manufacturing, construction sites are mostly unique and different which requires AI that can learn and adapt fast in changing environments. This means that AI technologies to be deployed in the construction industry must be useable in different construction projects or sites and tested thoroughly to convince construction contractors and businesses to use them. This may include taking advantage of other digital technologies like blockchain to improve trust and transparency.

Additionally, most machine learning systems follow a black-box approach, which means that they do not explain the reasons for reaching conclusions. To build trust in such systems, it is essential for construction practitioners to understand how the system makes its decisions. As such, this calls for the use of explainable AI (XAI) to produce explainable models and enable humans to understand, trust and manage the systems. Some popular approaches of XAI include local interpretable model-agnostic explanations (LIME) and layer-wise relevance propagation (LRP).

Security

Despite the capability of AI to strengthen security and detect intrusions, it is also a main target for exploitation by hackers, cybercrimes and privacy intrusion. This is a critical issue with huge economic and financial implications. Little mistakes in construction processes often lead to huge quality, cost and time implications, with a roll-over effect to the overall project plan (time, cost, supply chain and logistics, procurement, etc.). Most importantly, the safety of construction workers may be compromised which could lead to life-threatening accidents or loss of lives. For example, a computer vision system that identifies automated construction equipment may be tricked to mislabel a construction worker working at height. The use of AI to take full control of some processes or to augment the activities of construction workers needs to be done with minimal or no security risks. This calls for the mitigation strategies like the use of adversarial machine learning. Adversarial machine learning (ML) is the study of effective machine learning techniques against an adversarial opponent. The use of adversarial ML stems from a security defence perspective and the need to design algorithms that can resist high-level attacks. Also, further research needs to be carried out in this area especially for technologies that are emerging in use for construction research such as computer vision and robotics.

Talent Shortage

Currently, there is a global shortage of AI engineers with the requisite skills to spearhead serious developments across industries. It is quite difficult to get AI engineers with experience in the construction sector to build custom solutions targeted at solving the many problems in the industry. This can be mitigated by government spending more on science, technology, engineering, mathematics (STEM) education. Also, construction experts are needed to collaborate with researchers and industry experts in the AI field to merge ideas and birth new innovations that truly meets the needs of the construction industry.

High Initial Costs

The benefits of AI-driven solutions in the construction industry are indisputable. However, the initial costs required to invest in such AI solutions e.g. robotics are usually very high. The maintenance requirements of such solutions also need to be considered. This may be unaffordable for the vast majority of subcontractors and small firms that make up the bulk of the construction industry. Therefore, it is important for firms to determine the cost savings and return on investment of such technologies to determine whether to invest or not. Additionally, as these technologies become more accepted and prevalent in construction, prices are expected to drop, making it affordable for the smaller firms.

Ethics and Governance

One investigator elucidated that establishing and maintaining public trust in AI technologies is dependent on an inclusive, transparent and agile governance. This is a crucial issue that is of great importance to the society at large. The competences of AI technologies while promising great outputs, could also be dangerous if not properly regulated. For example, a big construction site robot that malfunctions and has to fall down in a busy construction site with plenty of workers. How does the robot make the decision of falling left or right, depending on the number of workers on each side, knowing fully well it could mean death for the workers? The use of some AI solutions could also lead to unfair advantage for some firms in the construction industry, which needs to be regulated.

Computing Power and Internet Connectivity

Construction sites are mostly remote and lacking power, telecommunications and internet connectivity. Sometimes, even construction activities lead to interruptions in power and internet connectivity. This poses a serious problem in the use of AI tools on construction sites whose operation relies mostly on good internet connectivity and power supply e.g. robots, site monitoring systems, etc. For example, sensors and actuators communicate information that needs to be computed in real-time during construction. It is pertinent to look for ways to solve this problem efficiently and effectively. The use of 4G (LTE/max) communication technologies have been able to solve this problem to a considerable extent. The emergence of 5G offers even greater reliability for construction sites due to its high data rate, reduced latency, energy saving, cost reduction, higher system capacity, and massive device connectivity.

Review Questions

1. Give an Overview of artificial intelligence and its subfields applicable to construction industry.
2. Explain the following terms:
   - Computer vision
   - Automated planning and scheduling
   - Robotics
   - Knowledge-based systems
   - Natural language processing
   - Optimisation
3. What are the potential benefits of expert systems?
4. What are the limitations of expert systems?
5. What is the needs of an expert system in construction industry?
6. With the help of a case study, explain the future opportunities of AI applications in the construction industry.

11

Information Technology and Automation

Construction industry has lagged behind other sectors in terms of applying Information Technologies. In the main, the boards of operating construction companies are dominated by engineers and surveyors whose knowledge of computing is very often limited to what systems have been implemented at head office, and whose interaction with computer is limited to occasional meetings with senior computer professionals rather than fellow engineers. Their general view is that the implementation of any new system takes a long time, is expensive and involves 'outsiders', and this is aggravated by with the degree of scepticism in terms that it's generally believed that the end product will in any case not achieve its initial specification.

Computing and communication technology, also commonly known as Information Technology (IT) have been radically transforming the way we live, learn, work and play.

Many companies in the construction industry do not generally appear to have appreciated the positive changes and advantages that the new technology was providing to companies in other sectors of the economy.

Nevertheless, many software is currently being used by the construction sector like CAD tools, Spreadsheets, Word Processing and so on. Information is used and should be managed from the conception stage, design, and execution, to the facilities management stage.

With the growth in the use of information technologies by construction companies, the increasing availability of electronic documents, manual classification becomes impractical.

With such amount of information, Document Management Systems which are considered as a glue or integration of all the other software for cost estimating, planning, scheduling, etc., generated in a specific stage of the project, are an indispensable tool for the current Project Management activities.

During the 1980s PCs were only used in few construction companies. Throughout the 1980s, although most building firms were using computer technology for many of their core functions such as accounting, wages, and salaries, very few of them evolved formal policies or strategies concerning the use of information technology.

By the latter part of the 1980s, about eight years after the introduction of reliable PC equipment, some companies had reached a situation in which their staffs on many of their larger projects were experiencing the advantages of the new technology through the use of planning, drawing, spreadsheet and word-processing software packages.

Today, a large number of software packages are available to all the disciplines of the construction team at every stage of the construction process. They provide support for a broad range of activities such as computer aided design and drafting, building visualization, design appraisal, project management, information storage and retrieval, cost estimation, structural analysis, on-site management, facilities management, etc.

The Development of Information and Communication Technology in Construction

The use of Information and Communication Technologies (ICTs) in the construction industry is generating new opportunities for collaboration, coordination and information exchange among organizations that work on a construction project.

Complex technical systems do not evolve fully formed, but rather as localized developments. In the development of automated systems for the transfer and transformation of materials, this unbalanced evolution leads to the problem of 'islands of automation', where highly automated materials flows are mixed in with completely manual ones. The same problem exists in the development of computerized information systems. The development of computing technology has meant that tools for analysis involving data manipulation have tended to develop earliest and in isolation. These tools play to the enormous strengths of computers in the rapid analysis of complex data sets–analysis which is frequently impossible if manually attempted. Thus stand-alone applications dependent on numerical analysis, ranging from finite element analysis to critical path analysis, had been developed by the 1960s. Information flows between these types of application continued to use traditional information technologies such as the paper-based engineering drawing.

During the 1970s, a new form of graphical manipulation developed to aid the creation of engineering drawings –computer aided design (CAD). Again, the output from these systems largely relied on traditional technologies for communication between different applications. The construction industry was at the forefront of these developments. By the fourteenth century, scaled technical drawings –probably the most important information technology of the last millennium after the printed book itself-were well established for use.

During the 1970s, large public sector projects –usually relying on extensive standardization and prefabrication-offered the opportunity to develop CAD systems. However, the demise of the large public sector construction programmes which have been essential to the development of ICT applications in every industrial sector meant that this initial momentum was lost.

The 1980s saw the development of the personal computer (PC) which dramatically reduced the cost of computing power, and enabled a much wider diffusion of computers within the industry, while the processing power of computers continued to grow exponentially. Most importantly, site offices could now be equipped with computers.

The development of communication technologies has taken an independent path. In comparison to computer technologies, developments were earlier and more profound. The telegraph and, more importantly, the telephone, greatly improved communication capabilities. The fax and photocopier are more recent innovations which have had a significant impact. Nonetheless, these communication technologies did not allow any further manipulation of the received data. It was not until the 1970s that they began to be connected to computers to provide integrated systems for the direct communication of information between computer systems. The development of Local and Wide Area Networks (LANs and WANs) proceeded steadily, but interconnectivity between computers was transformed by the breathless diffusion of the internet during the 1990s.

It is this rapid development of the interconnection between communication and information technologies over the last 20 years that has both opened up tremendous new opportunities and posed new technical challenges.

Current Information Technologies for the Construction Sector

Information technologies or Software applications are available to support most aspects of a construction project. They have been designed largely as solutions to specific problems. These applications can be grouped into the following categories:

- Computer Aided Design and Visualization
- Building Engineering Applications
- Computer Aided Cost Estimation
- Planning, Scheduling, Site Management
- Computer Aided Facilities Management
- Integration
- Business and Information Management

Figure 11.1 is a roadmap showing where and when these applications are being used along the construction process. The main purpose of the diagram is to indicate the main application areas for the existing discrete software packages in the construction supply chain.

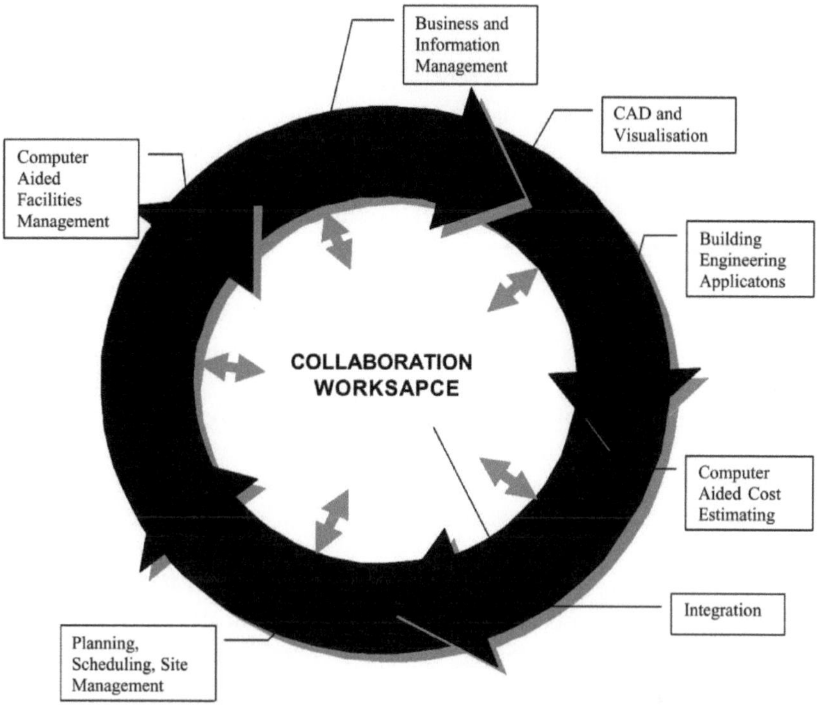

Fig. 11.1: Construction Process and IT application

Computer Aided Design and Visualization

Computer Aided Design (CAD) software is widely used by design professionals, and AutoCAD has the largest share of the CAD market (Howard 1998). Other popular CAD software include Microstation, ArchiCAD, MiniCAD, FastCAD, etc. These CAD programs have largely replaced the traditional drawing board at the production information stage.

The basic function of CAD tools is allowing the user to build up drawings by manipulating lines, circles, rectangles and texts interactively on the screen. The clear advantage of CAD software is the ability to allow 'editing', which means delete, move, copy, rotate, scale mirror, etc.

Furthermore, since the drawing can be saved at any stage, the designers are able to keep various versions of the building layout for later study. Once the geometrical information of the building design is stored in a CAD package, different views of the building can easily be produced.

Visualization and animation systems, like 3D studio, can produce photo-realistic, static, and moving images, so that the clients can view the final appearance of the building at the design stage. The emerging Virtual Reality technology even allows the user to interact with the design model and experience the building in simulated reality settings.

Building Engineering Applications

Nowadays, construction industry clients have ever-higher expectations. They want their buildings to look good, to be safe, to provide comfortable living environments for their occupants, to consume less energy in operation, etc. The ever more complex demands on the building design process have given rise to the need for a new approach to building engineering design based on computer software.

Once a building is constructed, it is very costly to correct any design defects. It is, therefore, important to simulate accurately the building's performance at the design stage so that problems can be identified and solved. Over the years, a variety of methods and algorithms have been developed to predict building performance in thermal, lighting, acoustics, and structural aspects.

Because of the complex and tedious calculations that involve these simulations, it was nearly impossible to carry them out before computers. During the last two decades, a range of building engineering applications have been developed for energy analysis, HVAC design, structural analysis, lighting simulation, etc.

The benefit of these applications is that they allow designers to evaluate alternative design solutions in order to reach optimum design.

Examples of this software are: ATEAN from Carrier, and CARGASW from Climasoft, that offer comprehensive range of software options for climatic energy design; CALCULUX for lighting and building services design, CYPE INGENIEROS S.A. for structural design, COSMOS for finite elements analysis, DUCTSIZE from Elite Software for electricity and water nets design.

Computer Aided Cost Estimation

Controlling costs is one of the most important requirements during a construction project. To achieve this, contractors and subcontractors must first make accurate cost estimation. Rigorous project accounting must then be used to control the spending. Today, there are sophisticated computer software packages, such as Esti-Mate, Manifest, FBS-Estimator, and PRESTO from Soft, GO from Star, ITEC and ARQ from AM2, which allow project managers to make estimations and to keep track of project spending.

Other software can help to measure, count, compute and tabulate quantities, lengths, areas, volumes, etc., of objects found in plans and specifications. Furthermore, most cost estimating programs can be integrated with databases of costs for labour, materials and equipment. The advantage is that cost data do not need to be re-entered, thus improving the celerity in estimating and avoiding errors. Computer based estimation of costs archives and retrieves large volumes of resource, cost and productivity information, makes fast and accurate calculations and presents results in an organized, neat and consistent manner.

Planning, Scheduling and Site Management

Construction works require careful planning and skilful management of human and physical resources. Computer systems can assist on-site managers to plan ahead, evaluate different options and adopt and execute the most efficient construction operation.

Besides the widespread use of planning packages such as Microsoft Project, Primavera, Power Project, etc., to plan and schedule detailed construction activities, some applications like JobMaster, ICON, GEST, and Presto Control, are designed to log and track internal processes during the construction phase.

Site operation simulation programs can emulate what happens in a real construction site by representing workers, machines, and materials, and computing the cycle of each step taking into consideration many uncertain factors.

Computer Aided Facilities Management

Facilities Management is a relatively new discipline that emerged in the early 80s. It is based on the operation, maintenance, and impact of the building operations on the life cycle costs.

The available software for Facilities Management is a combination of CAD and database management systems. Special routines enable blocking and stacking studies to be carried out to explore different layouts or to reflect organizational changes. Databases are the most important part of Facilities Management Software. It holds data from people and their services so that when they move, their services can follow them. ITE (Inspección Técnica de Edificios), for example, is a tool that makes possible to create a report of the state of the building.

Business and Information Management

The construction process is an information intensive one during which a huge amount of information in generated and consumed by all the professionals involved. The common type of information includes site survey, cost analysis, design drawings, documents, correspondence, fax, computer files or e-mails. Electronic Document Management Systems can create an environment in which disparate forms of information can be linked together, within the context of a project or organization, to achieve easy access and control.

All the previous tools and software for cost estimating, planning, scheduling, etc., are generated in a specific stage of the project. Sometimes this information is updated, modified, consulted, etc., at any moment of the project. Consequently, Document Management Systems are tools that comprise whatever information throughout the life cycle of the project, from the conception of the need to the maintenance.

Electronic Document Management Systems are applications that can be linked to Web Based Project Management Systems for to improve communication among partners and between them and the management of the project. In addition, Web Based Project Management Systems have other functionalities like project coordination, scheduling, etc.

Integration

From the first software applications, many different tools were developed. They use their own data formats, which are not compatible with each other. As a result, data cannot be electronically exchanged between them. In recent years, there is an increasing awareness of the need for integrated construction processes and many research projects are investigating related issues.

During the last two decades, advances in object oriented programming, database systems and product data modelling technologies have provided a solid platform for integration. Data standards are being developed first by the International Standards Organizations (ISO), and then by the International Alliance for Interoperability (IFC). At present, these standards are still evolving. An integrated project database covering the whole life cycle of construction projects remains a future prospect.

Current Communication Technologies for the Construction Sector

Communication technologies are the technologies dealing with the transmission of information. They support the process by which information is exchanged. Figure 11.2 shows the evolution of some information technologies (mainly, communication technologies) in construction. They are positioned according to their complexity (vertical axis) and construction specifics (horizontal axis). Notice that the most specific services tend to use or rely upon the generic ones, and that the most complex services tend to integrate simpler ones.

The first services were generic and included networked file archives, e-mail communications and text-based group conferencing. The web provided a much friendlier navigation and presentation of the files on remote machines. It was as that time when first construction related content appeared. Out of the first 1000 websites only two were related to construction.

The first construction specific services used the web to publish information such as scientific papers, building codes, product specifications, etc. The next generation services were staring to use the Internet as a collaboration platform for the companies involved in a construction project.

Since 1998, however, construction has been following trends of general service supplied on the Internet. A business model of the providers of various kind of Internet related software (e.g. for managing mailing lists, discussion forums, help desks, photo albums, etc.) has evolved from the sale of software for the users to install on their servers, to the offering of a service on their Website, that offers the same functionality to the end user. In the construction context there are now dozens of companies providing collaboration tools such as document managing, project coordination and scheduling.

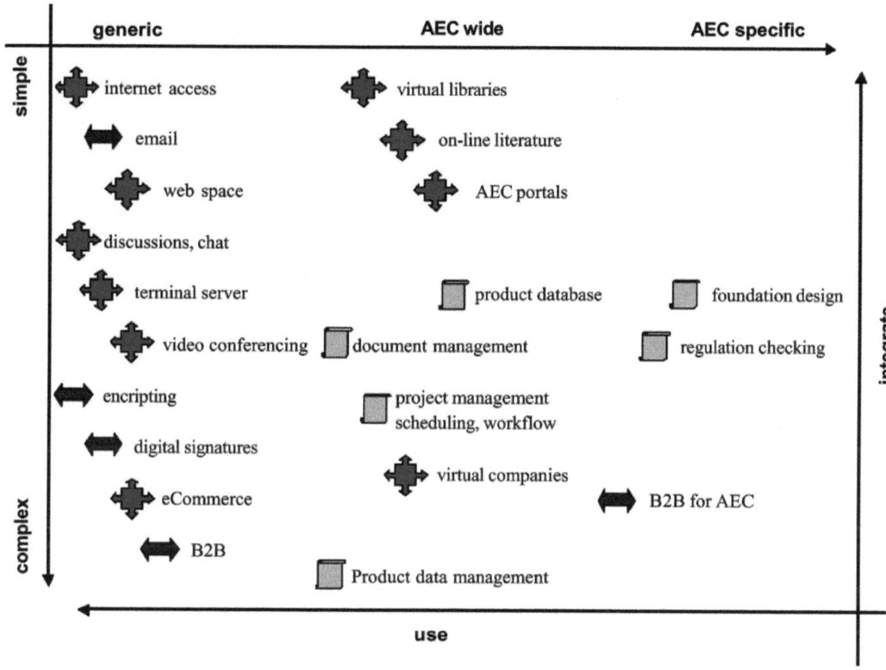

Fig. 11.2: Evolution of Internet related communication technologies

Computers in Construction

There is a growing library of packaged computer software as well as special-purpose computer programming languages available for small computers. Many of these can be profitably employed by the construction engineer. Some of the widely used software packages include word processors, electronics spreadsheets, data base programs, communications programs, graphics problems, and project management programs. Integration software packages that include several of these programs utilizing a common file structure are also available. Some of the software written specifically for the construction industry includes estimating programs, bidding programs, project management programs, and programs for maintaining cost and performance data for equipment and labor.

Word processors are used for general correspondence as well as for the preparation of memos, reports, training manuals, and procedures manuals. They are particularly useful for the preparation of repetitive document, such as contract specifications, where much of the material is standard but are modified somewhat for each specific project. Word processors often have associated spelling checkers which identify, word not contained in its standard dictionary. The user can either correct the word, accept it without adding it to the dictionary, or add it to the dictionary. Mailing-list programs are also available for many word processors. These enable the user to prepare form letter, and associated address files. Address files can be used to prepare mailing labels or envelope, as well to merge names, addresses, and other data into form letters.

An electronic spreadsheet is a more powerful form or the familiar row-and-column spreadsheet used for tabulating such data as quantity, unit cost, total cost, sale price, and profit. By allowing the user to specify mathematical relationship, between cells results for any input data can be quickly calcula1ed. The value for each row of column 4 may be specified as that obtained by multiplying column 2 by column 3. Many electronic spreadsheet programs abo contain built-in-function such as interest calculations, loan amortization, present value, future value, and internal rate of return. The use of such a program will enable the manager quickly to determine the effect produced by any change in the assumed or actual data.

Data base programs are used to organize, maintain and manipulate a collection of data. Special-purpose data base programs are written for a specific purpose such as inventory control. Although such specialized programs are relatively easy to learn and put into use, they can be used only for the specific purpose for which they were written. Therefore, general-purpose data base programs are much more widely used. General-purpose programs are very flexible but may be customized for each specific application.

Communications programs are used for communications between a small computer and another small computer or a large computer. They are also used to access such information services as electronic mail or electronic bulletin boards. Used with a computer and modem, they allow computer communications over an ordinary telephone line.

Graphics programs greatly speed up and facilitate the preparation of graphic material. They are widely, used for preparing charts and other illustrations for reports and presentations. Computer-aided design and drafting programs are becoming widely used for construction design.

Project management programs are usually built around the network planning technique. They often provide for maintaining and forecasting cost and resource data as well as time data. Some programs also contain functions capable of resource leveling.

Advanced Techniques

In addition to the quantitative management techniques and computer software described above, there are several more advanced techniques available to the progressive construction manager. For example, there are various optimization techniques available for the solution of optimization problems involving nonlinear functions. Computer simulation is a powerful tool for analyzing problems not easily solved by analytical methods. Construction operations such as earth-moving may be simulated by the use of packaged simulation programs or by writing a simulation program in a simulation language.

In this day of rapid technological advance, one can never predict the exact impact of new technology on construction. As we have seen, the wide availability of the microcomputer or personal computer has placed a powerful new tool at the disposal of all construction professionals. Computers have already begun to be applied to the control systems of earthmoving equipment. The use of robots in construction cannot be far away. Certainly, tomorrow's construction professional faces an exciting future.

Construction Automation

Construction has traditionally been resistant to technical innovation. Past efforts to industrialize construction in the U.S. were undertaken at the time when industrial automation technology was at its infancy. Additionally, engineering and economic analyses of prefabrication processes and systems were lacking. On the other hand, numerous construction tasks have or will become more attracted to automated technologies based upon the following characteristics: (1) repetitive, (2) tedious and boring, (3) hazardous to health, (4) physically dangerous, (5) unpleasant and dirty, (6) labor intensive, (7) vanishing skill area, (8) high skill requirement, (9) precision dexterity requirement, and (10) critical to productivity. For example, some construction tasks have been historically noted for their arduous, repetitive nature, with relatively little dynamic decision making required on the part of a human laborer. Such tasks may include placing of concrete, placing of drywall screws, finishing of concrete, and placing of masonry block, among others. The work involved in these tasks is rather unattractive for humans. Robots, however, are applicable to these types of work tasks provided that the technology and economics are feasible.

There have been increasing demands to enhance intelligence of construction equipment and systems. Many researchers have investigated the addition of sensors and control systems to existing construction equipment. A limited amount of research, however, has been conducted in developing intelligent construction equipment and systems. For semiautonomous and autonomous equipment with great potential for impact on the construction industry, artificial intelligence (AI) is required to generate instructions and plans necessary to perform tasks in dynamically changing environments on their own.

Construction automation refers to the use of a mechanical, electrical, and computer-based system to operate and control construction equipment and devices. There are two types of construction automation:

1. Fixed construction automation
2. Programmable construction automation

Fixed construction automation involves a sequence of operations performed by equipment fixed in their locations. In other words, an automated facility, whether it is permanently indoors or temporarily on the construction site, is set up specifically to perform only one function or produce one product. In programmable construction automation, equipment has the ability to change its sequence of operations easily to accommodate a wide variety of products.

Fixed Construction Automation

Fixed construction automation is useful in mass production or prefabrication of building components such as:

1. Reinforcing steel
2. Structural steel
3. Exterior building components (e.g., masonry, granite stone, precast concrete)

Examples of Fixed Construction Automation

In this section, selected examples of fixed construction automation are highlighted.

Automated Rebar Prefabrication System

The automated rebar prefabrication system places reinforcing bars for concrete slab construction. The system consists of a NEC PC98000XL high-resolution-mode personal computer that uses AutoCAD™, DBASE III Plus™, and BASIC™ software. The information regarding number, spacing, grade and dimension, and bending shapes of rebars is found from the database generated from an AutoCAD file. This information is used by an automatic assembly system to fabricate the rebar units.

The assembly system consists of two vehicles and a steel rebar arrangement support base. Of the two vehicles, one moves in the longitudinal direction and the other in the transverse direction. The longitudinally moving vehicle carries the rebars forward until it reaches the preset position. Then, it moves backward and places the rebars one by one at preset intervals on the support base. Upon completion of placement of the rebars by the longitudinally moving vehicle, the transversely moving vehicle places the rebars in a similar manner. The mesh unit formed by such a placement of rebars is tied together automatically.

Automated Brick Masonry

The automated brick masonry system, is designed to spread mortar and place bricks for masonry wall construction. The system consists of:

1. Mortar-spreading module
2. Brick-laying station

The controls of the system are centered around three personal computers responsible for:

1. Collecting and storing date in real time

2. Interfacing a stepping-motor controller and a robot controller
3. Controlling the mortar-spreading robot

A Lord 15/50 force-torque sensor is used to determine the placing force of each brick. The system is provided with an integrated control structure that includes a conveyor for handling the masonry bricks.

Fully Automated Masonry Plant

The fully automated masonry plant is designed to produce different brick types with the production capacity of 300 m wall elements per shift. The system consists of several components: a master computer, a database server, a file server, stone cutters, masonry robots, pallet rotation systems, refinement systems, storage systems, transversal platforms, a disposition management system, an inventory management system, and a CAD system.

Two individual brick types can be managed in parallel by unloading the gripper and the cutter-system consisting of two stone saws. By conveyer systems, stone units and fitting stones are transported to the masonry robot system. The masonry robots move two bricks at each cycle to the growing wall after a mortar robot puts a layer of mortar on it. A pallet rotation system carries the wall to the drying chamber.

After 48 hours, the wall is transported to destacking stations to group the wall elements of the same order. Finally, grouped wall elements are transported to the construction site.

Automated Stone Cutting

The purpose of the automated stone-cutting facility is to precut stone elements for exterior wall facings. The facility consists of the following subsystems:

1. Raw materials storage
2. Loading
3. Primary workstation
4. Detail workstation
5. Inspection station
6. End-product inventory

A special lifting device has been provided for automated materials handling. The boom's rigidity enables the computation of exact location and orientation of the hook. Designs for the pallets, the primary saw table, the vacuum lift assembly, and the detail workstation have also been proposed.

Programmable Construction Automation

Programmable construction automation includes the application of the construction robots and numerical control machines described below.

Construction Robots

The International Standards Organization (ISO) defines a robot as "an automatically controlled, reprogrammable, multi-purpose, manipulative machine with several reprogrammable axes, which may be either fixed in place or mobile for use in industrial automation applications" .

For construction applications, robots have been categorized into three types:

1. Tele-operated robots in hazardous or inaccessible environments
2. Programmed robots as commonly seen in industrial applications
3. Cognitive or intelligent robots that can sense, model the World, plan, and act to achieve working goals

The important attributes of robots from a construction point of view are their (1) manipulators, (2) end effectors, (3) electronic controls, (4) sensors, and (5) motion systems.

Applications of Construction Robots

John Deere 690C Excavator

The John Deere 690C excavator is a tele-operated machine; that is, it is fully controlled by a human operating from a remote site. It is equipped with a model 60466T, six-cylinder, four-stroke turbocharged diesel engine, producing a maximum net torque of 450 ft-lb (62.2 kgf-m) at 1300 revolutions per minute (rpm). The engine propels the excavator at traveling speeds ranging from 0 to 9.8 mph (15.8 km/h). The arm on the 690C excavator has a lifting capacity of 11,560 lb (5243 kgf) over side and 10,700 lb (4853 kgf) over end. The rated arm force is 15,900 lb (7211 kgf), and the bucket digging force is 25,230 lb (11,442 kgf).

The John Deere 690C excavator has been implemented in a cooperative development program with the U.S. Air Force within the Rapid Runway Repair (RRR) project. The major task of the RRR is the repair of runways damaged during bombing raids. The Air Force is currently investigating other areas in which the 690C could be implemented, including heavy construction work, combat earthmoving in forward areas, mine-field clearing, and hazardous-material handling.

Robot Excavator (REX)

The primary task of the robot excavator (REX) is to remove pipelines in areas where explosive gases may be present. This robot is an autonomous machine able to sense and adjust to its environment. REX achieves its autonomous functions by incorporating three elements into its programming:

1. Subsurface premapping of pipes, structures, and other objects is possible using available utility records and ground-penetrating sensors. Magnetic sensing is the leading candidate for premapping metallic pipes.
2. Primary excavation for gross access near target pipes is possible. Trenching and augering are the leading candidates for this operation.
3. Secondary excavation, the fine and benign digging that progresses from the primary excavation to clear piping, can be accomplished with the use of a supersonic air jet.

The hardware that REX uses for primary excavation is a conventional backhoe retrofitted with servo valves and joint resolvers that allow the computer to calculate arm positions within a three-dimensional space. The manipulator arm can lift a 300 lb (136 kg) payload at full extension and over 1000 lb (454 kg) in its optimal lifting position. REX uses two primary sensor modes: tactile and acoustic. The tactile sensor is an instrumented compliant nozzle. The instrumentation on the nozzle is an embedded tape switch that is activated when the nozzle is bent. The second sensor employed in excavation is an acoustical sensor, allowing for three dimensional imaging.

Haz-Trak

Haz-Trak, developed by Kraft Telerobotics, is a remotely controlled excavator that can be fitted with a bulldozer blade for grading, backfilling, and leveling operations. Haz-Trak uses force feedback technology, allowing the operator to actually feel objects held by the robot's manipulator. The operator controls the robot's arm, wrist, and grip movements through devices attached to his or her own arm. Thus, the robot arm instantly follows the operator's movements.

Pile-Driving Robot

The Hitachi RX2000 is a pile-driving machine directed by a computer-assisted guiding system. It consists of a piling attachment (such as an earth auger or a vibratory hammer) directly connected to the tip of a multijointed pile driver arm. The pile driver arm uses a computer-assisted guiding system called an "arm tip locus control." Coordinates of arm positions are calculated using feedback from angle sensors positioned at joints along the arm. A control lever operation system is provided to increase efficiency.

The compactness of the RX2000 and its leaderless front attachment enable efficient piling work even in congested locations with little ground stabilization. Further, the vibratory hammer has a center hole chuck that firmly chucks the middle part of a sheet pile or an H-steel pile. Hence, pile length is not limited by the base machine's dump height.

Laser-Aided Grading System

Spectra-Physics of Dayton, OH, developed a microcomputer-controlled, laser-guided soil-grading machine. A laser transmitter creates a plane of light over the job site. Laser light receptors mounted on the equipment measure the height of the blade relative to the laser plane. Data from the receiver are then sent to the microcomputer that controls the height of the blade through electronically activated valves installed in the machine's hydraulic system. A similar device has been developed by Agtek Company in cooperation with a construction contractor in California . An automated soil-grading process implemented by these machines relieves the operator from having to manually position and control the grading blades, thus increasing the speed and quality of grading, as well as work productivity.

Automatic Slipform Machines

Miller Formless Systems Company developed four automatic slip form machines — M1000, M7500, M8100, and M9000 — for sidewalk and curb and gutter construction.

All machines are able to pour concrete closer to obstacles than is possible with alternative forming techniques. They can be custom-assembled for the construction of bridge parapet walls, monolithic sidewalk, curb and gutter, barrier walls, and other continuously formed elements commonly used in road construction.

The M1000 machine is suitable for midrange jobs, such as the forming of standard curb and gutter, sidewalks to 4 ft, and cul-de-sacs. The M7500 is a side mount-design machine for pouring barrier walls, paved ditches, bridge parapets, bifurcated walls, and other types of light forming jobs. The M8100 is a midsize system with a side mount design combined with straddle-paving capabilities. The machine can be extended to 16-ft (4.88-m) slab widths with added bolt-on expansion sections. The M9000 multidirectional paver is designed for larger-volume construction projects. It can perform an 18-ft (5.49-m) wide paving in a straddle position. Options are available for wider pours, plus a variety of jobs from curbs to irrigation ditches, in its side mount mode.

Horizontal Concrete Distributor

The HCD, developed by Takenaka Company, is a hydraulically driven, three-boom telescopic arm that cantilevers from a steel column. The boom can extend 66 ft (20 m) in all directions over an 11,000-ft (1000-m) surface area. A cockpit located at the end of the distributor houses the controls for an operator to manipulate the boom direction and flow of concrete. The weight of the robot is 4.97 tons (4508 kgf), and it can be raised along the column by jacks for the next concrete pour. On average, the relocation procedure takes only 1.5 h.

Shotcrete Robot

Traditionally, in tunneling work, a skilled operator has been needed to regulate the amount of concrete to be sprayed on a tunnel surface and the quality of the hardening agent to be added, both of which depend on the consistency of the concrete. Kajima Construction Company of Japan developed and implemented a semiautonomous robotic applicator by which high-quality shotcrete placement can be achieved [Sagawa and Nakahara, 1985].

Slab-Finishing Robot

The robot designed for finishing cast-in-place concrete slabs by Kajima Construction Company, is mounted on a computer-controlled mobile platform and equipped with mechanical trowels that produce a smooth, flat surface. By means of a gyrocompass and a linear distance sensor, the machine navigates itself and automatically corrects any deviation from its prescheduled path. This mobile floor-finishing robot is able to work to within 1 m of walls. It is designed to perform the work of at least six skilled workers.

Auto-Claw and Auto-Clamp

Two robotic devices used for steel beam and column erection on construction sites have been developed by Obayashi Construction Company of Japan. Both construction robots have been developed to speed up erection time and to minimize the risks incurred by steelworkers. Both have been implemented on real job sites. The auto-claw consists of two steel clamps extended from a steel-encased unit containing a DC battery pack, electrical panel, and microprocessor unit, which is in turn suspended from a standard crane. The two clamps have a rated capacity of two tons (1.824 kgf) each and can be adjusted to fit beam flanges from 8 to 12 in. (203.2 to 304.8 mm). The clamps are automatically released by remote radio control once the beam is securely in place. Fail-safe electronic circuitry prevents the accidental release of the clamps during erection by keeping the circuit broken at such times. The steel beams require no special preparation for using this robot.

The auto-clamp's essential purpose and mechanics are the same as for the auto-claw, except that the auto-clamp uses a special electro steel cylinder tube to secure and erect columns. A steel appendage plate with a hole in the center must be welded to one end of the column. The steel cylinder is electrically inserted and locked into the hole by remote control, whereupon the column can be erected.

The autoclamp has a rated lifting capacity of 15 tons (13,605 kg). The appendage plates must be removed after the columns are erected. Like the auto-claw, the auto-clamp is equipped with a fail-safe system preventing the cylinder from retracting from the hole during erection.

Automated Pipe Construction

Research into automated pipe construction is under way at the University of Texas at Austin. Research efforts are focused on developing and integrating three pipe production technologies: bending, manipulation, and welding. The pipe manipulator, was adapted from a 20-ton rough-terrain hydraulic crane with an attachment to the main boom. The attachment includes an elevating, telescoping, auxiliary boom with a wrist and pipe-gripping jaws. Associated research has concentrated on improving productivity through automated lifting and manipulating of horizontal piping.

Blockbots

Another application involves the design, development, and testing of the "blockbot" robot intended to automate the placement of masonry blocks to form walls. The complete wall assembly consists of four major components:

1. A six-axis "head" that will actually place the blocks on the wall

2. A 20- to 30-ft (6- to 9.1-m) hydraulic scissors lift used to roughly position the placement head vertically and longitudinally.
3. A large-scale metrology system, sensors, and other related computer control equipment.
4. A block-feeding system/conveyor to continually supply the placement head. To facilitate construction, the blocks are stacked upon each other with no mortar between the levels.

The wall is then surface-bonded using Sure wall, a commercial fiberglass-reinforced bonding cement. This process produces a wall with strength comparable to that of a traditional mortar wall.

Wallbots

Researchers at the Massachusetts Institute of Technology (MIT) are engaged in the Integrated Construction Automation Design Methodology (ICADM) project. This work attempts to integrate the efforts of material suppliers, architects, contractors, and automated construction equipment designers.

The process of building interior wall partitions is divided between two separate robots: a trackbot and a studbot. Circumventing the need for complex navigational systems, the trackbot is guided by a laser beacon aligned manually by a construction worker. The trackbot is separated into two parallel workstations: an upper station for the ceiling track and a lower station for the floor track. Detectors are mounted on the ends of the effector arms to ensure that the laser guidance system achieves the necessary precision.

The placement of the track consists of four steps: (1) the effector arm grabs a piece of track, (2) the effector arm positions the track, (3) two pneumatic nail guns fasten the track, and (4) the trackbot moves forward, stopping twice to add additional fasteners.

Once the trackbot has completed a run of track, the studbot can begin placing studs. Location assessment is made by following the track and employing an encoding wheel or an electronic distance measuring (EDM) instrument. The studbot then references a previously sorted floor plan to ascertain locations of studs to be placed. The stud is removed from its bin and placed into position. The positioning arm then spot-welds the stud into place.

Interior Finishing Robot

An interior finishing robot, can execute the following tasks: (1) building walls and partitions, (2) plastering walls and ceilings, (3) painting walls and ceilings, and (4) tiling walls. The arm of the robot has six degrees of freedom with a nominal reach of 5.3 ft (1.6 m) and a lifting capacity of 66 lb (145 kgf). The robot is designed to perform interior finishing work in residential and commercial buildings with single or multiple floor levels and interior heights of 8.5 to 8.8 ft (2.60 to 2.70 m). A three wheel mobile carriage measuring 2.8 ft (0.85 m) enables motion of the robot between static workstations.

Fireproofing Spray Robot

Shimizu Company has developed two robot systems for spraying fireproofing material on structural steel. The first version, the SSR-1, was built to (1) use the same materials as in conventional fireproofing, (2) work sequentially and continuously with human help, (3) travel and position itself, and (4) have sufficient safety functions for the protection of human workers and of building components. The second robot version, the SSR-2, was developed to improve some of the job site functions of SSR-1. The SSR-2 can spray faster than a human worker but requires time for transportation and setup. The SSR-2 takes about 22 min for one work unit, whereas a human worker takes about 51 min.

The SSR-2 requires relatively little manpower for the spraying preparation — only some 2.1 person-days compared with 11.5 for the SSR-1. As the positional precision of the robot and supply of the rock wool feeder were improved, the SSR-2 could achieve the same quality of dispersion of spray thickness as for that applied by a human worker.

Exterior Wall Painting Robot

The exterior wall painting robot, paints walls of high-rise buildings, including walls with indentations and protrusions. The robot is mounted on mobile equipment that permits translational motion along the exterior wall of a building. The robot consists of the following:

1. Main body that sprays paint
2. Moving equipment to carry the robot main body to the proper work position
3. Paint supply equipment
4. A controller

The robot main body consists of the following:

1. Main frame
2. Painting gun
3. Gun driver
4. Control unit

The painting gun is driven in three principal translational directions (*x*,*y*, and *z*). The painting gun is also provided with two rotational degrees of freedom. The robot moving equipment consists of the following:

1. A transporter that propels the moving equipment along the outside of the building being painted
2. A work stage on which the robot main body is mounted
3. A mast that serves as a guide for raising and lowering the work stage

The top of the mast is attached to a travel fitting, and the fitting moves along a guide rail mounted on the top of the building.

Integrated Surface Patcher (ISP)

Secmar Company of France developed a prototype of the integrated surface patcher (ISP). The unit consists of the following components:

1. A 19-ton (17,234-kgf) carrier with rear-wheel steering
2. A 3.9-yd 3 (3-m 3) emulsion tank
3. A 5.2-yd 3 (4-m 3) aggregate container
4. A built-in spreader working from the tipper tailboard (a pneumatic chip spreader with 10 flaps and a 10-nozzle pressurized bar)
5. A compaction unit

The ISP unit has a compressor to pressurize the emulsion tank and operate the chip-spreading flaps. The machine uses a hydraulic system driven by an additional motor to operate its functional modules. The electronic valve controls are operated with power supplied by the vehicle battery. The ISP is used primarily for hot resurfacing repairs, including surface cutting, blowing and tack coating with emulsion, as well as for repairs requiring continuous treated or non-treated granular materials.

The unit is suitable for deep repairs using aggregate-bitumen mix, cement-bound granular materials, and untreated well-graded aggregate, as well as for sealing wearing courses with granulates.

The current design of the ISP allows only carriageway surface sealing. It is thus not well suited for surface reshaping or pothole filling. It is used only for routine maintenance tasks. In operational terms, ISP is not capable of on-line decision making on how to proceed in the case of an irregular crack or other non-predetermined task. However, automated patching can be started manually or automatically, depending on the presence of optical readers mounted on the equipment that read the delimiters of the work area, and on the mode of action chosen by the operator.

Autonomous Pipe Mapping

Another application is the development of an automated pipe-mapping system. Current manual methods are slow, inefficient, qualitative, and non-repetitive. The intention of the system is to autonomously establish size, depth, and orientation of buried pipes. This knowledge is extremely valuable in guiding excavation, validating as-built drawings, and building databases of piping details.

The system is composed of a computer-controlled Cartesian x-y table that allows various sensors to be swept across an arid area. The primary mapping is completed by a magnetic sensor that reads and records magnetic field intensities. These intensities are manipulated and interpreted, resulting in a line drawing representing the pipe locations. Higher-level processing estimates the depth of pipes and identifies interconnections such as elbows, tees, and crosses.

Terregator

A machine that may be used to transport the autonomous pipe mapping system is the terregator. Designed for autonomous outdoor navigation, it can be directly applied on a construction site. The terregator has been specifically designed to be extremely durable and powerful in order to prevent problems that inhibit machines designed for interior use. Its gearing is adjustable to allow it to be configured as a low-speed, high-torque machine or as a high-speed, low-torque machine. The terregator has a six-wheel-drive design to ensure mobility on rough terrain.

The terregator is also designed as a fully enclosed modular system to facilitate repairs, additions, or system improvements. The subsystems include locomotion, power, backup power, computer and controls, serial links, sensors, and a video link.

ODEX

ODEX III, developed by Odetics, Inc., is a six-legged, tele-operated, high-strength robot designed for inspection, surveillance, and material handling in nuclear power plants and outdoor hazardous environments

Numerical Control

Numerical control refers to control of construction equipment using numbers. Questions such as "What numbers are used to control a piece of equipment?" and "In what format are they presented to the equipment?" are basic to understanding numerical control. Numerically controlled equipment consists of a machine control unit (MCU) and a machine tool (such as an end effector). The MCU cannot think, judge, or reason in relation to the environment in which it works. The machine accepts and responds to commands from the control unit

For example, a numerically controlled pumped-concrete placement system may use numbers corresponding to (1) position (x, y, z) of the discharging end of the placement pipe, (2) pumping pressure, and (3) the speed at which the discharging end of the placement pipe travels.

Review Questions

1. Give an essay on development of Information and Communication Technology in Construction.
2. What are the Current Information Technologies available for the Construction Sector?
3. Explain the following terms:
   - Computer Aided Design and Visualization
   - Building Engineering Applications
   - Computer Aided Cost Estimation
   - Planning, Scheduling and Site Management
   - Computer Aided Facilities Management
   - Business and Information Management
4. What are the various applications of Computers in Construction?
5. What do you understand by Construction Automation? Describe various fields in construction where automation is being successful.

12

Ethics in Construction

The Importance of Ethics in Construction

As with any industry, integrating ethics into operations should be a top priority, and the construction industry is no different. Good relationships with partners, developers and subcontractors are vital when working on multi-million dollar projects. If any of these parties engage in unethical activities or building practices, it can be life threatening for both the workers and future residents. The construction industry generates a lot of revenue, which makes it an easy target for fraudulent or unethical activity. Some may be tempted to cut corners or commit environmental misconduct in order to speed up a project timeline and bring in more cash flow. We are committed to creating beautiful and safe housing for every resident in our communities, without compromising our values, to maintain our reputation, relationships with developers and ultimately our own safety.

According to the Construction Management Association of America (CMAA), there is a code of ethics that construction professionals should pledge to keep every project site a place of honest building. This includes obligations to the public, the client, the profession and the environment. When managers and everyone on the jobsite acknowledges and educates themselves on these obligations, chances of mistakes or unethical decision making severely decreases.

From the Top Down

Leaders should hold not only their team, but also themselves to the highest moral code. Employees from across all experience levels should be able to look to executive leadership as a model of how they should represent the company and conduct business. It is critical for leaders in the construction industry to be well versed in every law, trend and practice in which they operate. This gives them the resources they need to ensure they are abiding by law and leading a safe and ethical organization.

Ethically Educate the Team

Ethical decision making is something that can be taught. This is crucial in developing a company that values doing what is right. Construction firms can set specific processes in place to negate any issues that may arise. When training a team, construction professionals should provide channels of feedback. This includes maintaining an open door policy, reviews, internal audits and situational advice that will help guide your team members to make smart decisions.

Take Action

As much as you can prepare to handle ethical issues in the workplace, situations will still arise that need to be managed. When a member of your team comes to you with a concern, it is crucial to act on it and have a clearly defined plan of action when a report comes through. This can include party interviews and written statements. A situation may even require external legal counsel, but it's best to handle matters internally. This could include reprimand, an improvement plan or termination. By the same token, ethical decision making should be rewarded. This will act as an incentive and encourage the team to keep up the good work.

Need for more Ethical Leaders in Construction

According to the World Economic Forum, the construction industry accounts for about 6% of global GDP with annual revenues exceeding US$10 trillion dollars. For developing countries, construction accounts for more than 8% of GDP and employs more than 100 million persons worldwide. As of Q2 2020, construction accounted for about 5.3% of Trinidad and Tobago's GDP and employed some 80,000 persons. Despite the industry being a driver of economic growth and development, its complex nature, intricate stakeholder relationships, and prevailing culture have given way to numerous instances of unethical behaviour and the perceived inability of its leaders to openly acknowledge and confront these shortcomings.

According to Contractor, unethical conduct adversely affects the cost of construction. In a survey conducted in 2004, 61% of respondents estimated that for every US$1 million in project costs, $5,000 to $50,000 was lost to or "unaccounted for" in some sort of an unethical transaction. It is virtually safe to say, that most construction professionals have witnessed or encountered some form of unethical activity on their projects.

Within recent years, the prevalence of such conduct has increased. Some of the global industry's more prominent contractors and construction companies have openly faced fines, financial losses, and reputational damage arising from ethical breaches having engaged in such activities as front-end loading, bribery, corruption, collusion, unfair conduct, false claims, cover pricing, fraud, misrepresentation, and modern slavery.

In Trinidad and Tobago, the issue of unethical behaviour in construction appears to be of minimal concern to industry stakeholders. Apart from a few cases of improper conduct cited by investigative journalists, organizations and individuals involved in similar activities have not faced any consequences for their actions. A singular set of ethical standards or a code of conduct which governed industry behaviour could not be found. Noticeably absent were any published empirical research or studies which examined the prevalence and economic impact of unethical practices on the industry. Internationally, the issue of ethics in construction is of increasing concern, being more widely discussed and studied.

Findings of Research

A 2017 study conducted in the United States revealed that over 80% of architects, contractors, project managers, and other key industry stakeholders observed or experienced untrustworthy behaviour such as bribery, conflicts of interest, and the use of gifts in construction. The study concluded that the majority of unethical practices typically occurred prior to a project's award and the promotion of an ethical code among professionals might have helped to reduce such unethical practices.

In another study on the unethical behavior of professionals involved in projects in the Gaza Strip, the author concluded that its prevalence was high, could be largely attributed to the personal value systems of individuals, and recommended the development and implementation of ethical codes of conduct which promoted greater transparency and accountability amongst professionals and other industry stakeholders.

A comprehensive study on the ethical issues plaguing the Pakistani construction sector revealed that the industry was tainted by unethical practices with corruption and bribery, bid shopping, and fraud being the more prevalent. It concluded that these and other practices had adverse impacts on the quality management and performance of works, recommending that codes of ethical conduct be introduced throughout. Quite remarkable was the authors' call for more ethical leaders as industry role models within the industry.

Meaning to be Ethical

Ethics is a branch of moral philosophy that involves the systematizing, defending, and recommending of concepts of right and wrong. According to the Markkula Center for Applied Ethics, it also refers to consistent and well-founded standards that prescribe how we ought to conduct ourselves in various personal and professional situations, typically in terms of rights, obligations, and benefits to society, fairness, and natural justice.

When asked, many may equate ethics to certain perceptions, biases, or feelings held on an issue, what the law prescribes, or some other accepted societal norm. These rationalizations can at times, diverge from the principle of doing the right thing. For instance, a person following his or her feelings may withdraw from doing what is right, especially when the situation demands or necessitates it. Laws incorporate various ethical standards, but history has shown that they can deviate from what is ethical. Most individuals adhere to general standards and patterns of ethical behaviour, but organizations, industries, and even societies sometimes behave in an unethical or morally corrupt manner.

When one speaks of being ethical, it should be of those thoughts, decisions, and actions which are consistent with agreed-upon or mutually established codes of moral conduct and professional behaviour. These are foundational to who we are and what we represent. On construction projects, strong ethical principles and standards should therefore begin in and be established by its leadership and permeate throughout all aspects of the organization, thereby defining the team's overall thinking, culture, and operations. Embracing an ethically oriented organization has been shown to have a positive impact on the quality of people, relationships, and reputation that construction companies and projects both attract and enjoy.

In 2003, the Society for Construction Law considered the question of ethics in the construction industry, published a report which presented a multi-disciplinary ethical code covering the activities of all professionals involved in the construction industry, irrespective of their qualification or affiliation. The code prescribed that ethical behavior should comply with these seven principles:

- **Honesty.** Act with honesty and avoid conduct likely to result, directly or indirectly, in the deception of others.
- **Fairness.** Do not seek to obtain a benefit which arises directly or indirectly from the unfair treatment of other people.
- **Fair reward.** Avoid acts which are likely to result in another party being deprived of a fair reward for their work.
- **Reliability.** Maintain up to date skills and provide services only within your area of competence.
- **Integrity.** Having regard for the interests of the public, particularly people who will make use of or obtain an interest in the project in the future.
- **Objectivity.** Identity any potential conflicts of interest and disclose the conflict to any person who would be adversely affected by it.
- **Accountability.** Provide information and warning of matters within your knowledge which are of potential detriment to others who may be adversely affected by them. Warning must be given in sufficient time to allow the taking of effective action to avoid detriment.

In thinking about the future of our own industry, we too recognize that the introduction of ethical standards and codes of conduct at all levels would serve to improve project performance, outcomes, and the value created for clients and stakeholders. Any view that seeks to diminish its importance should not be accepted, as its proponents have historically promoted the misuse and mismanagement of capital and other resources, engendered inequality, enjoyed personal benefits and gains, and contributed to the construction of substandard buildings and infrastructure which may not have conformed to design specifications or be fit for use.

The industry has evolved to a point where the motives, decisions, integrity, and actions of its leaders can easily be called into question. The level of trust placed in our systems of project governance is waning. The current reality and imminent outlook demands the nurturing and development of the kind of moral leader who consistently does the right thing for the right reasons and empowers others to do the same, without fear or compromise. Leaders who seek not only to do what is right for the project and client, but all other stakeholders as well.

CMAA Code of Ethics

The CMAA Code of Professional Conduct represents the aspirations of the profession relative to clients, owners, colleagues, and the public.

"Construction Managers" as used in this Code is a term that applies to CMAA members in performance of their services as Construction and Program Managers. Construction Managers provide a broad range of services to their clients and employers that encompass program, project, and construction management and also serving as owner advisors or representatives.

The code is rooted in the historical values of the profession and linked directly to the CMAA Statement of Values and Policy Framework.

Four fundamental principles form the foundation of the code: Ethical Practice, Professional Excellence, Responsibility to the Public, and Client-Centered Practice.

As a member of CMAA, I/we adhere to the following Code of Professional Conduct:

Ethical Practice

1. Construction Managers should be guided in all their relationships by the highest standards of integrity and honesty.
2. Construction Managers should conduct themselves honourably, responsibly, ethically, and lawfully so as to enhance the honour, reputation and value of the profession.
3. Construction Managers should avoid conduct or practices that deceive the public or represent a real or perceived conflict of interest.
4. Construction Managers should respect the rights of others and should not discriminate on the basis of race, colour, gender, marital status, religion, national origin, age, disability, or sexual orientation nor knowingly violate any law, statute, or regulation in the performance of professional services. Construction managers should strive to create a diverse workforce.
5. Construction Managers should have a zero-tolerance policy for any form of harassment including sexual harassment and bullying.

Professional Excellence

6. Construction Managers should perform services only within their areas of competence and qualification.
7. Construction Managers should contribute to the advancement of the program, project, and construction management profession by using best practices, continuing their professional education, and contributing to the development of the future workforce.

Responsibility to the Public

8. Construction Managers should hold paramount the health, safety, and welfare of the workplace and the public.
9. Construction Managers should guide and aid in defining and meeting objectives for environmental sustainability and resiliency throughout a project's life cycle.

Unethical behaviours in Construction and Need to Care

(by Paul Netscher, 2017)

Are you ethical?

Unfortunately, many businesses carry out unethical practices. Contractors and the construction industry are often viewed as being unethical. Regrettably, there are many contractors who are unethical, but equally, there are many clients who also engage in unethical practices and don't pay contractors the monies they are due. But not all contractors are unethical, and nor are all clients unethical.

Importance of Ethics

Being ethical sets the tone of the relationship between the contracting parties. As soon as one party partakes in unethical practices we see the relationships begin to break down, often causing the construction project to suffer. When unethical behaviour is uncovered it leads to a breakdown in trust. The other party is left wondering what other unethical behaviours have been concealed and haven't been noticed.

But good ethical practices are an essential part of a construction company's reputation. Who wants to employ a contractor that engages in unethical practices? But also, who wants to work for a client that is unethical?

Unfortunately unethical behaviour is sometimes contagious. If management engages in unethical behaviour you can almost be certain that their employees also engage in unethical practices. When one party is unethical we sometimes see an unethical response from the other party. When someone engages in an unethical practice we often find they get caught-up with more unethical practices. Once someone has paid, or received, a bribe it's easier to repeat the offence in the future. If management are corrupt in their dealings with the client, or contractor, they sometimes engage in other unethical practices, even stealing from their own company.

Our ethics, or lack of ethics, express who we are, and define our dealings with others.

Behaviours that can be viewed as being unethical

1. **Not ensuring your project is safe**. This includes making sure neither your workers nor other people will be injured. To not care about a person's well-being or safety is unethical.
2. **Giving the client a poor-quality project**, or one that is defective. Frequently contractors knowingly hide mistakes and poor quality work. In some cases these defects have resulted in buildings collapsing, even killing people. Contractors have signed a contract which binds them to delivering particular quality requirements and specifications. To deliver less is, in essence, a form of theft.
3. **Deliberately omitting items**, or supplying products that do not conform to the project specifications. Some contractors try and save money by purchasing substandard materials. There are even reported cases of contractors not installing all of the reinforcing in concrete structures, sometimes even removing reinforcing steel after the work has been inspected by the client. In 2016 a building collapsed in an earthquake in Asia revealing the concrete foundations were filled with empty oilcans which were used by the contractor to save concrete.
4. **Paying bribes** or providing incentives to the client's representatives so they award the construction project to you, or so they accept work of a poor standard, or agree to approve variation claims which are inflated or bogus. Sometimes contractors or clients pay bribes to the authorities to 'smooth' the project approval process. Paying bribes leads to more problems when the person receiving the bribe accepts it as standard practice, demanding more and bigger bribes from other parties. Paying bribes reduces productivity and increases costs for everyone.
5. **Not paying employees their agreed wages** and salaries at the agreed time. These wages and salaries should be above the minimum wages for the country and state, and should also be a 'fair' wage. Employees who aren't paid fairly won't be productive. They won't give their best, which will ultimately cost the company more.
6. **Employing illegal migrants**. Some companies employ workers without the correct documentation. Often these workers are literally held hostage, unable to leave while paid wages below the minimum.
7. **Accommodating workers in cramped, dirty accommodation**. Workers deserve to be treated fairly. Happy employees are productive workers.
8. **Contractors that 'rig' or fix prices**. Usually clients invite a number of contractors to price their construction project, with the expectation that the bidding or tendering process will be competitive and the contractors will submit the cheapest possible price. Contractors have been known to 'organise' these bidding processes. This could be done in several ways, including all those pricing the project adding in a pre-

agreed 'losers fee', which is a fee that the winning contractor will pay to the other contractors that priced the project. A fee that the client actually, unbeknown to them, is paying. In other instances, contractors pre-arrange who will have the 'lowest' price. Often this price is inflated, and then the other contractors must ensure that their prices are more expensive than the contractor who has the 'lowest' price. On the next project another contractor is provided the opportunity to submit the 'lowest price', and so work is shared around amongst contractors. This means that clients pay more for their projects, but it also results in contractors becoming 'lazy', profits are easy to come by so there is little incentive to improve productivity, save costs and work smarter. These contractors can find it difficult to work in a competitive bidding environment.

9. **Clients awarding projects to contractors when the contractor's price is clearly too low**. Now many would say there is no harm done by this. I don't believe that clients should profit from contractor's mistakes. But, there is a real danger when clients award a project to a contractor whose price is too low. The contractor could become bankrupt part way through the project, leaving the client with delays and additional costs when they have to employ another contractor to finish the project. Contractors who are losing money on a construction project may be tempted to take short cuts, they may include substandard materials to save money. Contractors who are losing money probably won't put sufficient resources on the project which will jeopardise the construction schedule and quality on the project. There's nothing wrong with awarding a project to the lowest bidder providing they have the resources and the skills to successfully deliver the project. If the contractor has obviously made a mistake with their price then clients should have the integrity to point the error out, or select a contractor with a reasonable price.
10. **Clients that 'bid shop'.** When a number of contractors have competitively priced the project the client tells their preferred contractor what price they must provide to beat the other contractors. Contractors that submitted the lowest price have wasted their time since the client has rigged the bidding process and only ever intended to award the project to their favoured contractor. Eventually contractors won't submit prices to these clients, who will ultimately pay more for their projects in the future.
11. **Clients that don't pay their contractors in accordance with the contract and contractors who don't pay their subcontractors and suppliers on time.** Those that make unfair deductions, or withhold monies for no due reason. Who wants to work for these companies in the future? Those that do price work for these companies will add extra margin to cover for the inconvenience of working for the company.

12. **Clients that don't respond to variation claims that** deliberately ignore claims, or refuse to pay variation claims for no reason.
13. Contractors that submit **bogus or overly inflated variation claims**. This irritates clients who will be less inclined to deal fairly with legitimate claims. It negatively impacts the contractor's reputation.
14. **Vexatious legal disputes** raised by any of the parties. Usually contract documents are clear, but moreover, one expects all parties to enter a contract in a fair and honest way. Legal disputes take time and cost money, and the only winners are usually lawyers. I've found that most disputes can be resolved with clear and honest dialogue.
15. **Poor environmental practices**. Both clients and contractors are sometimes guilty of dumping hazardous waste in normal garbage. Some contractors dump construction waste in vacant lots or in parkland. Oil and dangerous chemicals are knowingly dumped on the ground or in storm water drains. Contractors who carry out illegal environmental practices probably engage in other unethical practices. Clients should ensure that they and their contractors engage in good environmental practices. If we care for our environment we are probably going to care for people. It's about respect.
16. **Theft** of any sort. This includes employees who don't work the full hours they claim, and the unauthorised use of company property.

Behaviours that shouldn't be viewed as being unethical

Unfortunately contractors are sometimes accused of being unethical and some may disagree with some of the points below.

1. **Submitting variation claims for delays, extra work or changes**, where the contractor is entitled to claim for these in terms of the contract, and providing these are a fair reflection of their entitlements. Some clients detest contractors who submit variation claims, even when these claims are correct.
2. **Making a large profit on a project**. Many clients believe that contractors aren't entitled to make a large profit on a project and the only reason they have made significant profits is because the contractor has 'ripped the client off'. They regard these profits as being ill-gotten gains. Construction is a risky business and sometimes contractors lose money on a project. On occasion projects do go well, the risks the contractor allowed for in their price didn't eventuate, sometimes the contractor developed innovative solutions which reduced costs, possibly productivities were better than allowed, or the contractor bought materials for less than they expected. Yes, things do sometimes go favourably – just as sometimes punters do win at the casino or on the lottery! Should we begrudge them their wins? Sometimes when the construction market is buoyant and there's plenty of work contractors

can lift their prices and increase their profit margins. Inevitably, due to the cyclical nature of construction, the volume of work will reduce and profit margins will fall again. Contractors are entitled to make big profits occasionally, providing these are made through legitimate means.

3. **Deducting monies** from contractors, subcontractors or suppliers, where these are legitimate charges for items which they should have supplied but didn't supply. However, it's good practice to timely notify the contractor, subcontractor or supplier of these charges, and where possible to agree the quantum of the charges beforehand.

Construction Ethics: 7 Examples of Ethical Practices

(by Christopher Bauer, 2011)

Here are some tips on things to do-and not do-to help keep your focus on doing the right thing through both the good times and the bad:

1. **Know your company's values, and let those values dictate each and every business decision you make**. Don't have a values statement? Don't worry-you're in good company as very few construction businesses do. Write one, and write it carefully. You want it to include a short list of the most persistent, most important priorities that should drive the behavior of every member of your company. Your values statement needs to be so clear so that any employee, at any moment and in any situation, can decide whether his/her intended actions are aligned with these essential values or not. Writing a great values statement is far more difficult than it sounds. However, taking the time to get it right is one of the best tools you can have for assuring that your entire company is doing what they are supposed to be doing, regardless of any pressure to do otherwise.
2. **Know the law**. That may sound obvious, but it's amazing how many construction business owners can't list many of the laws that apply to their business. Without knowing the relevant business laws and how to ethically apply them, you're not even in the game.
3. **Remember that you're not the only one going through tough financial times right now**. Why does that matter? It's not just because it might help you feel a little less alone-it's because there are probably more potential clients than ever out there who would be happy for you to cut ethical or legal corners to help their bottom line. Sweetheart deals are on the rise in many places and conflict of interest actions appear to be at an all-time high. Just because another business owner or a local official tells you that something wrong is really okay, that doesn't make it so. If you are approached about a contract that even feels shady-let alone one that you know to be shady-run, don't walk, away from that other person or company as fast as you can.

4. **Pay attention to your thinking and, in particular, to your rationalizations**. If you hear a voice in your head-or words from someone else-trying to make excuses for an action that doesn't seem right, pay attention; this is one of the easiest ethical "red flags" for ethics problems to identify. If you need to try to justify your actions-especially if it's before you've even done whatever it is-what are the chances that it's really something right? The chances are low. Of course, trying to rationalize something doesn't necessarily mean that the behavior is wrong. It's just the sign of a risk. Once you become aware of a potential legal or ethical problem, though, you can't ignore it. To do so simply leaves the door open to stepping over the ethical or legal line, however much you may say that doing so isn't your intention.
5. **Pay attention to your gut**. If you feel like something may not be right, it probably isn't. Mind you, our guts can steer us wrong, but it's usually the other way around. In other words, we can often convince ourselves that something wrong is really okay (that's why we need to pay such close attention to rationalizations), but we rarely tell ourselves that something okay is actually wrong. If it feels like you're heading down the wrong path, pay close attention. If your gut is correct, it's never too soon to keep yourself from stepping over the line, nor is it ever too late to stop something inappropriate that you've been doing.
6. **Not sure whether or not something is ethical or legal?** Seek out knowledgeable peers and ask them. There's never a problem with being unsure whether or not something is ethical or legal. However, there is a huge program with being unsure and then doing nothing to find out.
7. **Everyone is in this together, and the reputation of the construction industry depends on every member of the industry**. If you see another construction business owner or one of his/her employees doing something wrong, you can't simply turn a blind eye. To do so simply isn't holding up your end of the bargain. Be blunt with the other owner about your concerns-tactful, but honest. Let him/her know the nature of your concern, and give them a chance to explain. After all, you may have misunderstood what he/she is doing or, in fact, may actually be the one doing things incorrectly. You'll never know, though, unless you have that direct conversation. Will it feel tough to confront them? Maybe. But think of it this way-if someone else had concerns about you or one of your employees, would you rather they alerted you to it or simply ran down your reputation to others?

Ethics are funny. Everyone in business agrees that they (ethics) are critically important but believe that it's every other company who needs to pay closer attention to them. Most business owners believe that they somehow could not possibly have a problem and that it's always some other person or some other company doing things wrong. Obviously, though, it can't always be someone else.

The reality is that each one of us has the potential to cross the line; it's just a matter of whether or not we land in the perfect storm in which we have a need or wish (usually financial or interpersonal) and the opportunity to try to meet that need by doing something in spite of it being wrong. Especially in the current economy, many more of us are feeling financial and other pressures and that can be fertile ground for thinking about stepping over the line. None of that means that we are bad people-simply that we are human.

Fortunately, sooner or later the business climate will again improve. Like it or not, though, the risks we all carry for potential ethical problems will be with us all along. These seven tips will make it easier to stay on the right side of the ethical line no matter what shape the economy takes.

Code of Ethics for Corporate Members, Institution of Engineers (India)

(Effective from March 1, 2004)

Preamble

- The Corporate Members of The Institution of Engineers (India) are committed to promote and practice the profession of engineering for the common good of the community bearing in mind the following concerns:
  - Concern for ethical standard;
  - Concern for social justice, social order and human rights;
  - Concern for protection of the environment;
  - Concern for sustainable development;
  - Public safety and tranquillity.

The Tenets of the Code of Ethics

- A Corporate Member shall utilise his knowledge and expertise for the welfare, health and safety of the community without any discrimination for sectional or private interests.
- A Corporate Member shall maintain the honour, integrity and dignity in all his professional actions to be worthy of the trust of the community and the profession.
- A Corporate Member shall act only in the domains of his competence and with diligence, care, sincerity and honesty.
- A Corporate Member shall apply his knowledge and expertise in the interest of his employer or the clients for whom he shall work without compromising with other obligations to these Tenets.
- A Corporate Member shall not falsify or misrepresent his own or his associates' qualifications, experience, etc.
- A Corporate Member, wherever necessary and relevant, shall take all reasonable steps to inform himself, his employer or clients, of the environmental, economic, social and other possible consequences, which may arise out of his actions.

- A Corporate Member shall maintain utmost honesty and fairness in making statements or giving witness and shall do so on the basis of adequate knowledge.
- A Corporate Member shall not directly or indirectly injure the professional reputation of another member.
- A Corporate Member shall reject any kind of offer that may involve unfair practice or may cause avoidable damage to the ecosystem.
- A Corporate Member shall be concerned about and shall act in the best of his abilities for maintenance of sustainability of the process of development.
- A Corporate Member shall not act in any manner which may injure the reputation of the Institution or which may cause any damage to the Institution financially or otherwise.

General Guidance

The Tenets of the Code of Ethics are based on the recognition that –

- A common tie exists among the humanity and that The Institution of Engineers (India) derives its value from the people, so that the actions of its Corporate Members should indicate the member's highest regard for equality of opportunity, social justice and fairness;
- The Corporate Members of the Institution hold a privileged position in the community so as to make it a necessity for their not using the position for personal and sectional interests.

And, as such, a Corporate Member

- should keep his employer or client fully informed on all matters in respect of his assignment which are likely to lead to a conflict of interest or when, in his judgement, a project will not be viable on the basis of commercial, technical, environmental or any other risks;
- should maintain confidentiality of any information with utmost sincerity unless expressly permitted to disclose such information or unless such permission, if withheld, may adversely affect the welfare, health and safety of the community;
- should neither solicit nor accept financial or other considerations from anyone related to a project or assignment of which he is in the charge;
- should neither pay nor offer direct or indirect inducements to secure work;
- should compete on the basis of merit alone;
- should refrain from inducing a client to breach a contract entered into with another duly appointed engineer;
- should, if asked by the employer or a client, to review the work of another person or organisation, discuss the review with the other person or organisation to arrive at a balanced opinion;

- should make statements or give evidence before a tribunal or a court of law in an objective and accurate manner and express any opinion on the basis of adequate knowledge and competence; and
- should reveal the existence of any interest – pecuniary or otherwise – which may affect the judgement while giving an evidence or making a statement.

Any decision of the Council as per provisions of the relevant Bye-Laws of the Institution shall be final and binding on all Corporate Members.

Case Study

Ethical Issues in the Construction Industry: Contractor's Perspective (Hamimah Adnan et al., 2012)

Introduction

The construction industry is classified as the most fraudulent industry worldwide (Transparency International, 2005) providing the perfect environment for ethical dilemmas, with its low-price mentality, fierce competition and paper-thin margins (FMI, 2004). This industry is considered to be one of the most susceptible to unethical practices because it involves substantial capital investments, providing large scale opportunities for rent extraction as well as investments that usually cannot be redeployed after implementation. Unethical practices can take place at every phase of a construction project – during planning and design, pre-qualification and tender, project execution and operation and maintenance. Such practices can result in projects which when completed are considered unnecessary, unsuitable, overlay complex components, overpriced or delayed.

The effects of unethical practices have lasting impact detrimental to construction and engineering companies such as wasted tender expenses, tendering uncertainty, increased project costs, economic damage, blackmail, criminal prosecutions, fines, blacklisting and reputational risk. The issue of professional ethics has sparked enthusiastic concern and pragmatic discussion among the general public and there is growing demand by the current literature for good ethical practices and professional behavior in the construction industry. In the face of its size and universality, the construction industry is often cited as plagued with graft and malpractices. Common issues highlighted are tendering practice, substandard quality of construction work, safety culture, payment woes, corruption and most importantly, public accountability for money spent on public buildings and infrastructure. The scenario in the Malaysian construction industry is no exception, where graft and malpractices are numerously reported in the media and regarded as synonymous with construction players.

Ethical issues in the construction industry not only concern bribery or corruption but also conflict of interest and collusive tendering. Transparency International (2005) illustrates how corruption can add up to 25% to the cost of public contracting, generating waste of public resources, missed development opportunities, and unstable environment for businesses. The statistics also further indicate that the scale of corruption is higher in construction than any other of sector of the economy. For the Malaysian construction industry, reports indicate 17.3% of 417 government projects were considered 'sick' (Malaysia Ministry of Works, 2005) and of these sick projects, some of them were caused by unethical behaviours among the project participants.

One study found that poor management relating to poor inspection programmes, poor safety policies and lack of safety education programmes, and unsafe working methods as the main reasons behind these statistics. A lot of these companies are not even legally compliant, let alone going beyond compliance.

With the rise in unethical issues in the construction industry, contractors have always been perceived as the main culprits and blamed for the defects that happen in the industry. The thousands of successfully constructed projects that have stood the test of time without any defects are ignored by the media when reporting the defective projects. Further, it is generally forgotten that there are two other main players, the clients and the technical consultants, who may be contributory factors responsible for the defects in the project. Hence, fundamental questions related to unethical issues were raised; what is the common unethical behaviours evidenced by the contractors during the life cycle of the project? How seriously are the unethical practices affecting the industry and in what ways can they be reduced?

Literature Review

The Malaysian construction industry is generally classified into two categories namely, general construction and special trade. General construction comprises three subsectors which are residential construction, non-residential construction and civil engineering construction. The second category, special trade, involves activities such as metal works, electrical, plumbing, sewerage and sanitary, refrigeration and air-conditioning, painting, carpentry, tiling and flooring and glass.

Recent studies show that fraud in construction businesses is widespread with the manufacturing, construction and engineering sectors as those highly involved in fraudulent activities (Fraud Survey, 2009). At the same time, there is also evidence that poor governance disproportionately affects the poor, as they lack resources for alternatives. Since investment in infrastructure is a critical driver of economic growth, poor governance, mismanagement and corruption in construction projects can undermine social and economic development outcomes and serve as a structural brake on sustainable development.

There are many reasons why professionals are involved in unethical practices. This may be due to insufficient legislative enforcement, fierce competition, the economic downturn, insufficient ethical education from schools and professional institutions, cultural changes and high complexity of construction works. The Malaysian construction industry is no exception. Even though there are guidelines and acts to guide the procurement process, ethical issues still arise in projects, leading to lower project quality outcomes. In Malaysia, a number of high-profile public sector projects are muddied by problems; the headquarters of the Malaysian External Trade Development Corporation; delay in completion of school buildings; school computer labs; cracks in 31 pillars of the Middle Road Ring Two (MRR2) projects; and most recently, the collapsed roof of the Sultan Mizan Zainal Abidin Stadium in Terengganu.

The Malaysian government is taking bold steps to address these ethical issues through a number of redemptive measures. The reforms include the establishment of the Anti-Corruption Agency, followed by the establishment of the Anti-Corruption Commission Act, 2009. At the same time, the Construction Industry Development Board (CIDB) launched the Construction Industry Master Plan (CIMP) in 2004, followed by the establishment of the Code of Ethics for contractors in 2007. This code demands honesty and integrity in performing responsibilities, compliance to laws and regulations, respect for individuals and society and the realization of the importance of quality, skills, health, safety and environmental preservation.

In 2004, the Integrity Institute of Malaysia launched the National Integrity Plan (NIP) in a bid to reform integrity issues in Malaysia. The plan highlights the preventive approach to corruption by inculcating the values of honesty, integrity and ethics in individuals and society. The objectives of this plan are to establish a fully moral and ethical society whose citizens are strong in religious and spiritual values, with the highest ethical standards. With the current scenario where open competition based on equality in the procurement process is prevented, by an affirmative action policy that ensures opportunities for the Indigeneous Malay (Bumiputra) business community, there has been a consequential rise in unethical issues such as corruption and conflict of interest.

Various literatures have shown that most unethical behaviour in the construction industry takes the form of unfair conduct, negligence, conflict of interest, collusive tendering, fraud and bribery. Furthermore, in terms of professional conduct, it has been identified that the majority of professionals believe that obligations to the client and public are of equal. Vee and Skitmore (2003) have further classified unethical practices into four general types of actions which are unfair conduct, conflict of interest, collusion, fraud and bribery.

Unfair Conduct

Unfair conduct may occur in competition (unfair competition), in contracts (unfair contract terms), in staff promotion/dismissal/demotion (unfair labor practices) and in business practices (unfair business practice). For example, the following features of unfair conduct are been noted (Commonwealth of Australia, 1997):

- Little or no ability to negotiate terms of the contract (pro forma 'take it or leave it' contracts are used).
- Inadequate disclosure of relevant and important commercial information which the weaker party should be aware of before entering the transaction.
- Inadequate and unclear disclosure of important terms of the contract, particularly those which weighted against the weaker party.
- The dominant parties seek to vary the nature of a long – term relationship so that it is more favorable to them but which affects the viability of the weaker party.
- When disputes do arise there is often no quick, cheap and market – sensitive way of settling them; and even where such interventions do exist, there is a reluctance by weaker parties to access any remedial action through fear of reprisal.

In their survey, Vee and Skitmore (2003) combine unfairness with dishonesty. It could be argued that dishonesty constitutes illegal behavior and is thus more appropriately linked to fraud. Instances quoted by Vee and Skitmore (2003) as being 'unfair' include actions by government agencies, such as 'biased tendering evaluation system, the process of re-tendering and shopping for prices after tenders have closed', as well 'making it difficult for private businesses to compete'. Other instances relate to professional fees, such as 'non-payment of consultants' fees by clients and developers after engaging their services. The latter types of complaint is easier to prevent if consultants employ defensive contractual strategies, thereby helping to distinguish between business practices that are 'unfair' and those that are illegal.

Conflict of Interest

This is a situation in which someone in a position of trust, such as a lawyer, a politician or an executive or director of a corporation, has competing professional or personal interests which could make it difficult to fulfil his or her duties impartially. Even if there is no evidence of improper action, a conflict of interest can create an appearance of impropriety that can undermine confidence in the ability of that person to act properly in his/her position. Conflict of interest involves a personal interpretation of whether or not certain behavior is commonly acceptable, rather than if it is illegal. The onus should clearly be placed upon individuals to declare all possible instances of potential negative interpretation, before proceeding with projects. Vee and Skitmore's (2003) study suggests that there is a high incidence of unacceptable behavior of this type in the Australian construction industry.

Their study identified the internal and external (environmental and situational) factors that influence the moral decision- making process that leads to participation in collusive tendering agreements. The three most important factors are in the external environmental: whether I would be held legally liable for the action, whether there is a legal issue involved. Their research indicates that the professional's decision whether to become involved in collusion is determined primarily by the legal implications. Conscience (what my intuition told me to do) was rated fourth out of the 25 factors identified in the study.

Fraud and Bribery

Fraud and bribery are forms of corruption. Amundsen (2000) distinguished between political and bureaucratic corruption, 'individual' and 'collective' forms of corruption, and corruption as a mechanism of either 'upward extraction' or 'downward redistribution'. The main forms of corruption identified are bribery, embezzlement, fraud and extortion. These concepts are partly overlapping and at times interchangeable with other concepts. The Danish Ministry of Foreign Affairs (2006) identified the common forms of corruption involved in procurement process as: bid rigging, collusion by bidders, fraudulent bids, fraud in contract performance, and fraud in an audit inquiry, product substitution, defective pricing or parts, falsification/misrepresentation of costs, bribery and acceptance of gratuities, misuse of government funds, travel fraud and theft and embezzlement. Gordon and Miyake (2001) argue that it is not always easy to define exactly what constitutes bribery and other corrupt practices, but identified facilitation payment, gifts and hospitality, conflict of interest and the use of intermediaries as falling within these practices. They identified three difficulties associated with defining bribery, namely: (1) drawing the line between acceptable relationship building and corrupt practices or bribery; (2) a difference between extortion and facilitation payments; and (3) tolerance of different practices in different cultures as an excuse for bribery and corrupt practices.

Research Methodology

The objective of this study is to investigate the current status and practices of ethics in the Malaysian construction industry. Self-administered questionnaires were sent to obtain feedback from 300 Class A Contractors registered under the Construction Industry Development Board Malaysia (worth more than USD3.5 Million) on their experience of unethical practices in the construction industry. A total of 28 usable questionnaires were returned which is considered low. The collected data was then analysed and tabulated by using simple average mean and percentages. Based on the information collected, the problems related to ethics in the construction industry were identified and probable solutions were suggested to address the unethical practices.

Results and Discussion

Demographic Information

Of the 28 respondents, female respondents outnumbered male respondents. The majority of the respondents were between age 25 and 35. More than two third of all respondents self-reported having diploma or higher. Of the all the contractors surveyed, more than 50 percent of the respondents were professionally positioned at the medium to top level executives. The number of years the participants was employed in the construction field range 3 to 20 years with the average being 7 years. The survey generated twenty-eight (28) respondents, with female respondents outnumbering male respondents. The majority of the respondents were aged between 25 and 35. More than 50 percent of the respondents were professionally positioned as medium to top level executives. The number of years the participants was employed in the construction field ranged from 3 to 20 years with the average being 7 years.

Bribery

The majority of the respondents had experienced or witnessed bribery in the industry, in the form of cash inducement, gift, favours, trips and appointments in the construction industry.

Negligence

The majority of the respondents had witnessed various forms of negligence in the construction industry which includes late and short payments, poor quality and inadequate information, lack of supervision, subcontractors' lack of safety ethics, bad documentation and unfair treatment of contractors in tender/final account negotiations respectively.

Fraud

80% of the respondents had witnessed or experienced instances of fraud in the industry. 20% of the respondents affirmed that theft of materials and illogical request for time extensions happens quite often in their profession.

Dishonesty and Unfairness

88% of the respondents experienced unfair and dishonest conduct in the industry. These occur in the four major areas of tendering practices, bureaucratic or government policy, consultant fees and project costs and design and construction contractual agreements.

12% of the respondents experienced unfair and dishonest conduct quite often, with 41% stating that competitors quite often overstate their capacity and qualifications to secure work and falsify experience, capabilities and qualifications for tendering purposes. 38% of the respondents felt bias happens quite often in tender evaluations of the major contractors and 35% claim that clients pre-selected contractors before the calling of tenders which was conducted only to fulfil organisational and statutory requirements. Other unethical conduct highlighted as occurring quite often, as testified by more than 25% of respondents, are:

- Client divulging information to the preferred tenderer (Tendering Practices)
- Bureaucratic and Government Policy
- Main contractors not paying and deducting subcontractors' fees without proper justification

Ranking of Unethical Practices in Construction Industry

The highest ranked unethical practices as testified by the contractors are cover pricing (3.67), followed by bid cutting (3.30), late and short payments (3.29), subcontractor' lack of safety ethics (3.12), unfair treatment of contractors in tender/final account negotiations (3.10), competitors overstating their capacity and qualifications to secure work (3.08), competitors' falsification of experience, capabilities and qualifications (3.08), bureaucratic and government policy (3.05) and poor documentation (3.00). Respondents also suggested that changing the public's perceptions of the construction industry requires short, medium and long term measures and strategies. These were as follows:

Short Term Measures

- A complaints system in all departments should be enforced in the construction industry. Punitive measures, penalties or even cancellation of licenses on repeated violations may be introduced. At present, all complaints are submitted to the Malaysian Anti-Corruption Commission.
- Organizations should spearhead ethical practices through organizational leadership. Also, ethical conduct and values of employees should be the criteria for rewards and promotions besides work related performance.
- The introduction of 'the integrity pact' by the CIDB is imperative as a tool for clean bidding. The concept of agreement between the government and the bidder in the procurement process is to curb corruption in government procurement. This concept had been widely accepted in various countries worldwide.

Medium Term Measures

- An indigenous quality assurance group should be part of every project team to ensure quality along with ethical practices.
- Internship with approved organizations should be made regulatory/obligatory for young technical graduates for a period of six months to groom them on professional matters including ethical aspects.
- The media's role in promoting an ethical society is even more relevant today. It can play a very important role in overcoming ethical dilemmas and the formation of an ethical society by debating such issues publicly. In this context, training and increasing awareness is of paramount importance.

Long Term Measures

- The role of an independent judiciary in implementing the law cannot be overemphasized.
- Specific laws imposed by the government to 'check' unethical practices and the creation of an 'ombudsman' for the industry are necessary. Hence, the trend is towards a compliance model, whereby professionals and organizations will be regulated, due to perceived inability to regulate their own behaviors.
- Research and development should be undertaken in the public and private sectors to effectively address ethical matters.
- Tradesmen/technical manpower should not be allowed to work till they obtain proper training through recognized training institutions.
- Empowering ethics education is considered significant for setting the foundations of a professional, together with input on his roles in the industry from professional institutions later in his professional life, as a 'check and balance ethics structure'. Organizations should make a conscious decision to pursue a corporate code of ethics.
- An industry-wide code of ethics is considered necessary in order to foster an overall ethical environment. Crucially, its implementation requires the willingness and commitment by the legislative body to enforce it.

Review Questions

1. What is The Importance of Ethics in Construction?
2. Why there is a Need For More Ethical Leaders in Construction?
3. Give salient features of CMAA Code of Ethics.
4. What are the Unethical behaviours in Construction which Need to be Cared?
5. Give 7 Examples of Ethical Practices in Construction Ethics.
6. Describe the salient features of Code of Ethics for Corporate Members, Institution of Engineers (India)
7. Explain the Ethical Issues in the Construction Industry from Contractor's Perspective.

Bibliography

Adam, A., Josephson, P.E.B. and Lindahl, G., 2017. Aggregation of Factors Causing Cost overruns and Time delays in Large Public Construction Projects. *Engineering, Construction and Architectural Management,* 24, pp. 393-406.

Ahmed, S.M., Azhar, S., Castillo, M. and Kappagantula, P., 2002. Construction Delays in Florida: An Empirical Study. *Final Report. Department of Community Affairs, Florida, US.*

Aibinu, A.A. and Jagboro, G.O., 2002. The Effects of Construction Delays on Project Delivery in Nigerian Construction Industry. *International Journal of Project Management,* 20(8), pp. 593-599.

Akoa, B.B., 2011. *Cost Overruns and Time Delays in Highway and Beidge Projects in Developing Countries-Experiences from Cameroon* (Doctoral Dissertation, Michigan State University).

Al-Hazim, N., Abu Salem, Z., and Ahmed, H., 2017. Delay and Cost Overrun in Infrastructure Projects in Jordan. *Procedia Engineering,* 182, pp. 18-24.

Alinaitwe, H.M., Mwakali, J.A. and Hansson, B., 2007. Factors Affecting the Productivity of Building Craftsmen Studies of Uganda. *Journal of Civil Engineering and Management,* 13(3), pp. 169-176.

Alison Dykstra,. (2018). Construction Project Management: A Complete Introduction Hardcover – Import. Kirshner Pub Co Inc; 2nd Edition.

Allen, D. (1989), "Never the Twain shall meet?", Accountancy Age, January, p. 21. Andrews, K.R. (1971)

Al-Momani, A.H., 2000. Construction Delay: A Quantitative Analysis. *International Journal of Project Management,* 18(1), pp. 51-59.

Al-Sabah, R., Menassa, C.C., and Hanna, A., 2014. Evaluating Impact of Construction Risks in the Arabian Gulf Region from Perspective of Multinational Architecture, Engineering and Construction Firms. *Construction Management and Economics,* 32, p. 382-402.

AlSehaimi, A., Koskela, L., and Tzortzopoulos, P., 2013. Need for Alternative Research Approaches in Construction Management: Case of Delay Studies. *Journal of Management in Engineering,* 29, pp. 407-413.

Alzahrani, M W. Emsley (2013) The Impact of Contractors' Attributes on Construction Project Success: A Post Construction Evaluation, International Journal of Project Management 31 (2013) 313-322.

Amit Bijon Dutta,. (2020). Understanding Risk Management in Construction Projects Hardcover. Evincepub Publishing.

Amoatey, C.T., Ameyaw, Y.A., Adaku, E., and Famiyeh, S., 2015. Analysing Delay Causes and Effects in Ghanaian State Housing Construction Projects. *International Journal of Managing Projects in Business,* 8(1), pp. 198-214.

Ansar, A., Flyvbjerg, B., Budzier, A., and Lunn, D., 2014. Should we Build more Large Dams? The Actual Costs of Hydropower Megaprojects Development. *Energy Policy,* 69, pp. 43-56.

Anuradha Mansinghka & Namratha Mohan,. A Detailed Guide to Construction Management Plans-Last updated on June 14, 2021,. https://www.saviom.com/blog/detailed-guide-construction-project-management-plans/

Ashley, D.B., 10-12 March, 1986 "New Trends in Risk Management", Paper Presented at the Internet's 10th International Expert Seminar on New Approaches in Project Management, Zurich.

B.C. Punmia & K.K. Khandelwal,. (2022). Project Planning and Control with PERT AND CPM,. Laxmi Publications Private Limited; Fourth Edition.

B.S. Patil,. (2015). Civil Engineering Contracts & Estimates Paperback. Orient Blackswan Private Ltd; Fourth Edition.

Behnam and Joseph,. (2016). Identification of a Set of Appropriate Critical Success Factors for Successful TQM Implementation in Construction and other Industries. International Journal of Advanced Research. Vol. 4, No. 11, pp. 1581-1591.

Belout Adnane, Gauvereau Clothilde, 2003, "Factors Influencing Project Success: The Impact of Human Resource Management", International Journal of Project Management Vol. 22, page 1.

Berengueres, José (2007). *The Toyota Production System re-contextualized.* Tokyo: José Berengueres. p. 74. ISBN 978-1847534774. OCLC 906982187.

Bramble, B. and Callahan, M., 2000. Construction Delay Claims 3rd ed. *Gaithersburg, MD: Aspen Law and Business.*

Budget planning "https://opentextbc.ca/projectmanagement/chapter/chapter-12-budget- planning-project-management/" (24th Feb 2019).

Chan, A.P.C., Chan, A.P.L., (2004). Key Performance Indicators for measuring Construction Success. Benchmarking: An International Journal 11 (2), 203-221.

Cheng, Y.M., 2014. An Exploration into Cost-influencing Factors on Construction Projects. *International Journal of Project Management,* 32(5), pp. 850-860.

Collinge, W.H. and Harty, C.F., 2014. Stakeholder Interpretations of Design: Semiotic Insights into the Briefing Process. *Construction Management and Economics,* 32(7-8), pp. 760-772.

Construction Planning and Management "https://nptel.ac.in/courses/105103093/"

Cooke-Davies, T., 2002 "The 'real' Success Factors on Projects", International Journal of Project Management, Vol. 20 No. 3, pp. 185-190.

Corporate Strategy – An Analytic Approach to Business Policy for Growth and Expansion, Penguin, Harmondsworth, revised. Azzone, G., Masella, C. and Bertele', (1991)

Cost Control Problem https://www.cmu.edu/cee/projects/PMbook/12_Cost_Control,_Monitoring,_and_Accounting. html" (20th Feb 2019).

Cost Control Techniques "https://www.aboutcivil.org/cost-control-cost-management- techniques.html". (23th Feb 2019)

Cost Estimate "https://indianexpress.com/article/cities/mumbai/colaba-bandra-seepz- metro-underground-project-cost-set-to-escalate-further/" (24th Feb 2019)

Cost Estimation "https://www.aproplan.com/blog/efficiency/how-to-reduce-the-high- costs-in-your-construction-business" (25th Feb 2019)

D. Lal,. (2013). Construction Management and P.W.D. Accounts Paperback. Katson Publication. Delhi.

Day, D.A., & Benjamin, N.B.H. (1991). *Construction Equipment Guide* (2nd ed.). Canada: John Willey & Sons.

Design of Performance measures for Time-based Companies", International Journal of Operations & Production Management, Volume 11 No. 3 pp. 77-85.

Doloi, H., Sawhney, A., Iyer, K.C. and Rentala, S., 2012. Analysing Factors Affecting Delays in Indian Construction Projects. *International Journal of Project Management*, 30(4), pp. 479-489.

Dominic, A.D. D., and Smith, S.D., 2014. Rethinking Construction Cost Overruns: Cognition, Learning and Estimation. *Journal of Financial Management of Property and Construction,* 19, pp. 38-54.

Dvir Dov, Raz Tzvi, Shenhar Aaron J, 2002, "An Empirical Analysis of the Relationship between Project Planning and Project Success", International Journal of Project Management Vol. 21, page 89.

E M Rojas,. (2014). Construction Project Management: A Practical Guide for Building and Electrical Contractors Hardcover. Cengage Learning India.

Elmualim, A. and Gilder, J., 2014. BIM: Innovation in Design Management, Influence and Challenges of Implementation. *Architectural Engineering and Design Management*, 10(3-4), pp. 183-199.

Engineering Construction and Architectural Management 10(2): pp. 117-127. Eric Stokes and Salem Aram. "Project Management" 2008.

file:///C:/Users/user/Desktop/12%20Futuristic%20Construction%20Technology% 20Trends%20&% 20Ideas% 20[with%20Examples].htm

file:///C:/Users/user/Desktop/Code% 20of%20Ethics% 20Construction%20Management%20Association% 20of%20America.htm

file:///C:/Users/user/Desktop/Construction% 20Ethics% 207% 20Examples%20of% 20Ethical% 20Practices.htm file:///C:/Users/user/Desktop/Ethics% 20in%20Construction.%20What%20are% 20unethical%20behaviours% 20Why%20should%20we%20care.htm

file:///C:/Users/user/Desktop/Inventory% 20monitoring%20&% 20control%20in%20a% 20construction% 20company%20%E2% 80%93% 20SOURCING%20AND%20SUPPLY%20CHAIN.htm

file:///C:/Users/user/Desktop/Project% 20Scheduling_Network%20Techniques%20in%20Project% 20Management%20CPM%20and%20PERT%20Analysis.htm

Garold D. Oberlender,. (2014). Project Management for Engineering and Construction Paperback. McGraw Hill Education; Second Edition.

George Otim, Fiona Nakacwa & Michael Kyakula- "Cost Control Techniques Used On Building Construction Sites in Uganda".

Georges, P.G. & Vallance, G.V. (1986). Cost Management. Project Management Journal, Vol. 17 No. 3. pp. 47-54.

Hadidi, L., Assaf, S. and Alkhiami, A., 2017. A Systematic Approach for ERP Implementation in the Construction Industry. Journal of Civil Engineering and Management, 23(5), pp. 594-603.

Hailiang Huang & Chaofeng Zhang - "Cost Management-A Case Study of a Gardening Firm Project".

Hans Ottosson,. (2013). Practical Project Management for Building and Construction. Routledge and CRC Press.

Harding, J., 2012. Avoiding Project Failures. Chemical Engineering, 119(13), pp. 51-54.

Hemal, R.M.N., Waidyasekara, K.G.A.S., and Ekanayake, E.M.A.C., 2017. Integrating Design Freeze in to Large-scale Construction Projects in Sri Lanka. Moratuwa Engineering Research Conference, pp. 259-264.

Hira N. Ahuja, S.P. Dozzi & S.M. Abou Rizk. (1994). Project Management: Techniques in Planning and Controlling Construction Projects. Wiley; 2nd Edition.

http://apppm.man.dtu.dk/index.php/Financial_appraisal_in_construction-Financial appraisal in construction *Developed by Apostolos Bougas*

http://c.ymcdn.com/sites/www.nasfa.net/resource/resmgr/Learning_Series/Intro_t

http://www.jnkvv.org/PDF/17052020204513175201920.pdf

https://aip.scitation.org/doi/pdf/10.1063/1.4965067

https://civilengineeringnotes.com/brickwork-specification/

https://civilengineeringnotes.com/rate-analysis/

https://clc.gov.in/clc/acts-rules/building-and-other-construction-workers

https://cleartax.in/g/terms/depletion

https://cpwd.gov.in/Publication/GCC_Construction_2019.pdf— *General Conditions of Contract- 2019- Construction Works*

https://gobridgit.com/construction-resource-allocation-guide/ How to achieve effective resource allocation in construction management

https://gosmartbricks.com/construction-site-safety-7-tips-to-ensure-your-workers-are-safe/

https://hexagonxalt.com/resource-library/what-is-materials-management

https://ijesc.org/upload/de526df81fffa80ce22afba3cc305e71. Analysis%20of%20Equipment% 20Management% 20in%20Construction% 20Industry%20to%20Improve%20Productivity.pdf

https://kwaconstruction.com/2019/11/25/the-importance-of-ethics-in-construction/

https://sourcingandsupplychain.com/inventory-monitoring-control-in-a-construction-company/Inventory monitoring & control in a construction company by Balanarayanan N.K. Posted on March 31, 2020.

https://theconstructor.org/practical-guide/rate-analysis-of-plastering-with-cement-mortar/9620/

https://www.brainkart.com/article/Interest-rate-Formulas_5322/— Interest Formulae— Mechanical - Engineering Economics & Cost Analysis - Value Engineering

https://www.businessmanagementideas.com/project-management/pert-meaning-and-steps-network-analysis-project-ma

https://www.concreteshowindia.com/blog/quality-assurance-in-construction-industry/

https://www.educba.com/economic-order-quantity-formula/Madhuri Thakur

https://www.eresourcescheduler.com/blog/resource-allocation-tools-in-construction-challenges-and-recommendations

https://www.helplinelaw.com/real-estate-wills-probate-and-trust/BSIH/building-safety-in-india-how-builders-can-ensure-the-safety-of-workers.html

https://www.ijrte.org/wp-content/uploads/papers/v8i5/E6408018520.pdf

https://www.ikigaiconsulting.com/insights/we-need-more-ethical-leaders-in-construction

https://www.ipsnews.net/2020/08/getting-indias-construction-workers-entitlements/

https://www.irbnet.de/daten/iconda/CIB_DC26728.pdf

https://www.irjet.net/archives/V7/i5/IRJET-V7I51431.pdf

https://www.mondaq.com/india/construction-planning/616518/construction-engineering-law

https://www.profitbooks.net/what-is-depreciation/—Learn about depreciation.

https://www.projectmanager.com/blog/construction-quality-control

https://www.researchgate.net/publication/319359193,- Bar (Gantt)Charts—Saleh A. Mubarak-online material -Chapter · May 2010,.DOI: 10.1002/9780470912171.ch2.

https://www.researchgate.net/publication/311438258-Project Management Information System in Construction Industry: (By Pranavraj ambare and Arun Dhawale, 2017)

https://www.shrm.org/resourcesandtools/legal-and-compliance/employment-law/pages/india-construction.aspx

https://www.toolshero.com/management/materials-management/Materials Management-Toolshero.

https://www.vskills.in/certification/tutorial/selective-inventory-management/Role of Selective Inventory Control

Ibrahim, M., and Bruland, A., 2012. Cost Diviation in Road Construction Projects: The Case of Palestine. *Construction Economics and Building,* 12(1), pp. 58-71.

IS:125&-1967 Safety measures for fire.

IS:3606 (Part I & II) 1966; Safety measures for Scaffolding, Ladders, Formwork and Equipment.

IS:3764—1966 Safety measures for Excavation.

IS:4081-1960; Safety measures for Drilling and Blasting.

IS:4130-1976; Safety measures for Demolition.

IS:591 &-1970; Safety measures for Hot Bituminous Works.

Iyer, K.C. and Jha, K.N., 2005. Factors Affecting Cost Performance: Evidence from Indian Construction Projects. *International Journal of Project Management,* 23(4), pp. 283-295.

J F McCarthy,. (2011). Construction Project Management Hardcover – Import, Pareto Building Improvement; 1st Edition.

Jallow, A.K., Demian, P., Baldwin, A.N., and Anumba, C., 2014. An Empirical Study of the Complexity of Requirements Management in Construction Projects. *Engineering Construction & Architectural Management,* 21, pp. 505-531.

Jarkas, A.M, and Haupt, T.C., 2014. Major Construction Risk Factors Considered by General Contractors in Qatar. *Journal of Engineering, Design and Technology,* 13(1), pp. 165-194.

Jianbing LIU, Jianliang CAI "Construction Cost Control Based on PDCA Cycle"

Kaliba, C., Muya, M. and Mumba, K., 2009. Cost Escalation and Schedule Delays in Road Construction Projects in Zambia. *International Journal of Project Management,* 27(5), pp. 522-531.

Kenley, R., 2014. Productivity Improvement in the Construction Process.

Kissi, E., Mohammed, S.A. and Owusu-Diatuo, K.J., 2018. Challenges Facing Public Works Departments in Construction Project Delivery within Metropolitan Municipal and District Assemblies in Ghana. *UDS International Journal of Development,* 5(1), pp. 129-143.

Krishan Kumar Chitkara,. (2019). Construction Project Management | 4th Edition, McGraw-Hill, India.

Krishnamurthy K.G & Ravindra S.V,. (2018). Construction and Project Management 2ed (PB 2018) Paperback. CBS Publishers & Distributors, Delhi.

Kumar Neeraj Jha,. (2015). Construction Project Management, Theory & Practice. Pearson Education India; 2nd Edition.

Lappe, M. and Spang, K., 2014. Investments in Project Management are Profitable: A Case Study-based Analysis of the Relationship between the Costs and Benefits of Project Management. *International Journal of Project Management*, 32(4), pp. 603-612.

"Learning to See the Effects of Improved Workflow in Civil Engineering Projects", Lean Construction Journal, 9(4), pp. 390-398.

Len Holm,. (2018). Cost Accounting and Financial Management for Construction Project Managers -Paperback ,. Routledge; 1st Edition.

Lim C S and Mohamed M Z. (1999). Criteria of Project Success: An Exploratory re-examination. International Journal of Project Management, 1999; 17; 4; 243-248.

Mahanim Hanid, Mohan Siriwardena, Lauri Koskela "What are the Big Issues in Cost Management?"

Memon, A.H., 2013. *Structural Modelling of Cost Overrun Factors in Construction Industry* (Doctoral Dissertation, Universiti Tun Hussein Onn Malaysia).

Moen, Ronald; Norman, Clifford. "Evolution of the PDCA Cycle" (PDF). *westga.edu*. Paper Delivered to the Asian Network for Quality Conference in Tokyo on September 17, 2009. Retrieved 1 October 2011.

Mohammad Azim Eirgash,. (2020). https://www.researchgate.net/publication/348060444-Resource Allocation and Leveling in Construction Management Projects with Resource Histogram.

Muhammad Saqib, Rizwan U. Farooqui, Sarosh. H. Lodi, August 4-5, 2008, "Assessment of Critical Success Factors for Construction Projects in Pakistan", First International Conference on Construction in Developing Countries, Karachi, Pakistan.

Nassar, N.K. (2009). An Integrated Framework for Evaluation of Performance of Construction Projects. Paper Presented at PMI® Global Congress 2009—North America, Orlando, FL. Newtown Square, PA: Project Management Institute.

National Building Code of India, 2016 (BIS, Delhi).

Nazarko, L., 2015. Technology Assessment in Construction Sector as a Strategy towards Sustainability. *Procedia Engineering*, 122, pp. 290-295.

Nguyen, Long Duy, Ogunlana, Stephen O. And Lan, Do Thi Xuan, 2004 "A Study on Project Success Factors in Large Construction Projects in Vietnam", Engineering, Construction and Architectural Management, Vol. 11 No. 6, pp. 404-413.

Nikhila C,.Value Analysis: Meaning Steps and Advantages,. https://www.businessmanagementideas.com/production-management/cost-reduction/value-analysis-meaning-steps-and-advantages/7064—o_Lean_Construction_&.pdf (19-06-2017)

Ochieng, O.M., Shedrack, M., and Douglas, M., 2015. Role of Cost Estimation on Timelines in Completion of Kenya National Government Building Construction Projects. *International Journal of Management Research & Review,* 5, pp. 285-295.

Odeh, A.M. and Battaineh, H.T., 2002. Causes of Construction Delay: Traditional Contracts. *International Journal of Project Management,* 20(1), pp. 67-73.

Ogunlana, S.O., Promkuntong, K. and Jearkjirm, V., 1996. Construction Delays in a Fast-growing Economy: Comparing Thailand with other Economies. *International Journal of Project Management,* 14(1), pp. 37-45.

P. Strange, Gareth D. Vaughan,. (1993). TQM: A View from the Playing Field., Journal of Management in Engineering,

P.S. Gahlot & B.M. Dhir,. (2018). Construction Planning and Management Paperback. New Age International Publishers; Second Edition.

Parviz F. Rad,. (2003). Project Success Attributes. Journal of Cost Engineering, April, 2003; 45; 4; 23-28.

Paul Netscher,. (2017). Construction Management: From Project Concept to Completion Paperback. Createspace Independent Publisher.

Pawan Dua,. (2021). Risks & Deterrents in Construction Projects: For Customers, Contractors, Suppliers & Consultants ... Paperback. Notion Press; 1st Edition.

Peter Simonsson1, Anders Björnfot2, Jarkko Erikshammar3 & Thomas Olofsson4, "Learning to see the Effects of Improved Workflow in Civil Engineering Projects", Lean Construction Journal, 2012.

Piyush Bhandari,. (2021). Construction Project Management: From Project Conception to Completion.

Pourrostam, T. and Ismail, A., 2011. Significant Factors causing and effects of delay in Iranian Construction Projects. *Australian Journal of Basic and Applied Sciences,* 5(7), pp. 450-456.

Prem Vardhan,. (2020). Construction Business. Notion Press; 1st Edition.

Prof. Patil Dipak P., Prof. Pankaj P. Bhangale & Swapnil S. Kulkarni. "Study of Cost Control on Construction Project".

Purnima Bajpai,. (2019). Effective Materials Management in Construction Projects,. *J* -NBM & CW July 2019.

R. Jayaraman. "Project Cost Control: A New Method to Plan and Control Costs in Large Projects". Business Process Management Journal" Vol. 22 No. 6 pp. 1247-1268, November 2016.

R.P. Rethaliya,. (2020). Construction Management Paperback. Atul Prakashan.

Rangawala,. (2012). Construction of Structure and Management of Works. Charotar Publishing House. Pvt. Ltd., Anand, Gujarat.

Rahman, H., Wang, C., Takim, R. and Wong, S., 2011. Project Schedule Influenced by Financial Issues: Evidence in Construction Industry. *Scientific Research and Essays,* 6(1), pp. 205-212.

Rajiv Abhyankar,. (2015). Construction Management - Paperback. Vishwkarma Publications.

Rajiv Gupta,. (2008),. Construction Planning and Technology. CBS Publisher & Distributors Pvt. Ltd. Delhi.

Resource Analysis, "https://www.mak.ac.ug/documents/Makfiles/aet2011/Otim-2.pdf" (24th Feb. 2019)

Robert M Patty, End of Project Overruns, Nov. 2009

Robert Peurifoy, Clifford J. Schexnayder & Aviad Shapira,. (2010). Construction Planning, Equipment and Methods Paperback. McGraw Hill Education; 7th Edition.

S.C. Sharma., (2016). Construction Engineering and Management of Projects. Khanna Publishers, New Delhi.

S.P. Mahajan & Sanjay Mahajan,. (2018). Quantity Surveying & Valuation. Satya Prakashan, New Delhi.

Sadeh A, Dvir D and Shenhar A. (2000). The Role of Contract Type in the Success of R&D Defence Projects Under Increasing Uncertainty. Project Management Journal, September 2000; 31; 3; 14-21.

Saleh A. Mubarak,. (2019). Construction Project Scheduling and Control Hardcover. Wiley; 4th Edition.

Salleh, R. (2009), Critical Success Factors of Project Management for Brunei Construction Projects: Improving Project Performances. Page 35.

Sanders, D. and Eagles, W.D. (2001). *Delay, Disruption and Acceleration Claims*, Borden Ladner Gervais LLP, May, 3.

Sebastian Morris,. (2019), Financing Infrastructure in India – Issues and the Way Forward,. W.P. No. 2019-03-02, March 2019, Indian Institute of Management Ahmedabad, India.

Shahsavand, P., Marefat, A. and Parchamijalal, M., 2018. Causes of Delays in Construction Industry and Comparative Delay Analysis Techniques with SCL Protocol. *Engineering, Construction and Architectural Management*, 25(4), pp. 497-533.

Sidney M. Levy,. (2018). Project Management in Construction, Seventh Edition - Hardcover. McGraw Hill.

Siguroursan, S.F (2009). Critical Success Factor in Project Management: An Ethical Perspective.

Simonsson1, Anders Björnfot2, Jarkko Erikshammar3 & Thomas Olofsson4

Sindhuja S,. CPM: Meaning, Steps and Example | Network Analysis,. https://www.businessmanagementideas.com/project-management/cpm-meaning-steps-and-example-network-analysis/9

Sindhuja S,. PERT: Meaning and Steps | Network Analysis | Project Management,.

Singh, R., 2009. Delay and Cost Overrun in Infrastructure Projects: An Inquiry into Extent causes and Remedies. *Centre for Development Economics, Delhi.*

Sofiat O. Abioye, Lukumon O. Oyedele, Lukman Akanbi, Anuoluwapo Ajayi, Juan Manuel Davila Delgado, Muhammad Bilal, Olugbenga O. Akinade, Ashraf Ahmed,. (2021). Artificial Intelligence in the Construction Industry: A Review of Present Status, Opportunities and Future Challenges,. Journal of Building Engineering,. 44 (2021) 103299. www.elsevier.com/locate/jobe

Sohrab (1999),. Total Quality Management in Construction Industry. Journal of Indian Buildings Congress, Vol. 6, No. 1, pp. 1-8.

Steinfort, P and Walker (2007). Critical Success Factors in Project Management Globally and how they may be Applied to Aid Projects. In D. Baccarini (ed.) Proceedings of the PMOZ Achieving Excellence- 4th Annual Project Management Australia Conference, Brisbane, Australia, 28-31 August 2007.

Steven M. Bragg (2010). "Cost Reduction Analysis: Tools and Strategies", 1st Edition, Wiley, New York.

Suresh Kumar,. (2018). State-of –the- Art and Strategy of Implementation of Total Quality Management (TQM) in Building Construction Industry,. Ph.D Thesis. Punjab University, Chandigarh.

Syed Jamaluddin, S.Z.H., Mohammad, M.F., and Ahmad, K., 2014. Enhancing the Quality of Construction Environment by Minimizing the Cost Variance. *Procedia- Social and Behavioural Sciences,* 153, pp. 70-78.

Tarhini, A., Ammar, H., Tarhini, T. and Masa'deh, R.E., 2015. Analysis of the Critical Success Factors for Enterprise Resource Planning Implementation from Stakeholders' Perspective: A Systematic Review. *International Business Research,* 8(4), pp. 25-40.

The Concept of Corporate Strategy, Dow Jones-Irwin, Homewood, IL. Ansoff (1984)

Tracking of Man power material Machinery for Cost Budget "https:// www.cmu.edu/ cee/projects/PMbook/04 _Labor,_Material,_And_ Equipment_Utilizatio n.html" (22th Feb 2019).

U.K. Shrivastava,. (1999). Construction Planning and Management. Galgotia Publications Pvt. Ltd. Delhi.

Umesh Kohli & Krishan Kumar Chitkara,. (2012). Project Management Handbook,. I McGraw-Hill.India.

Urmila A Mahadik "Cost Reduction in Construction Projects", "International Journal of Engineering Technology, Management and Applied Sciences", September 2015, Volume 3, Special Issue, ISSN 2349-4476.

Varun Kumar Sharma,. (2020). Investigation of Risk Factors Affecting Highway Project Cost Escalation and Time Delay,. Ph.D Thesis. Punjab Engineering College (Deemed to be University), Chandigarh.

Vee, C. and Skitmore, R.M. (2003) Professional Ethics in the Construction Industry.

Vidalis, S.M. and Najafi, F.T., 2002, December. Cost and Time Overruns in Highway Construction. In *Canadian Society for Civil Engineering-30th Annual Conference: 2002 Chellenges Ahead* (pp. 2799-2808).

Walid Belassi, Tukel Oya Icmeli, 1996 "A New Framework for Determining Critical Success/Failure Factors in Projects", International Journal of Project Management Vol. 14, No. 3, page 141.

Work Break Down Structure "https://brainmass.com/business/management-tools-and- techniques/work-breakdown-structure-cost-control-456472" (24th Feb 2019).

Yadav, N. and Sagar, M., 2013. Performance Measurement and Management Frameworks. *Business Process Management Journal*, 19(6), pp. 947-970.

Yakubu Adisa Olawalen and Ming Sun- "Cost and Time Control of Construction Projects: Inhibiting Factors and Mitigating measures in Practice".

Youlin Li, 2018 IOP Conf. Ser.: Mater. Sci. Eng. 394 0320 "Research on Construction Projects Cost Management"

Index

❑ ❑ ❑ ❑ ❑ ❑

Printed by Libri Plureos GmbH in Hamburg,
Germany

9 789394 917002